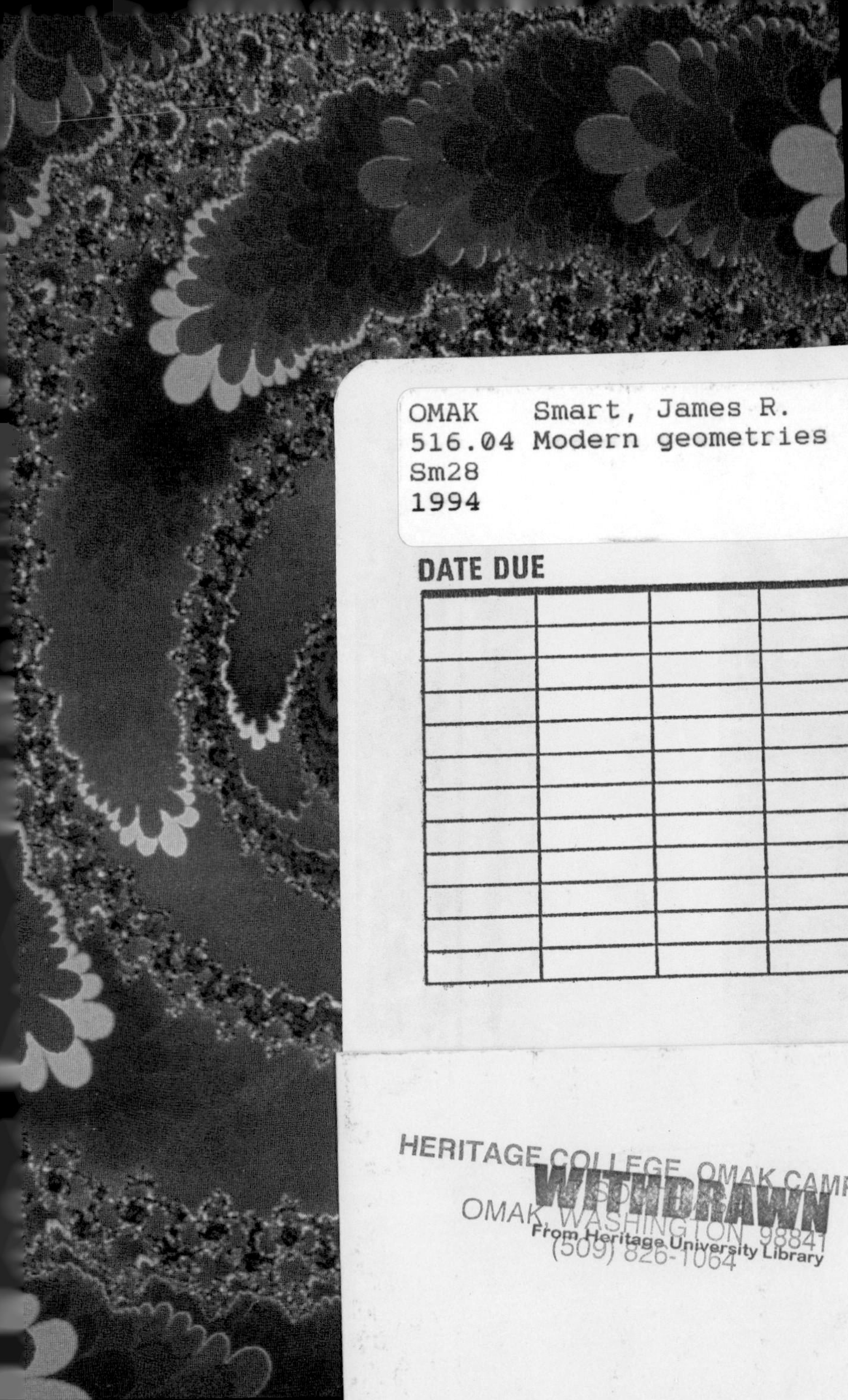

Modern Geometries

Fourth Edition

Modern Geometries

Fourth Edition

James R. Smart
San Jose State University

Brooks/Cole Publishing Company
Pacific Grove, California

THIS BOOK IS PRINTED ON ACID-FREE RECYCLED PAPER

CONTEMPORARY UNDERGRADUATE MATHEMATICS SERIES
Consulting Editor: Robert J. Wisner

Brooks/Cole Publishing Company
A Division of Wadsworth, Inc.

Printed in the United States of America
10 9 8 7 6 5 4 3 2 1

Library of Congress Cataloging-in-Publication Data
Smart, James R.
Modern geometries / James R. Smart.—4th ed.
Bibliography: p.
Includes index.
ISBN 0-534-21198-4
1. Geometry, Modern. I. Title.
QA473.S53 1993 93-29985
516′.04—dc20 CIP

- Sponsoring Editor: Jeremy Hayhurst
- Editorial Assistant: Elizabeth Barelli
- Production Services Manager: Joan Marsh
- Production Editor: Daryl Angney, TECH*arts*
- Manuscript Editor: Helen Berliner
- Permissions Editor: Carline Haga
- Cover Design: Lisa Berman
- Interior Illustration: TECH*arts*
- Design and Typesetting: Daryl Angney, TECH*arts*
- Printing and Binding: Arcata Graphics, Fairfield, Pennsylvania

This book is dedicated to Rachel

Preface

Since the publication of the third edition, the practice of including one or more courses in modern geometry at the junior–senior level in universities has continued. The use of both groups of transformations and sets of axioms to classify geometries remains important. The field of computer graphics has emerged as a major application of geometric ideas. The continuing emphasis on problem solving throughout the mathematics curriculum has shown geometry to be a fascinating source of problems, and has pointed up the relationship between skill in geometry and skill in problem-solving techniques. Increasingly, geometry is seen as an applied as well as an abstract study.

The general goals and recommendations from *Geometry's Future*, conference proceedings sponsored by COMAP, Inc., have been carefully considered in planning this new edition, and specific suggestions from the report have been incorporated.

This fourth edition also incorporates many improvements suggested by reviewers and by teachers and students who have used the previous editions. The exercise sets have been extended to include varying levels of difficulty, with more investigative-type problems appropriate for individuals or small groups. Other changes include additional explicit examples and figures, the improvement of many figures, clarification of wording, rewriting of some proofs, the inclusion of many new applications such as fractals and geometric probability, more historical notes, and a stronger bibliography. All of the desirable features of earlier editions have been retained, so that these changes result in a text that is both mathematically sound and highly teachable.

As the title indicates, the central theme of the text is the study of many different geometries, rather than a single geometry. The first two chapters give two ways of classifying some geometries: by means of sets of axioms or by the type of transformation defined. Both ways are used extensively in this text, and sometimes both are used for studying the same geometry. The finite geometries of Chapter 1; the convexity, Euclidean geometry, and constructions of Chapters 3, 4 and 5; the projective geometry of Chapter 7; and the non-Euclidean geometry of Chapter 9 are based on various sets of axioms. In addition to the examples of geometries presented from a transformational point of view in Chapter 2, there are separate chapters (6, 7, 8) on inversion, projective geometry, and topological transformations with the nature of the geometry depending on the transformation allowed.

Throughout, the text emphasizes practical and up-to-date applications of modern geometry. Students should be aware that many of these topics are discussed in current professional journals and that contemporary research mathematicians are seriously involved in the extension of geometric ideas. The major new applied area of computer graphics, with an increased emphasis on matrices for transformations, is covered in greater detail in the fourth edition.

This text is written for students who range widely in their mathematical abilities. While much of the material is appropriate for those with average or weak backgrounds in geometry, students with strong backgrounds will find many new extensions of ideas that will sharpen their problem-solving skills and encourage them to continue their investigations. The text is planned for both majors and minors in mathematics. It is appropriate for students interested in mathematics from the liberal arts standpoint and for those planning to be teachers of mathematics.

Many of the first exercises in each set can serve as the basis for classroom discussion; in that way, the instructor can make certain that fundamental concepts are understood. Later exercises allow extensive practice in providing independent proofs. The open-ended exercises provide extensive opportunities for further study. New in this edition is the designation of some of these investigative–type questions for use in cooperative learning situations. Chapter review exercises provide additional practice of many concepts and skills. Some instructors may also wish to use these as practice tests.

The various chapters of the text are largely independent. This and the arrangement of topics within each chapter allow for great flexibility in their use, according to the needs of the class and the desires of the instructor.

▲ Table Suggesting Some Possible Course Arrangements

I. Use of entire text for two-semester or three-quarter course

Plan A. (Semester or quarter system) Follow sequence of text

Plan B. (Semester system) One semester for geometry of transformations (Chapters 2, 6, 7, 8) and another semester for axiomatic systems (Chapters 1, 3, 4, 5, 9)

Plan C. (Quarter system) One quarter for geometry of transformations (Chapters 2, 7, 8), another quarter for axiomatic systems (Chapters 1, 6, 9), another quarter for extensions of Euclidean geometry (Chapters 3, 4, 5)

II. Use of portions of text for courses covering less than an academic year

Plan D. (Two-quarter) Any two quarters from plan C

Plan E. (Two-quarter) All of text except Chapters 5, 6, 9

Plan F. (Two-quarter) All of text except Chapters 3, 5, 8

Plan G. (One-semester) Either semester from plan B

Plan H. (One-semester) Chapters 1–4

Plan I. (One-semester) Chapters 4, 5, 7 and one other selected chapter.

Plan J. (One-quarter) Any one quarter from plan C

Plan K. Survey as many chapters as time permits by choosing first sections and omitting later sections of each chapter.

I would like to express my appreciation to James Moser of the University of Wisconsin, Bruce Partner of Ball State University, John Peterson of Brigham Young University, Demitrios Prekeges of Eastern Washington State College, Curtis Shaw of the University of Southwestern Louisiana, and Marvin Winzenread of California State University, Hayward, for reviewing the manuscript for the first edition; to Lewis Coon of Eastern Illinois University, Viggo Hansen of California State University, Northridge, Alan Hoffer of Boston University, John M. Lamb, Jr., of East Texas State University, and Alan Osborne of Ohio State University for reviewing the text in connection with preparations for the second edition; to Ione Boodt of the University of Indianapolis, George Cree of the University of Alberta, H. Howard Frisinger of Colorado State University, Jay Graening of the University of Arkansas, George C. Harrison of Norfolk State University, Alan Hoffer of Boston University, Wojciech Komornicki of Hamline University, Dave Logothetti of the University of Santa Clara, and Joanne Trimble of Marist College for reviewing the text and/or manuscript for the third edition; Fred Fiener of Northeastern Illinois University, Claudia Giamati of Northern Arizona University, Betty J. Krist of State University College, Buffalo, New York, Adrian Riskin of Northern Arizona University, and John Smashey of Southwest Baptist University, for reviewing the manuscript for the fourth edition; and to the many conscientious university faculty members and students who have made worthwhile suggestions for improvements.

I would also like to thank Robert Wisner, Jeremy Hayhurst, Elizabeth Barelli, Joan Marsh, Carline Haga, and all the other people at Brooks/Cole Publishing Company who helped in the preparation of this edition. Finally, let me thank Daryl Angney and the others at TECH*arts* for the great job of making work on the book such a pleasant experience.

James R. Smart

Contents

1,2,4,6,7,9

Modern Geometries

Fourth Edition

1

Sets of Axioms and Finite Geometries

1.1 Introduction

Today, the great increase in practical applications of geometry is leading to interest in many new topics. The idea that geometry consists only of the typical high school geometry course is completely inadequate. Your concept of the field of geometry should change dramatically as you study this text.

The word *geometry* literally means "earth (*geo*) measure (*metry*)." Although this literal meaning is far too narrow to include the various modern geometries explored in this text, the idea of earth measure was important in the ancient, pre-Greek development of geometry. These practical Egyptian and Babylonian applications of geometry involved measurement, to a great extent, and they were not complicated by formal proofs.

Hundreds of baked clay tablets from ancient Babylonia include mathematical ideas. The analysis of a tablet designated Plimpton 322 (catalog number 322 in the Plimpton collection at Columbia University) has shown the Babylonians, as early as about 1900–1600 B.C., had extended their knowledge of right triangles to a compilation of primitive Pythagorean triples. For example, the numbers 3, 4, 5 are primitive triples because they can be used as the lengths of the sides of a right triangle, and have no common integral factors other than one. The Babylonians used rules for finding areas and volumes of some figures, and are also given credit for dividing the circle into 360 parts.

Two ancient Egyptian papyruses dating from about the same period, the Moscow papyrus and the Rhind papyrus, contain a total of 26 geometric problems, most of which involve various mensuration formulas used for computing land area and the volume of granaries. Some problems concerned finding the slope of the face of a pyramid.

During the Greek period, the science of earth measure became more refined. About 230 B.C., Eratosthenes made a remarkably precise measurement of the size of the earth. According to the familiar story, Eratosthenes knew that at the summer solstice the sun shone directly into a well at Syene at noon. He found that at the same time, in

Alexandria, approximately 787 km due north of Syene, the rays of the sun were inclined about 7.2° from the vertical (Figure 1.1). With these measurements, Eratosthenes was able to find the diameter of the earth.

Example ➤ Use the measurements of Eratosthenes to find the approximate difference in angle of elevation of the sun in two places 476 km apart in a north-south line, both south of the equator.

$$\frac{787}{7.2} = \frac{476}{x}$$

$$x \approx 4.4°$$

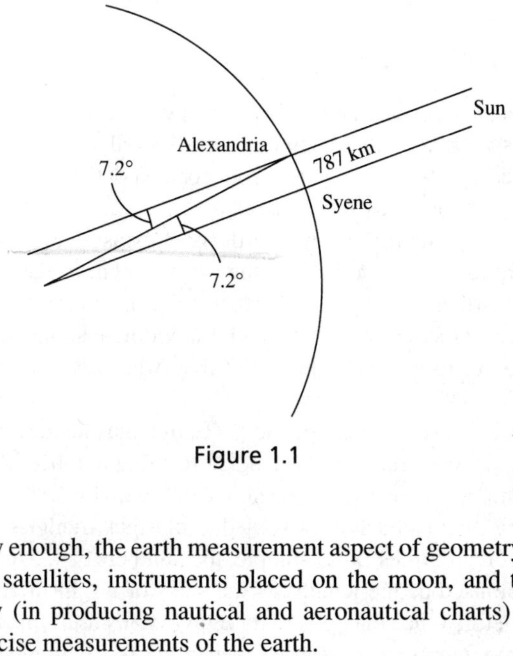

Figure 1.1

Interestingly enough, the earth measurement aspect of geometry has been of recent interest because satellites, instruments placed on the moon, and the U.S. Coast and Geodetic Survey (in producing nautical and aeronautical charts) have been able to provide very precise measurements of the earth.

Today, the science of geodesy includes making precise measurements of the earth's size and shape, as well as determining precise locations of points on the earth's surface. These measurements are used by cartographers in the production of maps. The earth is considered to be a geoidal surface whose composition is not uniform and which has lumps, or bulges, rather than being an exact sphere or even an exact oblate spheroid (with slightly different major and minor axes). Figure 1.2 illustrates two common map projections.

The first is the Mercator projection and the second is the Conic projection. A third type of map is discussed in Chapter 6. The concept of projection appears in Chapter 7.

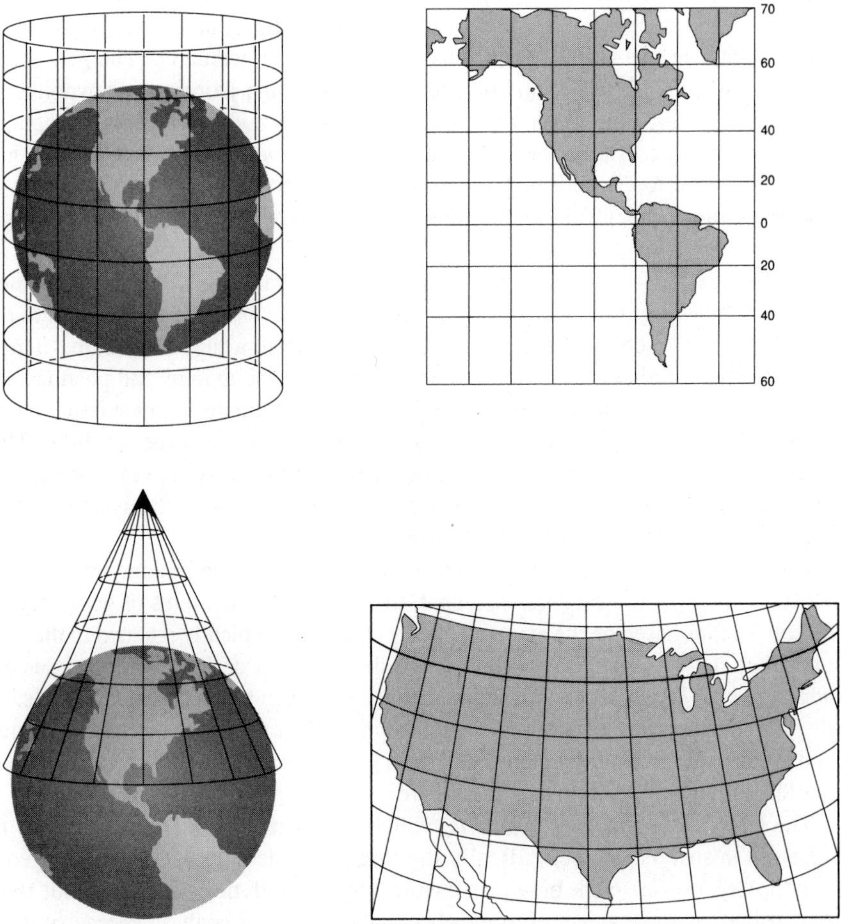

Figure 1.2

To many people today, however, the word *geometry* no longer suggests earth measure: mathematicians might describe the geometries included in this text as the study of the properties and relationships of sets of points. To the typical adult, though, the word *geometry* probably suggests the high school course they took in plane geometry or possibly the use of coordinates for points encountered in the study of algebra or calculus. The ancient Greeks of the period 500 B.C. to A.D. 100 receive much of the credit for developing the demonstrative geometry studied at the high school level. They recognized the beauty of geometry as a discipline with a structure and understood that the proof of a theorem could be even more exciting than the discovery of a practical application.

Greek geometry, called *Euclidean geometry* because of the monumental work of Euclid (300 B.C.), includes *undefined terms*, *defined terms*, *axioms or postulates*, and *theorems*. Other geometries studied in this text have the same sort of structure, so sets of axioms are one convenient means of classifying a geometry. The geometries in Chapters 1, 3, 4, 5, and 9 are all approached from the axiomatic viewpoint. In Euclidean geometry, undefined terms, which are arbitrary and could easily be replaced by other terms, normally include terms such as *points*, *lines*, and *planes*; it would also be possible to develop Euclidean geometry using such concepts as *distance* and *angle* and keeping them undefined. Definitions of new words involve using the undefined terms.

Today, the words *axiom* and *postulate* are used interchangeably. In the development of geometry, however, the word *postulate* was used for an assumption confined to one particular subject (such as geometry), while *axiom* denoted a "universal truth," a more general assumption that applied to all of mathematics. The axioms and postulates of Euclid are stated in Section 1.7. The *truth* of axioms or postulates is not at issue. These statements are beginning assumptions from which logical consequences follow. They are analogous to the rules for a game. Since the mathematical system to be developed depends on the axioms, changing the axioms can greatly change the system, just as changing the rules for a game would change the game.

Theorems are statements to be proved by using the axioms, definitions, and previous theorems as reasons for the logical steps in the proof. The theorems of geometry are valid conclusions based on the axioms. A simple theorem typically is stated in the form of an if-then statement such as, "If the sum of the measures of the opposite angles of a quadrilateral is 180 (in degrees), then the quadrilateral can be inscribed in a circle." In logic, this statement is a conditional. Selected ideas from logic are listed in Appendix 4, and selected theorems from plane Euclidean geometry are listed in Appendix 5. The first 28 theorems (propositions) of Euclid are listed in Appendix 6.

The Greek geometry was *synthetic*, which means it did not use coordinates for numbers as in analytic geometry. Really significant advances over the synthetic geometry of the Greeks were made only with the invention of analytic geometry (about 1637) and its subsequent use as a tool in modern analysis. While analytic geometry is not the dominant theme of this text, coordinates of points are used as an alternative to the synthetic approach when convenient.

The title, *Modern Geometries*, with emphasis on the plural, should prepare you for finding out much more about the existence of not just one, but an infinitude, of geometries. As the word *modern* in the title implies, the major emphasis is on newer geometries—those that have been developed since 1800. The emergence of modern algebra, with its theory of groups, and the introduction of axiomatics into algebra paved the way for Felix Klein's classification of geometries in 1872. The basic concept of transformations needed to understand this classification is discussed in Chapter 2. Many geometries can be explained using the basic idea of transformations. In each case, mathematicians are interested in properties that remain the same when sets of points are changed in some way. The geometries in Chapters 2, 6, 7, and 8 are all classified by means of transformations.

The latter part of the nineteenth century witnessed a revival of interest in the classical geometry of the circle and the triangle, with the result that the Greek geometry was extended by many significant additions (Chapters 4 and 5). Projective geometry (Chapter 7) was invented about 1822; material on non-Euclidean geometry (Chapter 9) was in print by about 1830. Inversive geometry (Chapter 6) was developed about the same time.

During the twentieth century, studies in the axiomatic foundations of geometry and the finite geometries (Chapter 1), the geometry of convexity (Chapter 3), and geometric topology (Chapter 8) have all been added to the great body of geometry that is relatively independent of analysis. A recent development, largely since 1975, has been the application of many ideas from geometry in the new applied area of computer graphics. Other recent developments include insertion of material on fractals, tessellations, and geometric probability in many geometry books.

Even this brief sketch of some of the major steps in the history of geometry should convince you that a discussion of modern geometries must deal with many different kinds of geometry. This dominant theme should be remembered each time a new geometry is encountered. It is an acknowledgment of the diversity of mathematical systems titled "geometry" that distinguishes a book on modern geometries from a traditional college geometry text of a quarter century ago, which concentrated only on a restudy and direct extension of the Euclidean geometry of the high school. Still other geometries might have been included. For example, in the early nineteenth century, Gauss, Riemann, and others applied calculus and analysis to geometry, spawning a specialty known as differential geometry. This approach was very fruitful, and led to results that have been crucial to the development of modern physics and cosmology.

Computer graphics is the use of a computer to produce pictorial output. Aspects of this production include defining the data points, manipulating the figure, and presenting it. The use of animation and the depiction of fractals are extensions of these ideas. Computer graphics solves the problem of creating a particular picture or pictures on the computer screen. Computer graphics may be studied using a small personal computer with graphics capability, but there are also complex and expensive computers especially designed for computer graphics use. Applications of computer graphics can enrich regular geometry texts and courses so that students see this important application at an earlier age.

The concepts of geometry such as points, lines, and slope are modified when used in computer graphics. For points, the use of integral coordinates on the computer screen means pictures composed of *pixels*, the smallest location on the screen that can be addressed. The computer graphics screen is a grid of pixels, though they often are called *dots* or *points* in computer graphics. A graphics picture is created by selecting and energizing some pattern of pixels.

The orientation and scale of axes are concepts that may be used differently in computer graphics than in high school geometry. For example, the positive y-axis may be directed downward on some computer screens, so that the origin is in the upper left hand corner. Because the horizontal–to–vertical ratio of pixels may be about 2 to 1 instead of 1 to 1, the effect on graphing is to change the looks of figures. For example, a line with slope of 1 and with positive y-axis pointed downward may look like Figure 1.3a. Circles may look like ellipses, and figures that are reflected about an axis may not look congruent. Although it is possible to make changes in the programs to compensate for this "distortion" on the graphics screen, geometry students should expect it and should learn to accept various representations as correct. One advantage of this application is to convince students of geometry that trying to prove something from a picture alone is dangerous.

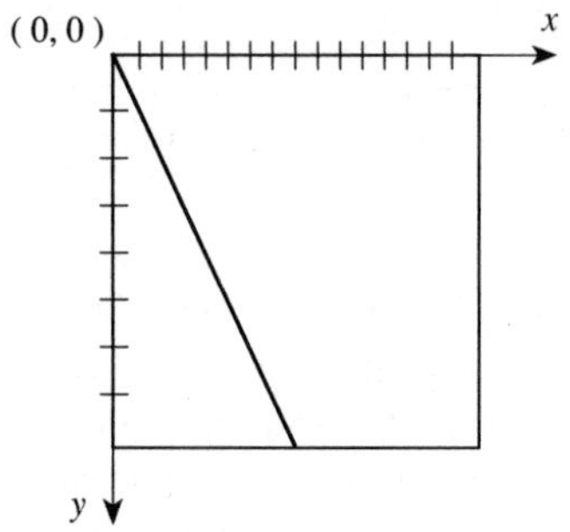

Figure 1.3a

Figure 1.3b

A specific application of computer graphics that takes advantage of the "ragged" appearance of lines and some curves in earlier inexpensive computers is the Moiré pattern, an example of which is shown in Figure 1.3b. These graphic designs simulate art produced by slight variations in position of a basic figure.

▲ EXERCISES 1.1 ▲

(Answers to selected exercises are given at the back of the text.)

1. Verify that the Pythagorean theorem $a^2 + b^2 = c^2$, holds for the primitive triple 3, 4, 5.
2. Find the perimeter of a right triangle with shorter sides measuring 5 and 12 units.
3. Find the perimeter of a right triangle with shorter sides measuring 9 and 12 units.
4. Why could a right triangle with integral lengths for sides not have shorter sides measuring 5 and 10 units?
5. Use the measurements of Eratosthenes to find the approximate diameter of the earth.

6. Use the measurements of Eratosthenes to find the approximate difference in angle of elevation of the sun at two places 1000 km apart in a north-south direction, both north of the equator.

For Exercises 7 and 8, assume that the method of Eratosthenes could be used for other planets.

7. If the distance were 300 miles and the angle difference 6°, what would be the circumference of the planet?
8. If the distance were 400 miles and the angle difference 8°, what would be the circumference of the planet?

For Exercises 9–14, answer true or false; then explain what is wrong with each false statement.

9. High school geometry owes more to the ancient Egyptians than to the ancient Greeks.
10. Euclid used the word *postulate* for an assumption confined to one particular subject.
11. Definitions must use only words that have previously been defined, not undefined terms.
12. Analytic geometry was invented before the development of finite geometrics.
13. The latter part of the nineteenth century witnessed a revival of interest in the classical geometry of the circle and the triangle.
14. Traditional college geometry of a quarter of a century ago included the study of more different geometries than are included today.

Throughout the text, exercises preceeded by the symbol ▷ are investigative-type projects that may require checking references, doing research, making generalizations, and other supplementary activities to extend the coverage of the section. Some investigative-type exercises are designated by ⏩ to show they are particularly appropriate for use with cooperative or small-group learning situations.

▷ 15. How could instruments on the moon be used to make precise measurements of the earth? Try to discover an answer before consulting a reference.

▷ 16. Read in a history of mathematics text about the geometry of ancient Babylonia and Egypt; then work several geometric problems found in the papyruses.

⏩ 17. Read more about the subject of geodesy and discuss it as an application of geometry.

▲ 1.2 Introduction to Finite Geometries

The Euclidean plane has an infinitude of points and lines in it, and a rich collection of theorems that continues to increase over the years. By contrast, "miniature" geometries have just a few axioms and theorems and a finite number of elements. These geometries are *finite geometries*, and the purpose of including them here is that they provide excellent opportunities for the study of geometries with a simple structure.

All the geometries studied in this text have a finite number of axioms and a finite number of undefined terms. Thus, those features do not make a geometry finite. Instead, a finite geometry has a finite number of elements—that is, points or lines or "things to work with." For the geometries studied in this chapter, these elements can be considered points and lines.

Historically, the first finite geometry was a three-dimensional geometry, each plane of which contained seven points and seven lines. The modernity of finite geometries is emphasized by the fact that Gino Fano explored this first finite geometry in 1892, although some ideas can be traced back to von Staudt (1856). It was not until 1906 that finite projective geometries were studied by Veblen and Bussey. Since that time, a great many finite geometries have been or are being studied. Many sets of points and lines that were already familiar figures in Euclidean geometry were investigated from this new point of view. Several of the finite geometries are an integral part of projective geometry, and a knowledge of the finite geometries of Chapter 1 will help in the study of some of the basic set of points and lines used in Chapter 7. However, at the present time it is quite possible for a mathematics major to graduate without ever encountering finite geometries, although it is also true that finite geometries are being used increasingly as enrichment topics and extension units at the high school level. Finite geometries also find a practical application in statistics.

All the finite geometries in this chapter have *point* and *line* as undefined terms. The connotation of *line* is not the same in finite geometry as in ordinary Euclidean geometry, however, since a line in finite geometry is assumed to have more than one, but only a finite number, of points.

The first simple finite geometry to be investigated, called a *three-point geometry* here for identification, has only four axioms:

Axioms for Three-Point Geometry

1. There exist exactly three distinct points in the geometry.
2. Two distinct points are on exactly one line. (Existence & Uniqueness)
3. Not all the points of the geometry are on the same line.
4. Two distinct lines are on at least one point.

Assume that the words *point*, *line*, and *on* are undefined terms. In Axiom 4, the two lines with a point in common are called *intersecting lines*.

Try to discover the answers to these questions before reading further:

- What kinds of figures or models could be drawn to represent the geometry?

- How many lines are in the geometry?
- Can any theorems be proved for the geometry?
- What representations are possible for the geometry, other than those with points and lines?
- Which properties of Euclidean geometry continue to hold in the three-point geometry, and which do not?

Not all of these questions can be answered completely for each finite geometry studied, but the questions do illustrate the nature of inquiry about a geometry based on an axiomatic system. It will be immediately helpful to give a partial answer to the first question.

Example ➤ Figure 1.4 shows four different specific representations for Three-Point Geometry. Verify the fact that all four axioms hold for each figure.

While the sets of points and lines in Figure 1.4 are such that all the axioms of the three-point geometry hold, there is still the possibility that the geometry might have additional lines not shown. This matter is settled by proving two theorems. First, comparing the wording of Axioms 2 and 4, we need to determine whether two distinct lines might be on more than one point.

We need to prove that two distinct lines are on exactly one point. This statement can be rewritten in the form of a conditional: If two lines are distinct, then they are on exactly one point. The proof will be accomplished by means of an *indirect argument*. Begin by considering two distinct lines. Then assume that these two lines are on more than one point. If a contradiction can be established, the assumption must be false, and the desired conclusion follows.

The proofs throughout this text are written in paragraph form. Each step in a proof can be supported by stating a previous axiom, theorem, or definition if necessary. When proving theorems as exercises, you should write the proofs in a style similar to that found in the text.

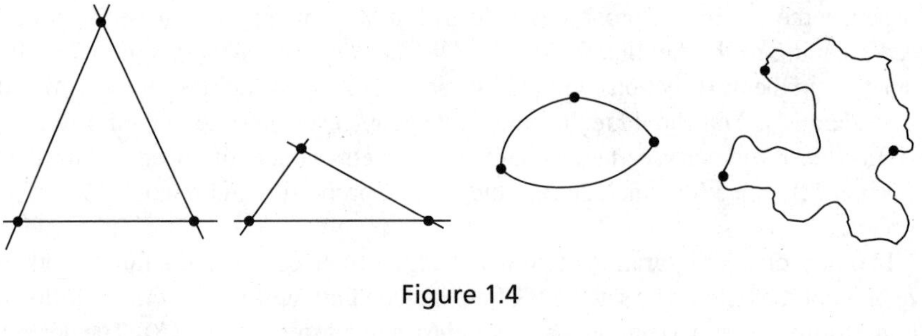

Figure 1.4

THEOREM 1.1 ➤ Two distinct lines are on exactly one point.

Proof: By Axiom 4, two distinct lines are on at least one point. Assume that two lines lie on more than one point. If lines l and m lie on points P and Q, then Axiom 2 is contradicted, since P and Q would be on two lines, l and m.

The exact number of lines in the three-point geometry can now be determined.

THEOREM 1.2 ➤ The three-point geometry has exactly three lines.

PROOF: From Axiom 2, each pair of points is on exactly one line. Each possible pair of points is on a distinct line, so the geometry has at least three lines. Suppose there is a fourth line. From Axiom 1, there are only the three points in the geometry. This fourth line must have a point in common with each of the other 3 lines, by Theorem 1.1, so that it must also be on two of the three points, which contradicts Axiom 2. Therefore, there can be no more than three lines in the geometry.

While *point* and *line* have been used as undefined terms in this first finite geometry, other words could be substituted to give an equally meaningful interpretation of the structure. For example, *tree* could be substituted for *point*, and *row* for *line*, so that the postulates would read as follows:

- ▲ There exist exactly three distinct trees.
- ▲ Two distinct trees are on exactly one row.
- ▲ Not all trees are on the same row.
- ▲ Two distinct rows have at least one tree in common.

Other interpretations could be found by using pairs of words such as *beads* and *wires*, *students* and *committees*, or *books* and *libraries*.

In the finite geometry under consideration, it should be evident from an examination of the axioms and Figure 1.4 that such Euclidean concepts as length of a segment, measure of an angle, and area—in fact all concepts concerning measurement—no longer apply. In this geometry, if a triangle is defined as three distinct lines meeting by pairs in three distinct points, not all collinear, then one and only one triangle exists. Another feature is that the concept of parallel lines, defined as two lines with no points in common, does not apply, since each two lines meet in a point. The familiar ideas of congruence also have no meaning in this geometry. Even though you are far more familiar with Euclidean geometry than with finite geometry, this consideration of whether the Euclidean properties hold in a new geometry will nevertheless give added meaning to the familiar concepts. The introduction of more significant finite geometries in the next sections will provide additional opportunities for the same kind of consideration. You should realize that finite geometries, in spite of their differences from Euclidean geometry, certainly qualify as geometries under the meaning introduced in Section 1.1, since they include the study of the properties and relationships of sets of points.

In order for a configuration of points and lines to be considered a finite geometry, several properties should be satisfied. These include finiteness (1 and 2 in the following list), uniformity (3 and 4), uniqueness (5 and 6), and existence (7 and 8). The desirable properties are:

1. The number of points is finite.
2. The number of lines is finite.
3. Each line is on the same number s of points, where $s \geq 2$.

4. Each point is on the same number t of lines, where $t \geq 2$.
5. Each pair of distinct points is on at most one line.
6. Each pair of distinct lines is on at most one point.
7. Not all points are on the same line.
8. There exists at least one line.

The simple geometry of this section does satisfy each of the properties (See Exercise 19I). The next sections contain many interesting finite geometries that should also be checked to see that each of these eight desirable properties is satisfied.

▲ EXERCISES 1.2 ▲

For the three-point geometry:

1. Draw a pictorial representation different from those in Figure 1.4.

For Exercises 2 and 3, tell why each representation in Figure 1.5 cannot be used for the three-point geometry.

2.

3.

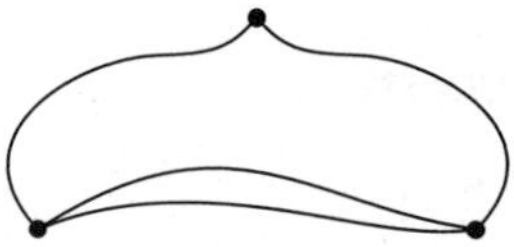

Figure 1.5

4. Rewrite the axioms, using the words *book* for *point* and *library* for *line*.
5. Rewrite the axioms, using the words *student* for *point* and *committee* for *line*.
6. Does each pair of lines in the geometry intersect in a point in the geometry?
7. Through a point not on a given line, is there at least one line not intersecting the given line?
8. For each two distinct points, does exactly one line exist on both of them?
9. Through a point not on a given line, how many lines are parallel to the given line?
10. Exactly how many points are on each line?
11. Must lines be straight in the Euclidean sense?
12. Could three lines all contain the same point?
13. Do any squares exist?

14. Prove that a line cannot contain three distinct points.

15. Prove that there exists a set of two lines containing all the points of the geometry.

16. Prove that there exists a set of two points such that each line of the geometry lies on at least one of these points.

▷ 17. Discuss whether or not each of the following concepts from Euclidean geometry also appears as a concept in the three-point geometry: angle, congruence, circle, area, intersection, similarity, hexagon.

▷▷ 18. Create other representations not shown in Figure 1.5, and explain why each of your representations *cannot* be used for the three-point geometry.

▷ 19. Check the three-point geometry to see that each of the eight desirable properties is satisfied.

▲ 1.3 Four-Line and Four-Point Geometries

The next finite geometry to be considered, like the geometry of Section 1.2, also has *point*, *line*, and *on* as undefined terms. The following three axioms completely characterize the geometry, called a *four-line geometry*.

Axioms for Four-Line Geometry

1. There exist exactly four lines.
2. Any two distinct lines have exactly one point on both of them.
3. Each point is on exactly two lines.

Axiom 1 is an *existence* axiom, because it guarantees that the geometry is not the empty set of points. The other axioms are *incidence* axioms, dealing with points on lines and lines on points. Before reading on, try to draw diagrams of points and lines that will satisfy all three axioms. Also, try to determine the total number of points.

THEOREM 1.3 ➤ The four-line geometry has exactly six points.

Proof: From Axiom 1, there are six pairs of lines. The number six is obtained as the combination of four things taken two at a time. A notation used for combinations is

$$\binom{4}{2}$$

The general formula for the combination of n things taken r at a time is

$$\binom{n}{r} = \frac{n!}{(n-r)!r!} \quad \text{and} \quad \binom{4}{2} = \frac{4!}{2!2!} = 6$$

By Axiom 2, each pair of lines has exactly one point on both of them. Suppose that two of these six points are not distinct. That would be a contradiction of Axiom 3 because each point would be on more than two lines. Also, by Axiom 3, no point could exist in the geometry other than those six on the pairs of lines.

THEOREM 1.4 ➤ Each line of the four-line geometry has exactly three points on it.

PROOF: By Axiom 2, each line of the geometry has a point in common with each of the other three lines, and all three of these distinct points are on the given line. Suppose there were a fourth point on one line. Then by Axiom 3, it must also be on one of the other lines. But this is impossible because the other three lines already determine exactly one point with the given line, and by Axiom 2, they can only determine one. Thus, each line of the geometry has exactly three points on it.

Figure 1.6 shows two diagrams that can be used to represent this finite geometry of four lines and six points. An examination of Figure 1.6 will lead to further inquiries about the four-line geometry. For example, the following questions are typical of those that should be answered:

- Do each two points of the geometry lie on a line?
- How many triangles exist in the geometry (all three sides must be lines of the geometry)?
- Does the geometry have examples of parallel lines (lines with no point in common)?

These questions are considered in the exercises at the end of the section.

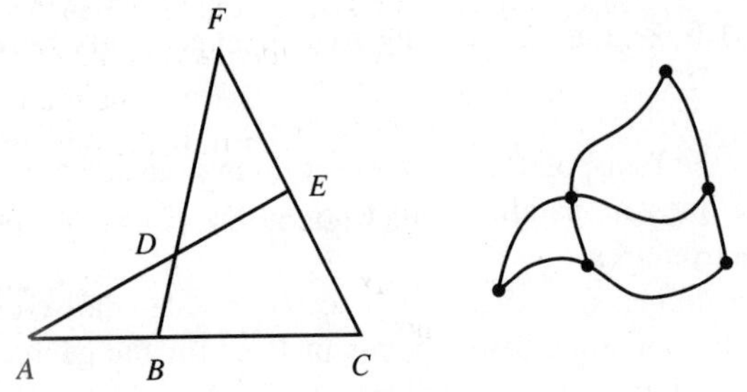

Figure 1.6

A new set of axioms can be generated using the axioms of the four-line geometry and the notion of plane dual. The *plane dual* of a statement is formed by exchanging the words *point* and *line* and making other necessary changes in the English. Writing the plane dual of each axiom for the previous four-line geometry results in a set of axioms for a *four-point geometry.*

Axioms for Four-Point Geometry

1. The total number of points in this geometry is four.
2. Any two distinct points have exactly one line on both of them.
3. Each line is on exactly two points.

Two possible representations for this geometry are shown in Figure 1.7.

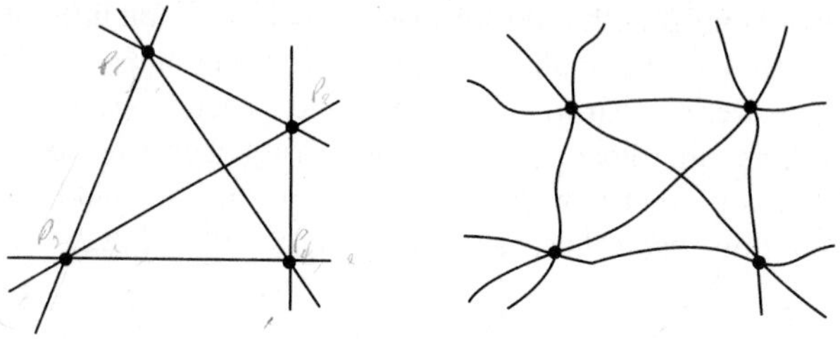

Figure 1.7

The lines meet only where points are indicated, not just where they appear to cross in the picture.

The concept of planar duality relates the four-point geometry and the four-line geometry. The plane dual of any valid theorem in one geometry becomes a theorem in the other, so that it is possible to gain more information about the new geometry rather easily. Thus, the plane duals of Theorems 1.3 and 1.4 become theorems for the four-point geometry.

THEOREM 1.5 ➤ The four-point geometry has exactly six lines.

THEOREM 1.6 ➤ Each point of the four-point geometry has exactly three lines on it.

Lines exist in the four-point geometry that do not have one of the four points in common, so these lines are parallel. Other properties of the four-point geometry are investigated in the exercises.

For each of the finite geometries so far, one axiom stated the exact number of points on a line or gave the total number of points or lines for the geometry. Without this limiting axiom, the set of axioms might have resulted in a geometry with an infinitude of points and lines. Indeed, most of the axioms for finite geometries are also axioms in Euclidean geometry. For example, Axioms 2 and 3 for the finite geometry of three points hold in ordinary Euclidean geometry, and even Axiom 4 holds when the lines are not parallel.

▲ EXERCISES 1.3 ▲

1. Write the plane dual of the axioms for the three-point geometry of Section 1.2.

Complete Exercises 2–9 for the four-line geometry:

2. Draw a representation for this geometry that is different from those shown in Figure 1.6.
3. Which axioms are also true statements in Euclidean geometry?
4. Rewrite the set of axioms for this geometry, using *student* for *point* and *committee* for *line*.
5. Do each two points of the geometry lie on a common line?
6. How many triangles exist in the geometry in which all three sides are lines of the geometry?
7. How many other lines are parallel to each line?
8. Use combinations to predict that there exist sets of three points in the geometry that do not all three lie on the same line of the geometry.
9. Prove that a set of two lines cannot contain all the points of the geometry.
10. Prove that there exists a pair of points in the geometry not joined by a line.

Complete Exercises 11–20 for the four-point geometry:

11. Draw another model for this geometry different from those shown in Figure 1.7.
12. Which axioms are also true statements in Euclidean geometry?
13. Rewrite the set of axioms for this geometry, using *tree* for *point* and *row* for *line*.
14. Do each two lines of this geometry determine a point?
15. If the points are A, B, C, D, name all sets of parallel lines.
16. How many other lines of the geometry are parallel to each line?
17. Prove, without using the idea of duality, that the geometry includes exactly six lines.
18. Prove, without using the idea of duality, that each point of the geometry is on exactly three lines.
19. Prove that a set of two lines exists that contains all the points of the geometry.
20. Without using duality, prove that there exists a pair of lines in the geometry without a point in common.
21. ▷ For each of these concepts from Euclidean geometry, explain why the concept appears, or does not appear, in (a) the four-line geometry and (b) the four-point geometry: angle, congruence, circle, area, intersection, similarity, hexagon, quadrilateral.

22. Draw representations that *cannot* be used for the geometries of this section and explain why each is inappropriate.

23. Check to see that each of the eight desirable properties is satisfied for the geometries of this section.

24. Let students stand in a group pattern to represent the points in the geometries of this section. Then they or others may check the various axioms and theorems.

1.4 Finite Geometries of Fano and Young

Fano originally studied a three-dimensional geometry, but the cross section formed by a plane passing through his configuration yields a plane finite geometry that is presented here. A complete set of axioms for this geometry follows.

Axioms for Fano's Geometry

1. There exists at least one line.
2. Every line of the geometry has exactly three points on it.
3. Not all points of the geometry are on the same line.
4. For two distinct points, there exists exactly one line on both of them.
5. Each two lines have at least one point on both of them.

For this geometry, *point* and *line* are undefined elements, and *on* is an undefined relation. In finite geometries as well as in ordinary Euclidean geometry, various expressions can be used for the same idea. For example, all of these state the same relationship:

A point is *on* a line.

The line *contains* the point.

The line *goes through* the point.

An almost immediate consequence of the axioms is:

THEOREM 1.7 ➤ Each two lines have exactly one point in common.

PROOF: By Axiom 5, two lines have at least one point on both of them. In a manner similar to the proof of Theorem 1.3, suppose that one pair of lines has two points in common. The assumption that they have two distinct points in common violates Axiom 4, because then the two distinct points would have two lines containing both of them.

It is more difficult to guess the number of points and lines for the geometry of Fano than for the previous geometries we have examined. Try to guess before reading the development leading to the next theorem.

From Axioms 1 and 2, there are at least three points in the geometry, while from Axiom 3 there is at least a fourth point, as symbolized in Figure 1.8a. By Axiom 4, there must be lines joining this fourth point and each of the existing points (Figure 1.8b), and by Axioms 4 and 5, there must be lines joining points 1, 6, 7; points 3, 6, 5; and points 5, 2, 7 (Figure 1.8c). Thus, the geometry of Fano contains at least seven points and seven lines.

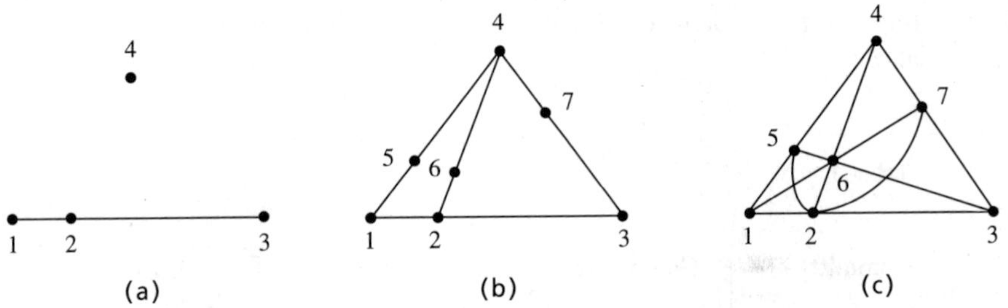

Figure 1.8

The fact that there are exactly seven points in Fano's geometry can be established by an indirect argument, since the assumption of an eighth point leads to a contradiction, as we will now show. Assume there is an eighth point, and consider, for example, as in Figure 1.9, the intersection of the line through points 1, 8 and the line 3, 7, 4. (The notation "line 3, 7, 4" means the line containing points 3, 7, and 4.) Axiom 5 requires that lines 1, 8 and 3, 7, 4 have a point of intersection. The point of intersection required by Axiom 5 cannot be point 3, 7, or 4, since that would violate Axiom 4. Thus, it must be a ninth point, but that violates Axiom 2. The assumption of an eighth point has led to a contradiction and must be rejected. The result is the following theorem:

THEOREM 1.8 ➤ Fano's geometry consists of exactly seven points and seven lines.

The models for Fano's geometry in Figures 1.8c and 1.9 show lines that appear to cross without actually having a point in common. They are not Euclidean lines, and the distinct points on them need to be clearly shown in drawings to avoid confusion.

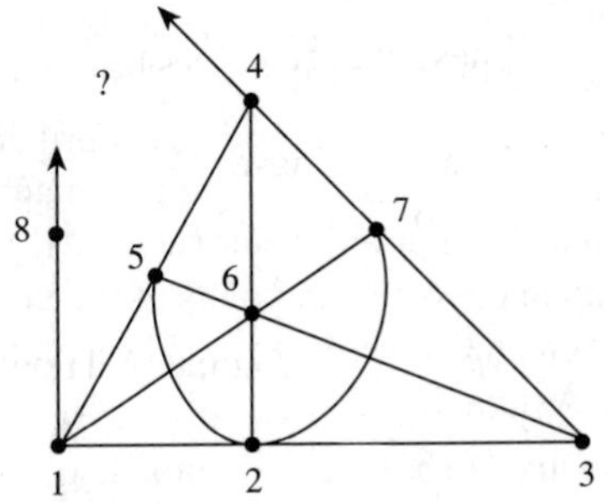

Figure 1.9

Consider a rewriting of the set of axioms for Fano's geometry with the word *point* replaced by *student* and the word *line* by *committee*. A table may be used to represent this finite system. For example, Table 1.1 shows committees in vertical columns. Substitution of numbers for the student names would show that Fano's geometry itself could be represented by the same sort of table. You should be able to set up a correspondence between each column in this table and each line in Figure 1.8c, noting that the students can be matched in alphabetical order with the points in numerical order.

Table 1.1

Committee 1	Committee 2	Committee 3	Committee 4
Alice Brad Cathy	Alice Dale Ellie	Alice Frank Greg	Brad Dale Frank
Committee 5	**Committee 6**	**Committee 7**	
Cathy Ellie Frank	Brad Ellie Greg	Cathy Dale Greg	

The connections between the theory of combinations and the number of points and lines in finite geometries are profitable to explore. For example, it might appear at first glance that a quick way to determine the number of lines in Fano's geometry would be to count all possible combinations of seven things taken three at a time, since each line has three points. However, there are 35 combinations of seven things taken three at a time, and there are only seven lines in the geometry. A study of Figure 1.9 (or 1.8c) will make it possible to reconcile this difference. Consider, for example, points 1 and 5 and the line they determine. There are five other possible points—2, 3, 4, 6, 7—that could be matched with the two given points to form lines, and all of these possibilities are counted in the 35. But only one of these points, point 4, is actually on the line through points 1 and 5; hence, only $\frac{1}{5}$ of the actual number of possibilities results in lines. Since $\frac{35}{5} = 7$, there are a total of seven lines. Further insight can be gained by asking why the other four points—2, 3, 6, 7—are not matched with points 1 and 5 to determine lines of the geometry. For any point such as 1, there are three lines of the geometry on it. These contain the other six points of the geometry (two on each line). None of the pairs of points on one of these lines can be matched with a point on another line without contradicting Axiom 4.

A different finite geometry can be obtained from Fano's geometry by a modification of the last axiom. The new geometry, called *Young's geometry*, has the first four axioms of Fano's geometry along with the following substitute axiom for Axiom 5:

5. If a point does not lie on a given line, then there exists exactly one line on that point that does not intersect the given line.

Axiom 5 shows that the ordinary Euclidean concept of parallel lines applies. Young's geometry is a finite geometry of nine points and twelve lines.

▲ EXERCISES 1.4 ▲

For Exercises 1–5, tell which statements are true for the geometry of Fano.

1. The geometry has exactly the same number of points as it has lines.
2. Each pair of lines in the geometry intersects in a point in the geometry.
3. Each two distinct points of the geometry determine a unique line in the geometry.
4. Changing one axiom results in Young's geometry.
5. The geometry contains at least eight lines.
6. Rewrite the set of axioms for the geometry of Fano, using *book* for *point* and *library* for *line*.
7. Which axioms in the geometry of Fano are also true statements in Euclidean geometry?
8. Using Figure 1.8c, name all the triangles in the geometry of Fano having point 4 as one vertex.
9. Use combinations to predict that there exist sets of three points in the geometry that do not all three lie on the same line of the geometry.
10. Explain why $\binom{7}{2}$ does not give the number of lines in Fano's geometry, but can be used to derive the correct number of lines.
11. For Fano's geometry, prove that each point lies on exactly three lines.
12. For Fano's geometry, prove that the lines through any one point of the geometry contain all the points of the geometry.
13. For Fano's geometry, prove that, for any pair of points in the geometry, there exist exactly two lines not containing either point.
14. For Fano's geometry, prove that, for a set of three lines not all containing the same point, there exists exactly one point in the geometry not on any of the three lines.
15. Sketch a model for a geometry that satisfies Axioms 1 and 2 of Fano's geometry but not Axiom 3.
16. Sketch a model for a geometry that satisfies Axioms 1 and 3 of Fano's geometry but not Axiom 2.
17. Draw a model for Young's geometry.
18. Prove that Young's geometry includes at least nine points.

19. For Young's geometry, prove that two lines parallel to a third line are parallel to each other. *Hint*: Use Axiom 5 and prove by contradiction.

▷ **20.** Write the plane dual for each axiom of Fano's geometry; then draw a model for this new set of axioms. Compare it with the model for the original set. Then read the beginning paragraphs of Section 1.6.

▷ **21.** Check to see that each of the eight desirable properties is satisfied by the geometries of this section.

▲ 1.5 Finite Geometries of Pappus and Desargues

The finite geometry of Pappus arises from a Euclidean geometry theorem called the theorem of Pappus. Figure 1.10 illustrates the theorem, which was discovered and proved by Pappus of Alexandria about A.D. 340. The theorem is stated here without proof.

THEOREM 1.9. ➤ *Theorem of Pappus.* If A, B, and C are three distinct points on one line and if A', B', and C' are three different distinct points on a second line, then the intersections of $\overleftrightarrow{AC'}$ and $\overleftrightarrow{CA'}$, $\overleftrightarrow{AB'}$ and $\overleftrightarrow{BA'}$, and $\overleftrightarrow{BC'}$ and $\overleftrightarrow{CB'}$ are collinear.

The notation $\overleftrightarrow{AC'}$ means "the line containing points A and C'." Points are called collinear if they lie on the same line. There are exceptions to the theorem if some of the lines are parallel (in the ordinary Euclidean sense), but attention will be restricted here to those circumstances in which the lines actually intersect. The theorem of Pappus involves nine distinct points, lying by threes on three lines. There are nine points and nine lines in Figure 1.10, and these may be studied as a finite geometry with the familiar Euclidean properties no longer evident and with only the following axioms.

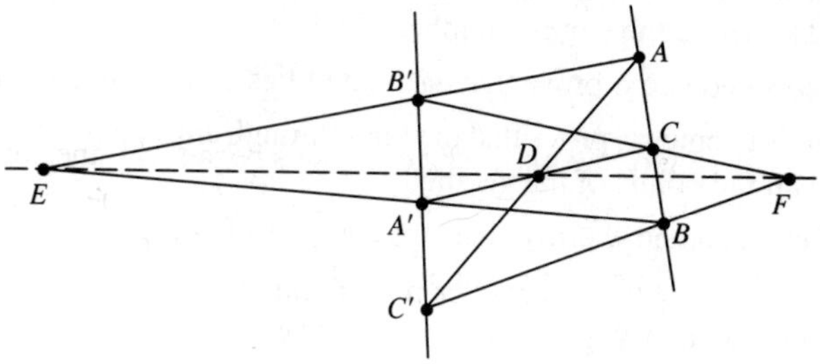

Figure 1.10

Axioms for the Finite Geometry of Pappus

1. There exists at least one line.
2. Every line has exactly three points.
3. Not all points are on the same line.

4. If a point is not on a given line, then there exists exactly one line on the point that is parallel to the given line.
5. If P is a point not on a line, there exists exactly one point P' on the line such that no line joins P and P'.
6. With the exception in Axiom 5, if P and Q are distinct points, then exactly one line contains both of them.

Study Figure 1.10 to see that each line of the finite geometry seems to have exactly two lines parallel to it. Of course, this observation does not constitute a proof.

The next theorem gives an additional property that can be proved from the axioms.

THEOREM 1.10 ➤ Each point in the geometry of Pappus lies on exactly three lines.

PROOF: Let X be any point (Figure 1.11). By Axiom 3, there is a line not containing X. From Axiom 2, this line contains points A, B, and C. By Axiom 5, X lies on lines meeting two of the points on the given line, say B and C. By Axiom 4, there is exactly one line through X parallel to $\overleftrightarrow{BC}$ so that there are at least three lines on X. However, there cannot be a fourth line through X. By Axiom 5, there is no line connecting X and A, and by Axiom 4, there is no other line through X not meeting $\overleftrightarrow{BC}$.

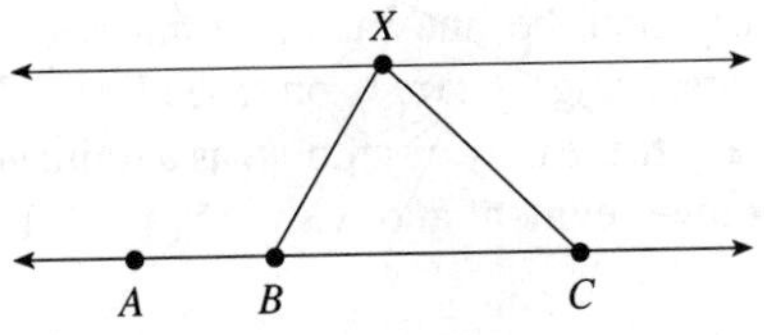

Figure 1.11

Although discussion of the geometry of Pappus began with the completed figure, it can be proved from the axioms alone (see Exercises 1.5) that the geometry of Pappus does in fact consist of exactly nine points and nine lines. Other versions of the axioms can be found using substitute words for points and lines, but it is also instructive to arrange the information in the form of a table. In Table 1.2, the notation from Figure 1.10 is used, with each vertical column representing a line of the geometry.

Table 1.2

A	B'	D	A	A	F	F	B	C
B	A'	E	D	B'	C	B	A'	D
C	C'	F	C'	E	B'	C'	E	A'

It is important to understand that a table such as Table 1.2 could be used, rather than a set of axioms, to give the initial representation of a geometry. In this case, most proofs are very simple, since they depend just on checking all possible cases in the table. For example, the table can be checked directly to see that each point lies on exactly three lines. One of the major reasons for studying finite geometries is that every possible case can be listed and counted.

Another example of a modern finite geometry that is actually a study of a famous set of points from Euclidean geometry is the *finite geometry of Desargues*. Several concepts must be introduced so they can be used in the study of this new geometry. Triangles ABC and $A'B'C'$ of Figure 1.12 are *perspective from point* P. This means that point P is the common point (the point of concurrency) of the three lines joining corresponding vertices. Thus, P lies on $\overleftrightarrow{AA'}$, $\overleftrightarrow{BB'}$, and $\overleftrightarrow{CC'}$. According to Desargues's theorem, studied in more detail in Chapter 7, two triangles perspective from a point are also *perspective from a line*. If triangles are perspective from a line, corresponding sides of the triangles meet at points on this line. The line of perspectivity in Figure 1.12 contains points R, S, T. For example, corresponding sides $\overleftrightarrow{AB}$ and $\overleftrightarrow{A'B'}$ meet at T, a point on this line.

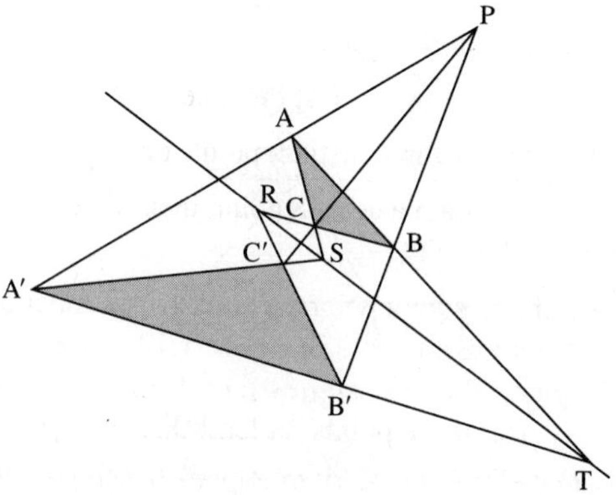

Figure 1.12

Further study of Figure 1.12 shows a total of ten labeled points on ten lines, with three points on each line and three lines on each point. These ten points and ten lines are the elements in the finite geometry of Desargues. One further concept is necessary to understand the axioms: a method of setting up a correspondence between points and lines is needed. If a point is the point of perspectivity for two triangles, and if a line is the line of perspectivity for the same two triangles, then the point is called the *pole* of the line and the line is called the *polar* of the point. In the finite geometry of Desargues, no line joins a pole and a point on a polar. For example, in Figure 1.12 the point P is the pole of the line containing points R, S, T. This fact leads to the following formal definitions, which do not use the concepts of point or line of perspectivity:

DEFINITION ➤ The line l in the finite geometry of Desargues is a polar of the point P if there is no line connecting P and a point on l.

DEFINITION ➤ The point P in the finite geometry of Desargues is a pole of the line l if there is no point common to l and any line on P.

The intuitive development has led to the axioms, but now the axioms become the beginning ideas; nothing can be proved from Figure 1.12—proof must come from the axioms themselves.

See if you can identify various poles and polars in Figure 1.12. For example, if point T is taken as the pole, where are the polars? This question can be answered by considering the lines through T and noting that there is exactly one line in the figure that has no points in common with these lines. That line is $\overleftrightarrow{PC'} = \{P, C, C'\}$; hence, $\overleftrightarrow{PC'}$ is the polar.

Axioms for the Finite Geometry of Desargues

1. There exists at least one point.
2. Each point has at least one polar.
3. Every line has at most one pole.
4. Two distinct points are on at most one line.
5. Every line has exactly three distinct points on it.
6. If a line does not contain a certain point, then there is a point on both the line and any polar of the point.

The exact number of poles for a polar and the exact number of polars for a pole are not stated explicitly in the axioms. Study Figure 1.13. Assume that line p is a polar of P, since Axiom 2 guarantees the existence of at least one. By definition, no line through P contains any one of the points (at least three) on p, but this also says there is no point common to p and any line on P, so that P satisfies the definition of a pole of p. This information, along with Axioms 2 and 3 and Exercise 16 in Exercises 1.5, leads to the following theorems.

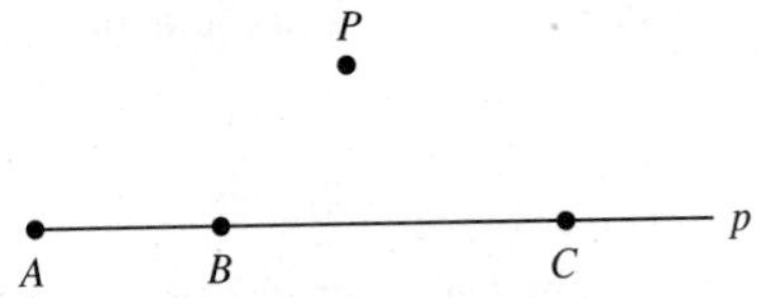

Figure 1.13

THEOREM 1.11 ➤ Every line in the geometry of Desargues has exactly one pole.

THEOREM 1.12 ➤ Every point in the geometry of Desargues has exactly one polar.

Parallel lines exist in the geometry of Desargues, but their properties are different from those of ordinary Euclidean parallels. For example, note in Figure 1.12 that three different lines can be drawn parallel to line R, C, B through point A' but that only one line can be drawn through A' parallel to line A, B, T.

It is worthwhile to study the axioms and the drawings for a geometry in order to make additional conjectures. Proofs of additional theorems for the finite geometry of Desargues appear as exercises. The theories that gave rise to the finite geometries of Pappus and Desargues are encountered as important concepts in projective geometry in Chapter 7.

▲ EXERCISES 1.5 ▲

1. Rewrite the set of axioms for the geometry of Pappus, using the words *tree* for *point* and *row* for *line*.
2. Which axioms in the geometry of Pappus are also true statements in Euclidean geometry?
3. For each point in the geometry of Pappus, how many other points in the geometry do not lie on a line through the given point?
4. In Figure 1.10, let each of the points A, B, and C represent point P in Axiom 5 for the geometry of Pappus. For each, name all the points corresponding to P' in the axiom.
5. Prove that there are at least two lines in the geometry of Pappus parallel to a given line.
6. Prove that the geometry of Pappus contains exactly nine points.
7. Prove that the geometry of Pappus contains exactly nine lines.
8. For each point in the geometry of Pappus, there are exactly how many points in the geometry that do not lie on a line through the given point? Prove that your answer is correct.
9. Prove that there are exactly two lines in the geometry of Pappus parallel to a given line.

Exercises 10–18 concern the finite geometry of Desargues.

10. Prepare a table to represent the geometry of Desargues, using the points as named in Figure 1.12 and letting each column of the table represent a line in the geometry.
11. In Figure 1.12, name the pole of:

    a. $\overleftrightarrow{RB}$ **b.** $\overleftrightarrow{AS}$

12. In Figure 1.12, name the polar of:

a. Point R **b.** Point S

13. Using Figure 1.12, name the pair of triangles in the Euclidean figure perspective from point T and from $\overleftrightarrow{PC'}$.

14. Which axioms in the geometry of Desargues are also true statements in Euclidean geometry?

15. In Figure 1.12, name all the lines parallel to $\overleftrightarrow{AS}$.

16. Prove that if point P is on the polar of point Q, then point Q is on the polar of point P.

17. Prove that two lines parallel to the same line are not parallel to each other.

18. Prove there is a line through two distinct points if and only if their polars intersect.

For Exercises 19I–22I, prepare a table like Table 1.2 for the geometry indicated.

▷ **19.** Three-point geometry

▷ **20.** Four-line geometry

▷ **21.** Four-point geometry

▷ **22.** Fano's geometry

▷ **23.** Check to see that each of the eight desirable properties is satisfied by the geometries of this section.

▲ 1.6 Other Finite Geometries

Fano's finite geometry has the special property of being *self-dual*, which means that the plane dual of each true statement is a true statement for the geometry. Actually, Fano discussed a particular set of finite geometries, each of which was self-dual. The general symbol for geometries of this special type is $PG(n, q)$. The letters PG stand for projective geometry. The number n is the dimension ($n = 2$ here/for plane geometries) and q must be a positive integral power of a prime number. The theory of projective geometry can be used to establish the following theorem, assumed here.

THEOREM 1.13 ➤ The general formula for the total number of points in $PG(n, q)$ is

$$\frac{q^{n+1} - 1}{q - 1}$$

For example, $PG(2, 25)$ is a projective geometry, but $PG(2, 15)$ is not. The geometry $PG(2, q)$ has $q + 1$ points on each line. Here, the number $q + 1$ is called the *order* of the finite plane. By duality, it also has $q+1$ lines on each point. The geometry $PG(2, 2)$ is Fano's geometry.

For the geometry of Fano,

$$\frac{q^{n+1}-1}{q-1}=\frac{2^3-1}{1}=7$$

If $q = 3$, then $PG(2, 3)$ is a new finite geometry that is self-dual. From Theorem 1.13, the total number of points is 13. $PG(2, 3)$ is a finite geometry of 13 points and 13 lines. This geometry has the same axioms as Fano's geometry, except that there are four points rather than three on every line. The geometry of Desargues is also an example of a self-dual geometry, but its structure is such that it is not a member of the family of geometries associated with Fano.

Additional development of the concept of duality is found in Chapter 7, and the reader will discover additional figures with sets of points and lines that may be used as the elements for finite geometries. In Chapter 7, the introduction of homogeneous coordinates for projective geometry will make it possible to investigate some finite geometries by assigning coordinates to the points. A different sort of finite geometry is introduced in Section 9.1, one that has a postulate characteristic of a non-Euclidean geometry.

The interested student can make up finite geometries of his or her own. Various examples of simple drawings that can be used for a finite geometry are shown in Figure 1.14. A few of the interesting features of each of these figures are mentioned in the exercises that follow.

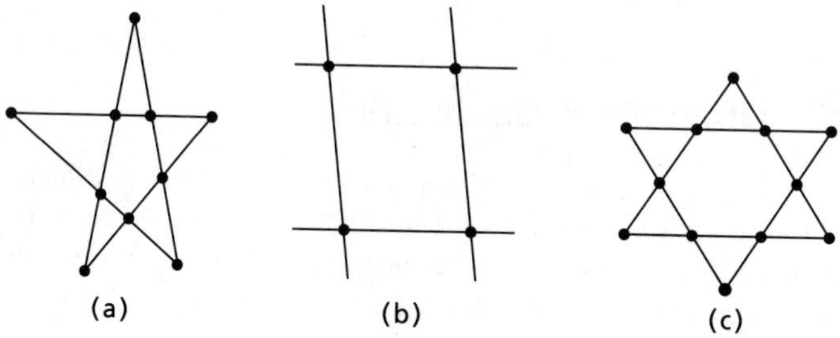

Figure 1.14

▲ EXERCISES 1.6 ▲

1. Prepare a table for the finite geometry $PG(2, 3)$, using the numbers 1 to 13 for the points and showing the points on a line as one column of the table.

Use the table in Exercise 1 for Exercises 2–5. For Exercises 3–5, answer true or false.

2. There are how many lines on each point?

3. For a set of four distinct points of the geometry, there is exactly one line of the geometry not containing any of those points.
4. All the lines through one point of the geometry contain each point of the geometry.
5. There exists a set of three distinct points in the geometry that do not have a line in common.
6. For the geometry $PG(2, 3)$, use combinations to show that there exist sets of four points in the geometry that do not all four lie on the same line of the geometry.
7. Which of these symbols represent self-dual geometries?

   a. $PG(2, 4)$ **b.** $PG(2, 5)$

   c. $PG(2, 6)$ **d.** $PG(2, 7)$
8. Find the number of points and lines, where possible, for the geometries of Exercise 7.

For the geometry of Figure 1.14a:

9. Each line has how many points on it?
10. The geometry consists of how many points and how many lines?
11. Each line has how many other lines parallel to it?
12. Does each pair of lines in the geometry intersect in a point in the geometry?

For the geometry of Figure 1.14b:

13. Each line has how many points on it?
14. The geometry consists of how many points and how many lines?
15. Each line has how many other lines parallel to it?
16. For each two distinct points, does there exist exactly one line on both of them?

For the geometry of Figure 1.14c:

17. The geometry consists of how many points and how many lines?
18. Each line has how many other lines parallel to it?
19. For each two distinct points, does there exist exactly one line on both of them?
20. Create a finite geometry of your own by representing it as in Figure 1.14. Be sure that it satisfies each of the general properties noted in Section 1.2. Answer each of Exercises 9–12 for your geometry. Write some statements that might be used as axioms in your geometry.
21. Check to see that each of the eight desirable properties is satisfied by the geometries of this section.
22. Find out about the application of finite geometry in error–detecting and correcting codes in a textbook on modern applied algebra.

▲ 1.7 Sets of Axioms for Euclidean Geometry

Now that you have studied sets of axioms for several finite geometries, let us look at sets of axioms for Euclidean geometry and at some ideas involving all sets of axioms.

A major problem in geometry for some 2000 years was how to provide an adequate set of axioms for ordinary Euclidean geometry. Several such sets of axioms are discussed in this section, and you should be familiar with the set of axioms contained in the high school geometry book you studied.

Euclid's *Elements* (about 300 B.C.) consisted of 13 books containing 465 propositions. The word *proposition* means a statement suggested for acceptance, so Euclid's statements really were theorems followed by the proofs. In addition to geometry, the *Elements* included number theory and geometric algebra. A famous set of five axioms and five postulates were given as follows:

Axioms (or common notions)

1. Things that are equal to the same thing are also equal to one another.
2. If equals are added to equals, the wholes are equal.
3. If equals are subtracted from equals, the remainders are equal.
4. Things that coincide with one another are equal to one another.
5. The whole is greater than the part.

Postulates

1. A straight line can be drawn from any point to any point.
2. A finite straight line can be produced continuously in a straight line.
3. A circle may be described with any point as center and any distance as radius.
4. All right angles are equal to one another.
5. If a transversal falls on two lines in such a way that the interior angles on one side of the transversal are less than two right angles, then the lines meet on that side on which the angles are less than two right angles.

Postulate 2 means in essence that a segment can be extended indefinitely to form a line. The *fifth postulate* of Euclid (the parallel postulate) will be considered in more detail later in this section. It also plays a fundamental role in the discussion of non-Euclidean geometry in Chapter 9.

Within the past hundred years in particular, mathematicians studying the foundations of mathematics have pointed out various flaws in the assumptions of Euclid. Euclid actually used other, tacit assumptions (assumptions not stated explicitly). Logical problems pointed out have included:

- ▲ The need for a definite statement about the continuity of lines and circles.
- ▲ The need for a statement about the infinite extent of a straight line.
- ▲ The need to state the fact that if a straight line enters a triangle at a vertex, it must intersect the opposite side.
- ▲ The need for statements about the order of points on a line.

- The need for a statement about the concept of betweenness.
- The need for a statement guaranteeing the uniqueness of a line joining two distinct points.
- The need for a list of undefined terms.
- The need for a more logical approach that does not depend on the concept of *superposition*. Euclid assumed that a triangle could be picked up and put down in another place with all the properties remaining invariant, yet no statement to this effect was made. The geometry of transformations, a major part of modern geometries today (Chapter 2 and elsewhere), originated because of this need for a sounder replacement for the concept of superposition.

Many modern sets of axioms for Euclidean geometry have been introduced to remedy the defects in Euclid. In a course on the foundations of geometry, or on the foundations of mathematics, these are studied extensively. In general, these newer sets of axioms are more comprehensive than that of Euclid, and for this reason they appear more complex. It has been difficult to incorporate them into the beginning high school geometry course without introducing still more axioms in order to avoid proving very difficult theorems at the beginning of the course.

One of the first modern sets of axioms for Euclidean geometry was devised by Moritz Pasch in 1882. He is given credit for what is called *Pasch's axiom*: A line entering a triangle at a vertex intersects the opposite side, and a line that intersects one side of a triangle at a point other than the vertex also intersects a second side.

Guiseppi Peano provided another new approach in 1889. Probably the most famous set of axioms for Euclidean geometry, however, was given by David Hilbert and was published in English in 1902. These axioms are given in Appendix 1. Hilbert used six undefined terms: *point*, *line*, *plane*, *between*, *congruent*, and *on*. He gave his axioms in five groups, and the grouping helps show how the logical difficulties of Euclid were overcome. For example, Hilbert's first three axioms clarify Euclid's postulate about drawing a straight line from any point to any point, and his axioms of congruence take care of the logical defect of superposition.

Since the time of Hilbert there have been many other sets of modern postulates for Euclidean geometry. Some of these are by Oswald Veblen (in 1904 and 1911), E. V. Huntington (1913), Henry Forder (1927), G. D. Birkhoff (1932), and others. The axioms of Birkhoff (Appendix 2) are of special significance because (1) they emphasize the connections between geometry and the real numbers, (2) they include *distance* and *angle* as undefined terms, and (3) they have been modified for incorporation into geometry courses.

One of the main reasons why Birkhoff's system is so brief is the power of the first postulate. Since this assumption has the effect of assuming all the properties of the real numbers, the order relations for points depend only on theorems and definitions rather than on additional postulates.

Because modern geometry programs depend on the coordination of analytic geometry with synthetic geometry more than was the case even at the time of Hilbert, axioms for the real numbers and the logical assumptions of algebra for relations are needed in geometry. Sets of axioms for modern geometry texts ordinarily contain specific axioms needed for each of the following purposes not readily apparent as a result of the axioms of Hilbert:

- ▲ To show the existence of a correspondence that associates a unique number with every pair of distinct points.
- ▲ To establish the measure of the distance between any two points in the line as the absolute value of the difference of their corresponding numbers.
- ▲ To state the existence of a unique coordinate system for a line that assigns to two distinct points two given distinct real numbers.
- ▲ To formulate the logical assumption necessary for the development of the theory of convexity (Chapter 3) by stating that a line in a plane partitions the points of the plane not on the line into two convex sets such that every segment that joins a point of one set to a point of the other intersects the line.
- ▲ To include additional assumptions about congruence of triangles—assumptions that eliminate the need for lengthy proofs at an early stage in the text. The two most common additional axioms are the assumptions that (1) congruence of triangles follows from congruence of two angles and the included side and that (2) congruence of triangles follows from congruence of the three sides.
- ▲ To postulate additional assumptions about fundamental concepts of area. These include the existence of a correspondence that associates the number 1 with a certain polygonal region and a unique positive real number with every convex polygonal region; a statement that if two triangles are congruent, the respective triangular regions have the same area; and the assumption that the measure of area of a rectangular region is the product of the measures of the lengths of its base and altitude.

Besides the axioms of Euclidean geometry and finite geometries in Chapter 1, you will encounter axioms for projective geometry in Chapter 7 and axioms for non-Euclidean geometries in Chapter 9. A brief consideration of desirable properties for any set of axioms for a geometry concludes this section.

Sets of axioms for a geometry should have at least two important properties.

1. The set should be *consistent*. In a consistent set of axioms, it is not possible to use them to prove a theorem that contradicts any axiom or other theorem that has already been proved. All of the sets of axioms in this text are examples of consistent systems. The method of proving a set of axioms consistent is discussed in Chapter 9 on non-Euclidean geometry. An example of an inconsistent system could be one that included *both* of these axioms: (a) Two distinct points determine exactly one line. (b) Two distinct points determine exactly two lines.

2. The set should be *complete*. Although completeness may be explained in several ways, it is used here to mean that for any two contradictory statements involving concepts of the set of axioms, at least one of the statements can be proved by using the axioms of the set.

A well-known example of the importance of completeness concerns Euclid's fifth postulate. Without a parallel postulate (or the equivalent), the set of axioms for Euclidean geometry could not be complete. The geometry of the remaining set of axioms is called absolute geometry, and it is not possible in absolute geometry to prove the theorem that the sum of the measures of the angles of a triangle either is or is not 180 degrees.

In the study of foundations of mathematics, it is often important that the set of axioms have at least one additional property: that none of the axioms can be proved from the remaining set of axioms. Sets of axioms having this property are called *independent sets* of axioms. A course in foundations of geometry includes proofs of independence.

To show that one axiom of a set is independent of the rest, one can show a *model* in which all the other axioms hold, but the remaining axiom is contradicted. Two simple examples follow from two different finite geometries. Figure 1.15 is a model for the first two axioms of the four-line geometry. Axiom 3 does not hold and, hence, is independent. Figure 1.16 shows a model for the first two axioms of the four-point geometry. Here, too, Axiom 3 does not hold and, hence, is independent.

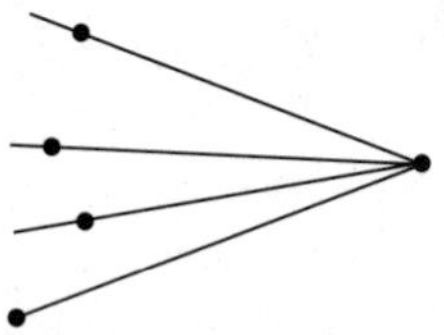

Figure 1.15

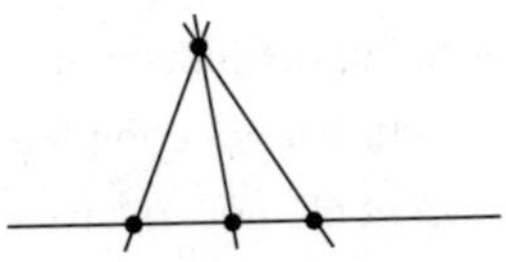

Figure 1.16

A well-known example of the significance of independence involves the fifth postulate. The fifth postulate is independent of the rest. Non-Euclidean geometries, studied in Chapter 9, are the result of substituting other parallel postulates that contradict the one in Euclid's set. In that chapter, models are used to show that geometries based on alternatives to the fifth postulate are as consistent as ordinary Euclidean geometry.

Euclid himself seemed reluctant to use the fifth postulate in his proofs. In his *Elements*, the first 28 propositions are proved without reference to the fifth postulate or to any alternate parallel postulate. Because of this, these 28 propositions form a significant part of what is called *absolute geometry*, the geometry based on all the axioms and the postulates of Euclid except the fifth postulate (see Appendix 6).

The requirement of independence is not always desirable at a more elementary level. Therefore, secondary geometry texts ordinarily have additional axioms that can be proved from the others. These axioms are included because they are convenient to state and to use early in the course, or because the proofs are too difficult for that level. These axioms may become theorems in a more advanced course. The assumptions of Euclidean geometry, along with some theorems proved in any introductory course, are the basic assumptions used in Chapters 2 through 5 of this text.

▲ EXERCISES 1.7 ▲

1. Draw a figure to explain the precise wording of Euclid's fifth postulate.

For Exercises 2–7, draw a figure to explain the following axioms of Hilbert.

2. Order Axiom 2
3. Order Axiom 4
4. Order Axiom 5
5. Congruence Axiom 3
6. Congruence Axiom 4
7. The axiom of continuity

For Exercises 8–10, name the axioms of Hilbert that:

8. State Pasch's axiom.
9. Guarantee the uniqueness of a line joining two distinct points.
10. Deal with betweenness for points on a line.

For Exercises 11–14, state whether a set of axioms could be:

11. Complete but not independent
12. Independent but not complete
13. Independent but not consistent
14. Consistent but not independent

For Exercises 15–22, state an axiom that could be added to the set of axioms given for each geometry so that the set would no longer be consistent.

15. Euclid's axioms and postulates
16. Birkhoff's postulates
17. Axioms for three-point geometry
18. Axioms for four-line geometry
19. Axioms for four-point geometry
20. Axioms for Fano's geometry
21. Axioms for the geometry of Pappus
22. Axioms for the geometry of Desargues

▷ 23. Find, in a recent high school geometry text, a set of axioms that is not an independent set. Point out at least one "axiom" that could be proved as a theorem, using the remaining axioms.

▷▷ 24. Try to discover additional models, besides those shown in Figures 1.15 and 1.16, that show the independence of other axioms in the geometries introduced in this chapter.

▷ **25.** Study the proofs of the first 28 propositions in Euclid's *Elements* to verify that each is a theorem in absolute geometry.

▲ CHAPTER 1 REVIEW EXERCISES ▲

1. A finite geometry is called finite because it has a finite number of what?

2. For measurements of distance of 700 km and difference in angle elevation of 8.3°, use the method of Eratosthenes to find the circumference of a planet.

For Exercises 3–58, tell whether or not the statement is always true for the particular finite geometry. The columns represent the following geometries: a, three-point; b, four-line; c, four-point; d, Fano; e, Pappus; f, Desargues; g, Figure 1.14c.

	a	b	c	d	e	f	g
For each two distinct points, there exists exactly one line on both of them.	3.	4.	5.	6.	7.	8.	9.
Each line has two or fewer other lines of the geometry parallel to it.	10.	11.	12.	13.	14.	15.	16.
Each point of the geometry lies on exactly three lines of the geometry.	17.	18.	19.	20.	21.	22.	23.
All of the axioms are also true statements in ordinary Euclidean geometry.	24.	25.	26.	27.	28.	29.	30.
Changing the last axiom can result in Young's geometry.	31.	32.	33.	34.	35.	36.	37.
For each set of three points, there is exactly one line containing them.	38.	39.	40.	41.	42.	43.	44.
If a point is not on a given line, then there is at least one point on the line not lying on any line through the given point.	45.	46.	47.	48.	49.	50.	51.
The lines through each point in the geometry contain every point of the geometry.	52.	53.	54.	55.	56.	57.	58.

Additional chapter review exercises may be selected from any exercises not already assigned from previous exercise sets in this chapter.

2 Geometric Transformations

▲ 2.1 Introduction to Transformations

In the first chapter, sets of axioms provided the basis for the study of geometries. In this chapter, the concept of transformation provides a different way of studying some geometries. In early experiences in geometry, students may use flips, turns, and slides to study geometric relations informally. In high school geometry, students encounter rotation and translation. Some triangles are congruent, and others are similar. Recall that the ratio of lengths of two sides of a triangle remains invariant for the corresponding sides of another triangle that is similar to the first. In geometries to be studied later in this text, one talks of the ideas of inversion and projection. All these ideas are related to a very basic concept in geometry–that of a *geometric transformation*.

Here are two examples of transformations:

1. A pairing indicated by $x \rightarrow 2x + 3$ of integral points on a number line. The arrow notation simply designates that the second element is the image of the first under the transformation. For example, point 1 would be paired with 5 and 2 with 7. See Figure 2.1a.
2. $(x, y) \rightarrow (x + 2, y - 1)$ is an example of a translation. As shown in Figure 2.1b, each point with real coordinates in the Euclidean plane is paired with a point 2 units to the right and 1 unit below the original point.

In this section, geometric transformations are defined, and several concepts common to all transformations are introduced. Section 2.2 discusses special sets of transformations called *groups*, which play an important role in geometries that are defined in terms of transformations. The section includes an example of a finite group of transformations. Sections 2.3–2.7 present many ideas from the Euclidean geometry of two and three dimensions, using the concept of transformations to show familiar ideas from a new point of view. Section 2.5 uses matrix multiplication in the application of transformations to computer graphics. The next-to-last section, Section 2.8, enlarges the concept of transformation by introducing the group of similarities, a somewhat more general group that includes Euclidean transformations as a subgroup. This section is a crucial step toward the much more general geometry of Chapter 7. Section 2.9 introduces the popular enrichment topic of fractals as an application of similarities.

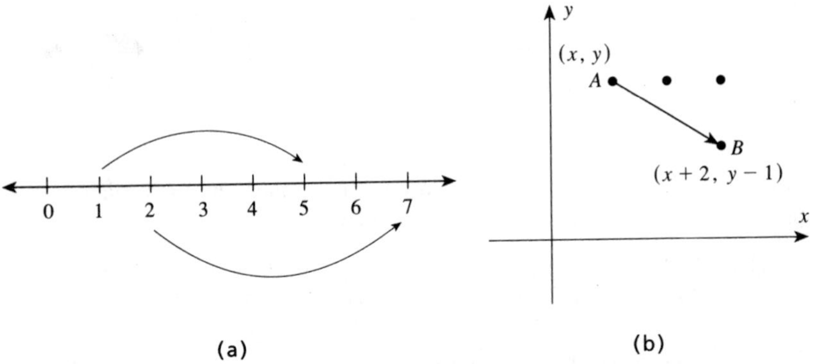

Figure 2.1

Before giving a precise definition of transformation, it is necessary to explain the idea of mapping.

DEFINITION ➤ A *mapping* of a set A into a set B is a pairing of elements of A and a subset of B so that each element of A is paired with *exactly one* element of the subset of B, and each element of that subset of B is paired with *at least one* element of A.

The special case of a mapping of set A into set B for which *each element* of B is paired with at least one element of A is called a mapping of set A *onto* set B.

DEFINITION ➤ F is a mapping of a set A into a set B if for every element a of A there exists a unique element b of B that is paired with a; this pairing is denoted by $f(a) = b$. The set A is called the domain of f, and the set B is called the codomain of f.

DEFINITION ➤ Given a mapping f from A to B, the range of f is the set of points b such that $b = f(a)$ for some a in A.

DEFINITION ➤ A mapping f from A to B is onto if the range of f is equal to B.

DEFINITION ➤ A mapping f from A to B is one-to-one if each element in the range of f is the image of exactly one element of A. That is, if $f(a) = f(c)$, then $a = c$.

Figure 2.2 shows an example of a mapping whose ordered pairs are (a_1, b_1), (a_2, b_1), and (a_3, b_2). A mapping of set A into set B may also be indicated by the notation $f(a) = b$. Here, an element b of B is the image of an element a of A under the mapping f.

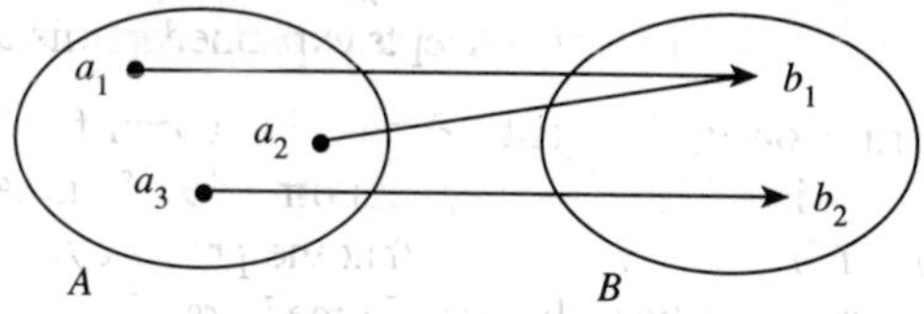

Figure 2.2

A special kind of mapping is of particular importance in mathematics; this is a mapping that is both one-to-one and onto, or a *transformation*.

DEFINITION ➤ A transformation is a mapping f of A onto B such that each element of B is the image of exactly one element of A.

In other words, a one-to-one correspondence exists between the sets of elements of A and B. A simple example of a transformation is shown in Figure 2.3.

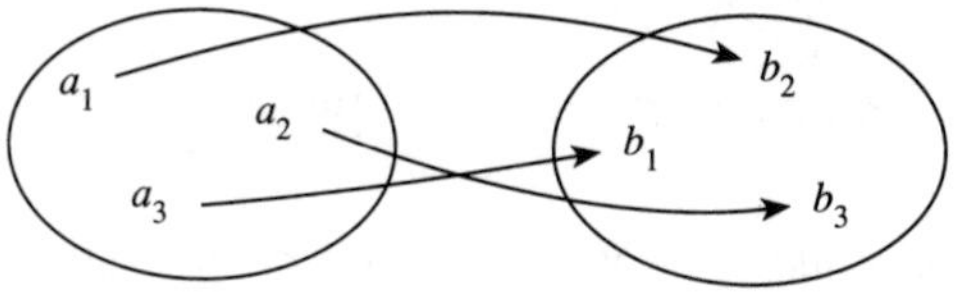

Figure 2.3

A *function* can be defined as a set of ordered pairs with no two different pairs having the same first element. It can be seen that the definition of mapping is equivalent to a typical definition of function, since in a mapping, having two different pairs with the same first element would mean that an element of A is paired with more than one element of B. It is customary to use the term *mapping* rather than *function* in geometry, however. One of the essential differences between a mapping and a transformation is that reversing the elements in the pairs of a mapping does not necessarily result in a mapping, while reversing the elements in the pairs of a transformation does also result in a transformation.

Two of the earliest mathematicians to write about transformation theory were Arthur Cayley (1821–1895), a prolific English mathematician, and James Sylvester (1814–1897), also an English mathematician, who taught at Johns Hopkins University and was founding editor of the *American Journal of Mathematics*.

The concept of transformation is important because of the fact that sets of transformations can be used to classify geometries. In high school geometry, for example, the transformations allowed are rotation, translation, and sometimes similarity. Allowing other, more general types of transformations results in other geometries. In a particular geometry, the student studies *properties* of figures and their images under a set of transformations. *Invariant properties* are those that do not change. In the study of geometry through transformations, the student is asked to notice close relationships between abstract algebra and modern geometry. Readers who have not yet studied abstract algebra will find the necessary concepts explained in this text.

If f is a transformation from A onto B and g is a transformation from B onto C, the product $h = gf$ is defined as the transformation from A onto C such that $h(P) = g[f(P)]$ for each point of A. Note that the product gf is defined in such a way that the transformation on the right is performed first. In Figure 2.4, the pairs in h are $(a_1, c_3), (a_2, c_2), (a_3, c_1)$.

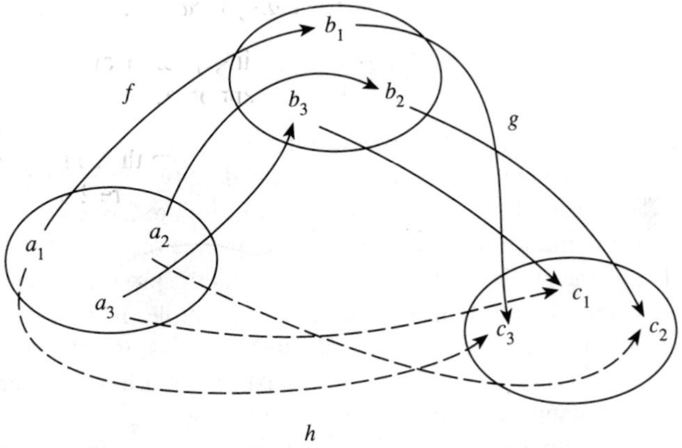

Figure 2.4

Example ➤ Let $f = \{(3, 4), (2, 6), (7, 1)\}$ and $g = \{(4, 5), (6, 3), (1, 2)\}$ be transformations of points on a line. Considering the first ordered pair of f, this notation means the point with coordinate 4 is the image of 3 under the transformation f. Find gf and fg. $gf = \{(3, 5), (2, 3), (7, 2)\}$ and fg is not defined. For example, in fg, g takes 4 to 5, but f does not give the image of 5.

Example ➤ As an example of the product of two transformations, consider the two defined as follows for all points (x, y).

$$f(x, y) \to (x + 1, y + 1)$$

$$g(x, y) \to (x + 2, y - 2)$$

The results of f followed by g, gf, for the sample points (2, 2) and (1, 5) are shown in Figure 2.5. The combined effect of gf is to cause (x, y) to have the image $(x + 3, y - 1)$.

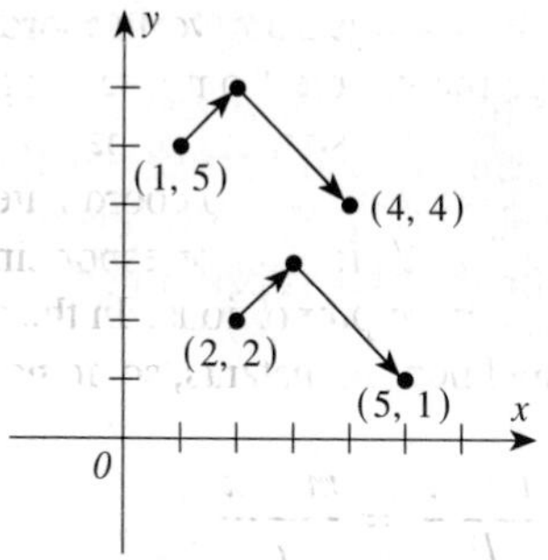

Figure 2.5

In special cases the product of two transformations is denoted by I. For the identity transformation, each element is its own image. The identity transformation leaves each point fixed. In this case, if $I = gf$, as illustrated in Figure 2.6, then g is the *inverse transformation* of f, indicated by $g = f^{-1}$. The inverse of a transformation has the effect of "undoing" the original transformation, so that the product of a transformation and its inverse is the identity transformation. For example, if f is a transformation such that (x, y) has the image $(x + 4, y - 2)$, then f^{-1} is a transformation such that (x, y) has the image $(x - 4, y + 2)$. Every transformation has an inverse. If a pair (a, b) occurs in the mapping f, then (b, a) is in the mapping f^{-1}. The fact that f is one-to-one and onto implies that f^{-1}, defined in this way, is a mapping and is onto. If g is any other transformation satisfying gf = identity, then g must equal f^{-1}. Thus, inverses of transformations are unique.

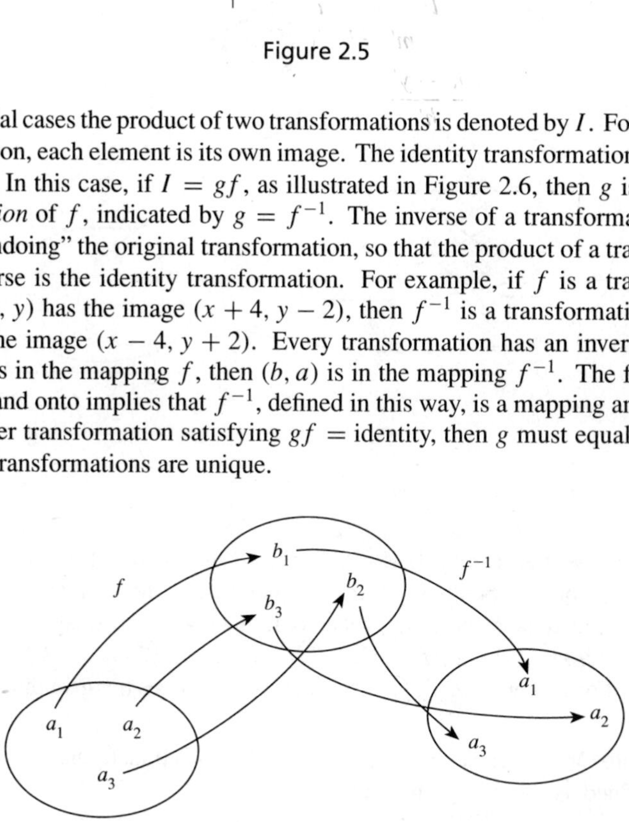

Figure 2.6

A somewhat elementary but modern and practical application of a mapping is the change from world coordinates to screen coordinates in computer graphics. A world coordinate system refers to measurements of actual physical objects. For example, a house might be measured in feet. That part of a real life scene to be shown on the screen is a rectangular region called a "window." The rectangular portion of the computer screen showing the contents of the window is called the "viewport." The basic application uses a window-to-viewport mapping to change from world coordinates to screen coordinates.

The mapping must change the picture from a rectangular region of one size and shape to a rectangular region of another size and perhaps another shape. The equations for the mapping must take any point with world coordinates (m, n) in the window and produce the screen coordinates (m', n') for the corresponding point in the viewport. See Figure 2.7. The equations come from proportions. In the resulting equations, observe that the ratios l'/l and h'/h are known constants, so do not have to be recalculated for each individual point.

$$\frac{m' - x'}{l'} = \frac{m - x}{l}$$
$$m' = l'/l(m - x) + x'$$
$$\frac{n' - y'}{h'} = \frac{n - y}{h}$$
$$n' = h'/h(n - y) + y'$$

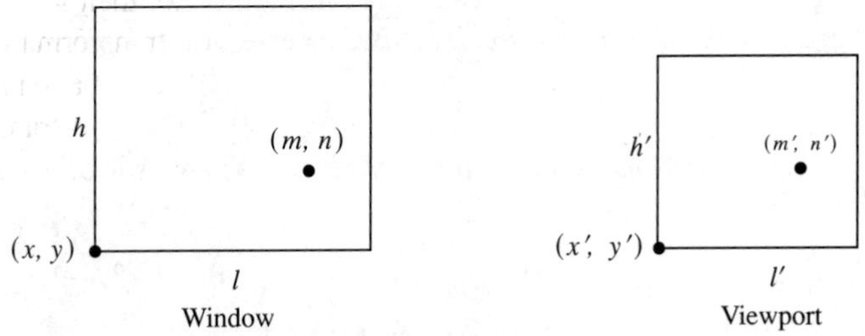

Figure 2.7

Example ➤ Find the corresponding point in the viewport for (20, 30) in the window if $l = 30, l' = 20, h = 40, h' = 45$. Also, let $x = y = 0$ and $x' = 10, y' = 20$.

Here, $m = 20$ and $n = 30$. Then

$$m' = 20/30(20 - 0) + 10$$
$$= 70/3$$

and

$$n' = 45/40(30 - 0) + 20$$
$$= 215/4.$$

The corresponding point is (70/3, 215/4).

▲ EXERCISES 2.1 ▲

1. Draw a diagram showing two sets, one with six elements and one with eight; then indicate a mapping that is not a transformation.

2. Which of the mappings shown in Figure 2.8 are transformations?

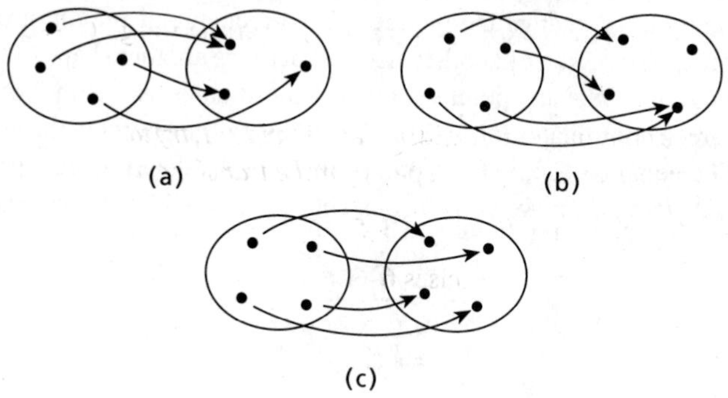

Figure 2.8

In Exercises 3–5, for the transformation indicated by $x \to 2x - 1$*, give the image of:*

3. 7 **4.** -2 **5.** $-\frac{3}{4}$

In Exercises 6–8, for the transformation indicated by $(x, y) \to (x+5, y-3)$*, give the image of each point:*

6. $(2, 1)$ **7.** $(0, 0)$ **8.** $(-3, -2)$

For Exercises 9–12, let $f = \{(a, b), (c, d), (e, h)\}$ *and* $g = \{(b, i), (d, j), (h, k)\}$ *be transformations of points on a line.*

9. Find the product gf. **10.** Find the product fg.

11. Find f^{-1}. **12.** Find g^{-1}.

For Exercises 13–16, let f *be the transformation such that* (x, y) *has the image* $(x-5, y+2)$ *and* g *be the transformation such that* (x, y) *has the image* $(x+2, y-3)$

13. Find the product fg. **14.** Find the product gf.

15. Find f^{-1}. **16.** Find g^{-1}.

For Exercises 17–18, find the screen coordinates for the corresponding point for the point with given world coordinates if:

$$l = 25, l' = 30, h = 20, h' = 18, x = y = 5, x' = y' = 20.$$

17. $(10, 15)$ **18.** $(32, 18)$

19. If the product T_2T_1 exists, then the transformation T_2 is said to be compatible with T_1. If T_2 is compatible with T_1, is T_1 necessarily compatible with T_2?

20. State and prove a theorem about compatibility if T_1 and T_2 are transformations of a set onto itself.

21. A transformation f such that $ff = I$, the identity, is called an *involutory* transformation. If f is an involutory transformation, prove that $f = f^{-1}$.

22. If f is the transformation such that (x, y) has the image $(x + a, y + b)$, find the values of a and b such that $ff = I$.

For the transformations indicated in Exercises 23l and 24l, try to discover if the distance between two points is an invariant property for the transformation.

▷ **23.** The transformation in Exercises 3–5

▷ **24.** The transformation in Exercises 6–8

▲ 2.2 Groups of Transformations

Modern geometries ordinarily involve a set of transformations rather than a single transformation. The type of set of transformations most commonly encountered is a *group*.

DEFINITION ➤ A *group of transformations* of a set A onto itself is a nonempty set S of transformations together with an operation such that:

1. If f and g are in S, then fg and gf are in S. (Closure)
2. If f, g, and h are in S, then $(fg)h = f(gh)$. (Associativity)
3. There is a unique element in S satisfying $If = fI = f$ for all f in S. (Identity)
4. Given f in S, there exists a unique element f^{-1} satisfying $f^{-1}f = ff^{-1} = I$. (Inverses)

The first of the properties is the *closure* property. This means the product of any two transformations in the set is also a transformation in the set. Stated another way, the operation of multiplication can always be performed within a group of transformations without going outside the group.

The second of the properties is the *associative* property. The notation $f(gh)$ and $(fg)h$ indicates the product of the same three transformations in the same order.

Example ➤ Consider the three transformations that have been defined as follows:

$$\begin{aligned} f &: x \to x + 3 \\ g &: x \to x - 2 \\ h &: x \to 2x \\ f(gh) &: x \to ([2x] - 2) + 3 = 2x + 1 \\ (fg)h &: x \to 2x + (-2 + 3) = 2x + 1 \end{aligned}$$

To show that the associative property $(T_3T_2)T_1 = T_3(T_2T_1)$ always holds for transformations, we can consider what happens to any member s_1 of a set S. Assume that the image of s_1 under T_1 is s_2, the image of s_2 under T_2 is s_3, and the image of s_3 under T_3 is s_4. This implies the image of s_1 under T_2T_1 is s_3, and then the image of s_3 under T_3 is s_4. In the same way, the image of s_1 under T_1 is s_2, and the image of s_2 under T_3T_2 is s_4. Thus $(T_3T_2)T_1$ and $T_3(T_2T_1)$ are the same mapping of a set onto inself.

The third property states the existence of a unique *identity*. Since the product of a transformation and the identity is the transformation, the effect of the identity is to leave each point invariant.

The fourth property states the existence of a unique *inverse* transformation for each transformation of the set. Since the product of a transformation and its inverse is the identity, the effect of the inverse is to undo the transformation, returning each point to the original position. The commutative property $fg = gf$ does not hold for transformations. If the commutative property does hold for all pairs of elements in a group, that group is called a *commutative* or *Abelian* group (after the Norwegian mathematician N. H. Abel, 1802–1829).

Many of the groups of transformations in geometry are infinite groups–that is, groups with an infinitude of members. However, studying examples of finite groups will help you to understand more about the concept of a group of transformations. It is important to remember that the elements of a group of transformations are transformations, not points.

Consider the set of all *symmetries of an equilateral triangle*, as shown in Figure 2.9. "Symmetries of the triangle" designates the reflections about the axes of symmetry or rotation about the center of gravity such that the new figure coincides with the old. The result of each symmetry can be represented by a renaming of the three vertices. Check each of the figures in Figure 2.9 to see that it corresponds to the following definitions of the elements of the set of symmetries of an equilateral triangle:

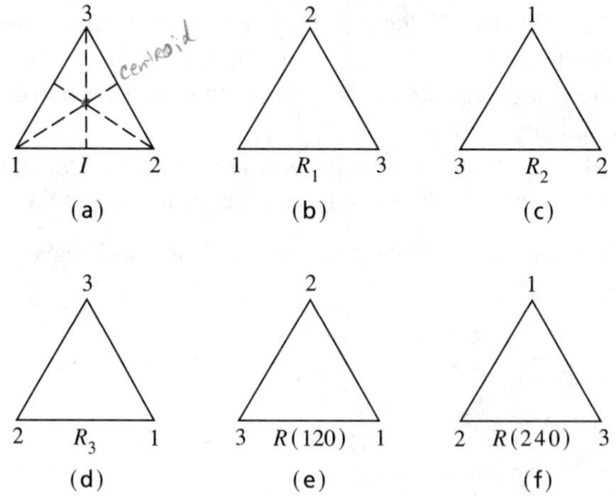

Figure 2.9

I	Identity. Rotation through an angle of $0° + (n)(360°)$, $(n = 0, 1, 2, \ldots)$ counterclockwise
R_1	Reflection about the axis through vertex 1
R_2	Reflection about the axis through vertex 2
R_3	Reflection about the axis through vertex 3
$R(120)$	Rotation through an angle of 120° counterclockwise
$R(240)$	Rotation through an angle of 240° counterclockwise

To verify that the set of symmetries of an equilateral triangle constitutes a group, it is necessary to verify that $f \in S$ implies that $f^{-1} \in S$ and that $f \in S$ and $g \in S$ imply fg and $gf \in S$. The inverse element for each transformation is as follows:

f	f^{-1}
I	I
R_1	R_1
R_2	R_2
R_3	R_3
$R(120)$	$R(240)$
$R(240)$	$R(120)$

The verification of both facts can be accomplished by completing an operation table for the elements. The entries in the table are found by performing the transformations in the order indicated. For example, Table 2.1 shows $R_2[R(120)] = R_3$. Recall that $R(120)$ comes first in this multiplication.

Table 2.1

Second Transformation Performed (columns); **First Transformation Performed** (rows)

	I	R_1	R_2	R_3	$R(120)$	$R(240)$
I	I	R_1	R_2	R_3	$R(120)$	$R(240)$
R_1	R_1	I	$R(240)$	$R(120)$	R_3	R_2
R_2	R_2	$R(120)$	I	$R(240)$	R_1	R_3
R_3	R_3	$R(240)$	$R(120)$	I	R_2	R_1
$R(120)$	$R(120)$	R_2	R_3	R_1	$R(240)$	I
$R(240)$	$R(240)$	R_3	R_1	R_2	I	$R(120)$

Introduction of the *permutation group symbols* will illustrate a common agreement that ties the work on transformation groups more closely to that found in modern algebra texts. Two rows of numbers are used to define a transformation. The first row shows the original vertices, and the second row shows the new position of those vertices.

$$R_1 = \begin{pmatrix} 1 & 2 & 3 \\ 1 & 3 & 2 \end{pmatrix}$$

Explanation for R_1. Vertex 1 remains fixed, vertex 2 moves to the place where vertex 3 was originally, and vertex 3 moves to the place where vertex 2 was originally.

$$R_2 = \begin{pmatrix} 1 & 2 & 3 \\ 3 & 2 & 1 \end{pmatrix}$$

$$R_3 = \begin{pmatrix} 1 & 2 & 3 \\ 2 & 1 & 3 \end{pmatrix}$$

$$R(120) = \begin{pmatrix} 1 & 2 & 3 \\ 2 & 3 & 1 \end{pmatrix}$$

Explanation for $R(120)$. Vertex 1 moves to the place where vertex 2 was originally, vertex 2 moves to the place where 3 was originally, and vertex 3 moves to the place where 1 was originally.

$$R(240) = \begin{pmatrix} 1 & 2 & 3 \\ 3 & 1 & 2 \end{pmatrix}$$

There are five other equivalent forms for each of the permutation group symbols. This is true because the elements on the first row can be arranged in any of six ways. For example, two other equivalent forms for R_1 are

$$\begin{pmatrix} 3 & 2 & 1 \\ 2 & 3 & 1 \end{pmatrix} \quad \text{and} \quad \begin{pmatrix} 3 & 1 & 2 \\ 2 & 1 & 3 \end{pmatrix}$$

Either the permutation symbols, or a series of pictures, can be used to check the entries in the multiplication table. The following three examples illustrate both methods.

Example ➤

$$R_1R_2 = \begin{pmatrix} 3 & 2 & 1 \\ 2 & 3 & 1 \end{pmatrix}\begin{pmatrix} 1 & 2 & 3 \\ 3 & 2 & 1 \end{pmatrix} = \begin{pmatrix} 1 & 2 & 3 \\ 2 & 3 & 1 \end{pmatrix} = R(120)$$

The notation means that R_2 is first. R_2 is indicated by

$$\begin{pmatrix} 1 & 2 & 3 \\ 3 & 2 & 1 \end{pmatrix}$$

Then the result on the bottom line becomes the top line of one of the equivalent forms for R_1.

$$\begin{pmatrix} 3 & 2 & 1 \\ 2 & 3 & 1 \end{pmatrix}$$

The product has the top row from R_2 and the bottom row from the form of R_1, and can be identified as the definition of $R(120)$.

Finding the product of the two transformations can also be illustrated, as shown in Figure 2.10.

Note the convention observed in the first row of Figure 2.10. The second transformation is performed using the original positions for the vertices. Thus, the vertex that was where vertex 3 was originally, (1), is moved to where vertex 2 was originally. The vertex that was where vertex 2 was originally, (2), is moved to where vertex 3 was originally, and the vertex that was where vertex 1 was originally, (3), remains unchanged.

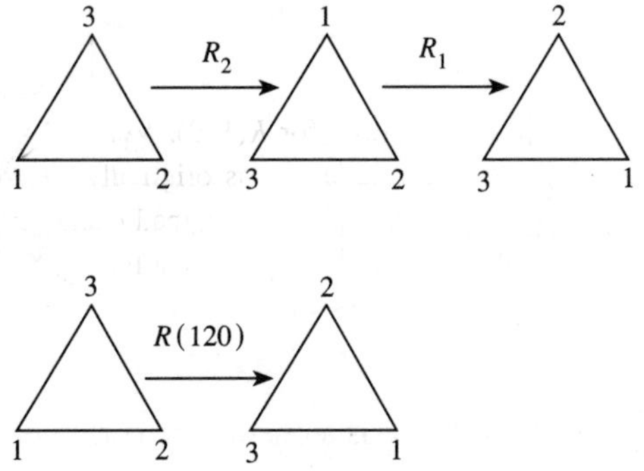

Figure 2.10

Example ➤

$$R(120)R_2 = \begin{pmatrix} 3 & 2 & 1 \\ 1 & 3 & 2 \end{pmatrix}\begin{pmatrix} 1 & 2 & 3 \\ 3 & 2 & 1 \end{pmatrix} = \begin{pmatrix} 1 & 2 & 3 \\ 1 & 3 & 2 \end{pmatrix} = R_1$$

Example ➤

$$R(240)R(120) = \begin{pmatrix} 2 & 3 & 1 \\ 1 & 2 & 3 \end{pmatrix}\begin{pmatrix} 1 & 2 & 3 \\ 2 & 3 & 1 \end{pmatrix} = \begin{pmatrix} 1 & 2 & 3 \\ 1 & 2 & 3 \end{pmatrix} = I$$

The last two examples are illustrated in Figure 2.11.

The conclusion from the discussion of the symmetries of an equilateral triangle is stated as a theorem.

THEOREM 2.1 ➤ The set of symmetries of an equilateral triangle is a group of transformations.

The group of symmetries of an equilateral triangle is not a commutative group. As a specific example, $R_1R(240) = R_3$, but $R(240)R_1 = R_2$. A consequence of this lack of commutativity is that, when the elements are listed in the same horizontal and vertical order, the operation table is not symmetric about the diagonal from upper left to lower right.

Within a group of transformations, there may be subgroups of transformations. A *subgroup* is a subset of a group of transformations that is itself a group. For example, one subgroup of the symmetries of an equilateral triangle has elements I, $R(120)$, and $R(240)$. Verify from the operation table that the products are elements of a subgroup. Furthermore, observe that this subgroup is a commutative group.

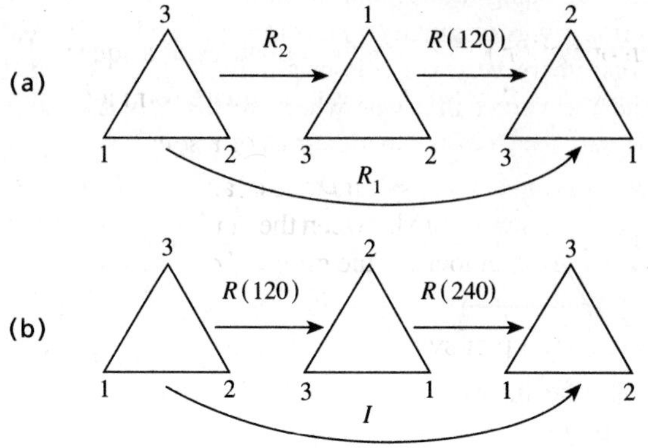

Figure 2.11

In general, a symmetry of a figure is a one-to-one correspondence that preserves distances. Other examples of finite groups of transformations are the symmetries of a square, the symmetries of an isosceles triangle, and the symmetries of a nonsquare rectangle. This group of symmetries of the non-square rectangle is sometimes called the *Klein 4-group*, after Felix Klein. These are explored in Exercises 2.2.

▲ EXERCISES 2.2 ▲

1. Using only the other three properties, prove that a group of transformations includes the identity element.
2. Write an explanation for the permutation group symbols for R_2, R_3, and $R(240)$.

For Exercises 3–6, find the products from the multiplication table for the symmetries of an equilateral triangle; then use the permutation notation to verify each answer.

3. $R(240)R_3$
4. $R_3R(240)$
5. R_1R_3
6. R_3R_1
7. Is the set $\{I, R_1, R_2, R_3\}$ of symmetries for the equilateral triangle a subgroup?
8. Prepare an operation table for the symmetries of an isosceles triangle.
9. Verify that the symmetries of an isosceles triangle form a group.
10. Is the group of symmetries of an isosceles triangle a commutative group?
11. List all the subgroups of the symmetries of an isosceles triangle.
12. Prepare an operation table for the symmetries of a nonsquare rectangle.

13. Verify that the symmetries of a nonsquare rectangle form a group.

14. List all subgroups of the group of symmetries of a nonsquare rectangle.

15. Which subgroups of the symmetries of a nonsquare rectangle are commutative groups?

16. Prepare an operation table for the symmetries of a square.

17. Verify the fact that the symmetries of a square form a group.

18. List all the subgroups of the symmetries of a square.

▷ **19.** List all the subgroups of the symmetries of an equilateral triangle and see if you can discover the relationship between the number of members of each subgroup and the number of members of the group. Try your conjecture for other groups of symmetries.

▶▷ **20.** Investigate the group of symmetries of a regular hexagon and its subgroups.

▷ **21.** Explain how the instrument called a kaleidoscope works, and why it is an application of groups of symmetries of regular polygons.

▲ 2.3 Euclidean Motions of the Plane

Unlike the finite groups of transformations in the previous section, the sets of transformations in this section have infinitely many members. This section and the next involve looking at Euclidean geometry from a point of view different from the typical secondary school approach. The transformations studied are transformations of the Euclidean plane onto itself.

The essential characteristic of the transformations of Euclidean geometry is that *distance* is preserved. That is, distance must be an invariant property.

DEFINITION ➤ A transformation f is an *isometry* of A onto B if it preserves distances. For any two points P_1, P_2 of A, the distance from P_1 to P_2 is equal to the distance from $f(P_1)$ to $f(P_2)$. The word *isometry* is derived from *iso* (same) and *metry* (distance).

This idea is illustrated in Figure 2.12. In Euclidean geometry, the distance d between points (x_1, y_1) and (x_2, y_2) is defined by

$$d = \sqrt{(x_2 - x_1)^2 + (y_2 - y_1)^2}$$

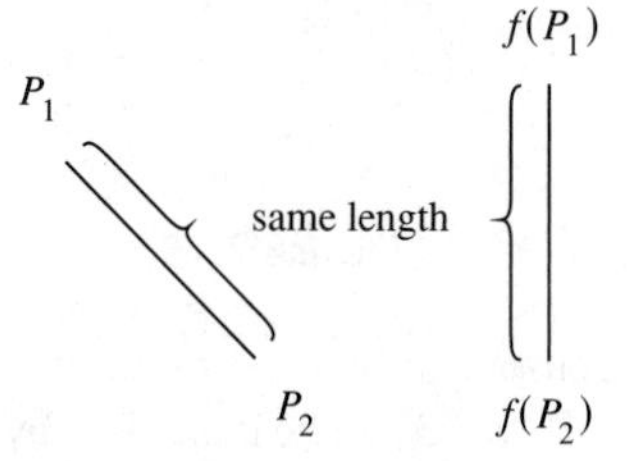

Figure 2.12

If an isometry exists between two segments, we also say the segments are *congruent*, or that they have the same length.

In ordinary Euclidean geometry, the isometries studied are isometries of a set of points onto itself. Isometries of this kind are called *motions*. Euclidean geometry involves the study of motions of the Cartesian plane. Correspondence between figures, rather than any physical motion, is the essential idea.

In 1872, Felix Klein classified geometries by applying this definition: A geometry is the study of invariant properties of a set of points under a group of transformations. Thus, Euclidean geometry is the study of invariant properties, such as angle measure, area, and parallelism of sets of points under the group of motions.

What types of transformations can be applied to a set of points in the plane so that the distance between any two points is always preserved? The first type of plane motion we will consider is a *translation*. Intuitively, a translation is a correspondence between points and their image points so that each image is the same distance in the same direction from the original point.

The mathematical concept of a vector is useful in describing a translation. A vector is a directed line segment. The notation $\boldsymbol{AB}$ (using boldface) indicates the vector from A to B. The length (or magnitude) of the vector can be indicated by the absolute value, written $|\boldsymbol{AB}|$.

Josiah Willard Gibbs (1839–1903), an American physicist, developed vector analysis in 1881. Vectors are used in physics and elsewhere because they have both a magnitude and a direction associated with them. Two vectors are defined as equal if they have the same magnitude and the same direction. For example, the vectors in Figure 2.13a are equal. Because of this, we can identify this set of equal vectors by giving the endpoint coordinates, assuming the beginning point is at the origin. Thus, the vectors in Figure 2.13a might be designated [2, 3].

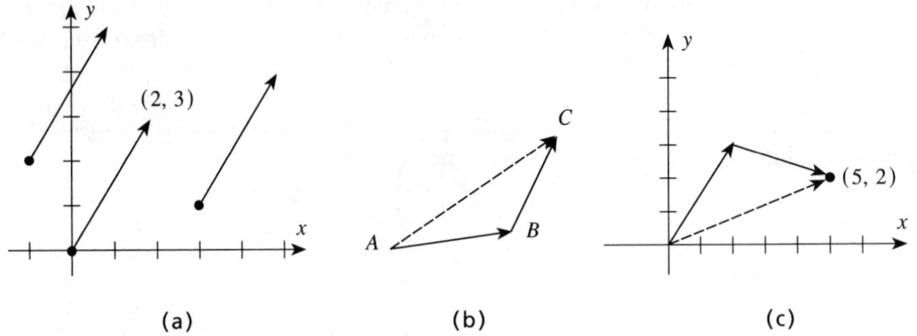

Figure 2.13

This idea shows why vectors can be used to describe translations. For example, the translation $x' = x + 2$, $y' = y + 3$ can be described by the vector [2, 3], interpreted to mean that the x coordinate of each point is increased by 2 and the y coordinate of each point is increased by 3.

The vector addition $\boldsymbol{AB} + \boldsymbol{BC} = \boldsymbol{AC}$ is shown in Figure 2.13b. Vector addition can also be shown by adding x and y values: $[2, 3] + [3, -1] = [5, 2]$. See Figure 2.13c. Vector addition can be used to find the product of two translations. For example, the product of a translation represented by [2, 3] and a translation represented by [3, −1] is a translation represented by [5, 2].

Figure 2.14a shows a translation of segment $\overline{AB}$ in a direction represented by the vector $\boldsymbol{AA'}$. Figure 2.14b shows a translation of triangle ABC in a direction represented by the vector $\boldsymbol{BB'}$. Check each of the following observations about a translation, and list any additional discoveries you make during this activity.

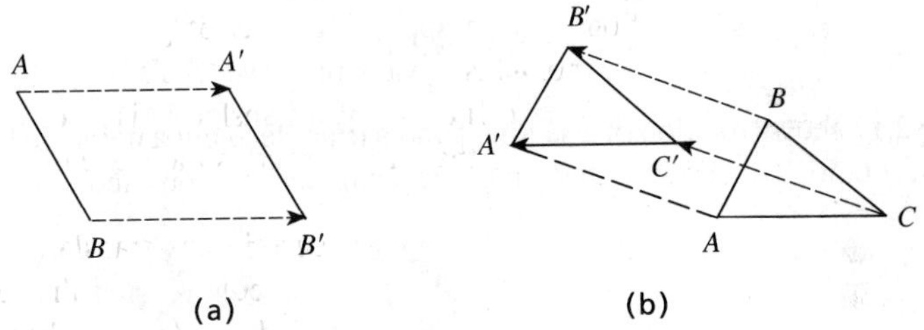

Figure 2.14

- ▲ A segment is translated into a parallel segment.
- ▲ All the vectors connecting corresponding points are equal. (Equal vectors have the same length and direction.)
- ▲ The inverse of a translation is another translation the same distance in the opposite direction.
- ▲ As illustrated in Figure 2.15, the product of two translations is a translation. The vector for the product is the sum of the vectors for the two translations.

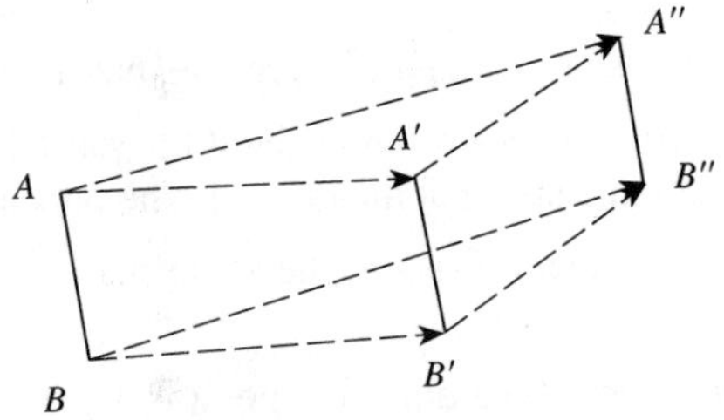

Figure 2.15

- ▲ The set of all translations forms a group. The identity for this group is a translation with zero vector, for which each point is its own image.

Additional information about translations can be found in the next sections, where the discussion includes the use of coordinates.

The second basic type of motion of the plane is a *rotation*. The symbol $R(O, \alpha)$ indicates a rotation through an angle of α about the point O, as in Figure 2.16. By convention, a counterclockwise rotation is associated with a positive angle.

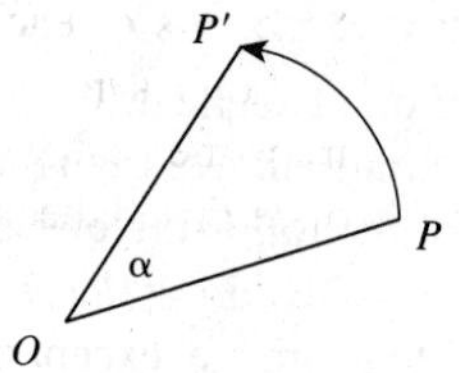

Figure 2.16

Figure 2.17 shows rotation of a segment and a triangle about a point. In both cases, the angle of rotation is $\angle AOA'$. Check the following observations about the motion of rotation:

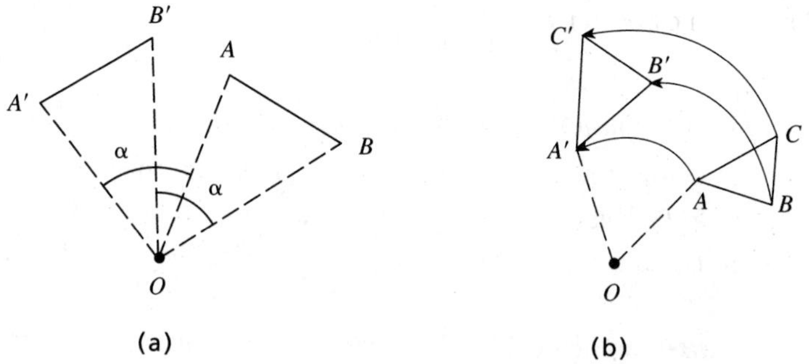

Figure 2.17

- A segment usually is not parallel to its image, but it may be.
- The inverse of a rotation is a rotation about the same point and with an angle of rotation of the same size, but measured in the opposite sense.
- The product of two rotations about the same point is another rotation about that point.
- The set of all rotations about one fixed point is a group of transformations.
- Rotations with center O and angles of α and $\alpha + 360°n$ are the same rotation (for $n = 1, 2, 3, \ldots$).

The set of all translations and rotations is called the set of *rigid motions* or *displacements*. Rigid motions can be illustrated by using a cardboard model of a triangular region and sliding or turning the model from one position to another in a rigid way without changing its size or shape.

The third basic example of motion of a plane is a *reflection*. A reflection R_1 utilizes a fixed line l, as in Figure 2.18. A point on l is its own image. Any other point P is mapped into a point P' such that l is the perpendicular bisector of $\overline{PP'}$. (The notation $\overline{PP'}$ means the segment with P and P' as endpoints.) As the name *reflection* implies, a set of points and their images are reflections of each other; it is as if line l were a mirror. A set of points having l as a line of symmetry is mapped onto itself by a reflection about l. One half of the figure is the image of the other half—the two sets of points are symmetric with respect to the line. Check these additional observations about reflections:

- A segment is not parallel to its image, except in some special cases.
- The inverse of a reflection is the same reflection.
- The product of two reflections about the same line is not a reflection.
- A reflection cannot be considered a sliding in the plane. It is necessary to flip over the cardboard model of the triangular region to have it correspond to the image. The clockwise orientation of $\triangle ABC$ is reversed in its image $\triangle A'B'C'$. Clockwise orientation means the path from A to B to C is in a direction around the figure corresponding to the direction in which the hands of a clock move around a circle. See Figure 2.19.

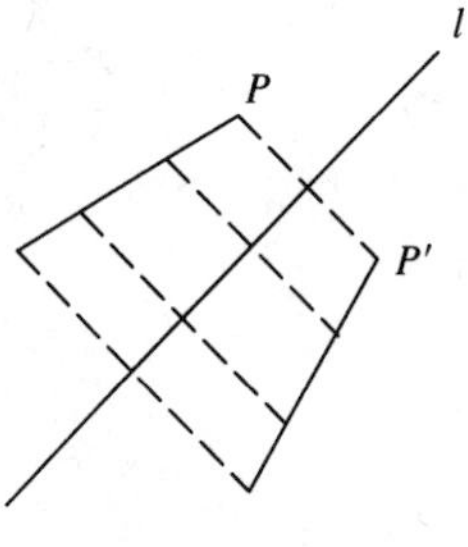

Figure 2.18

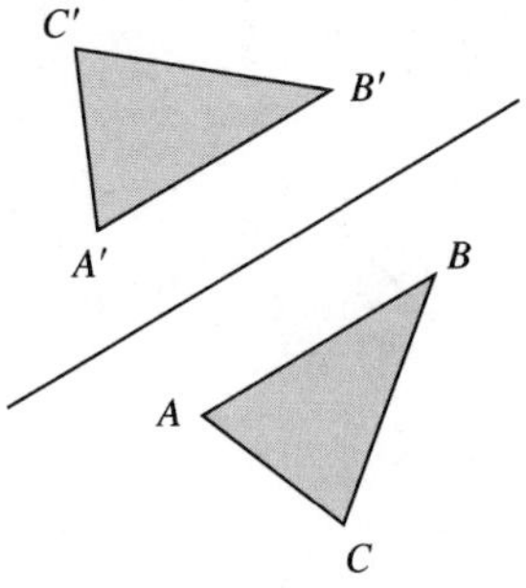

Figure 2.19

The three transformations given so far—translation, rotation, and reflection—may be considered the three basic motions of the plane. But one more type of transformation is necessary for closure. What this means is that some products of two of these transformations are not a translation, a rotation, or a reflection, but yet another type of transformation. Specifically, the product of a reflection and one of the other two transformations may not be an element of the set.

The fourth type of motion of a plane is a *glide reflection*. A glide reflection is the product of a reflection and a translation parallel to the fixed line of reflection, as in Figure 2.20, in which P is mapped into P''. The same glide reflection arises if the translation is done first. (See Exercise 10.)

The exercises include questions that will require experimentation with drawings or cardboard models to provide additional information about glide reflections.

Example ➤ For $\triangle BCD$, Figure 2.21a, sketch the image under: (1) a translation represented by the vector shown; (2) a rotation $R(A, 60°)$; (3) a reflection about the line shown; and (4) a glide reflection about the line shown using a translation vector of the same length as the one given, parallel to the line, and pointing downward. The answers are given in Figures 2.21b–e.

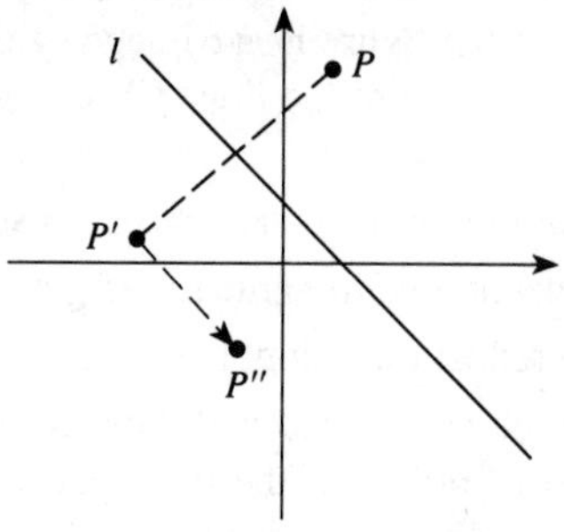

Figure 2.20

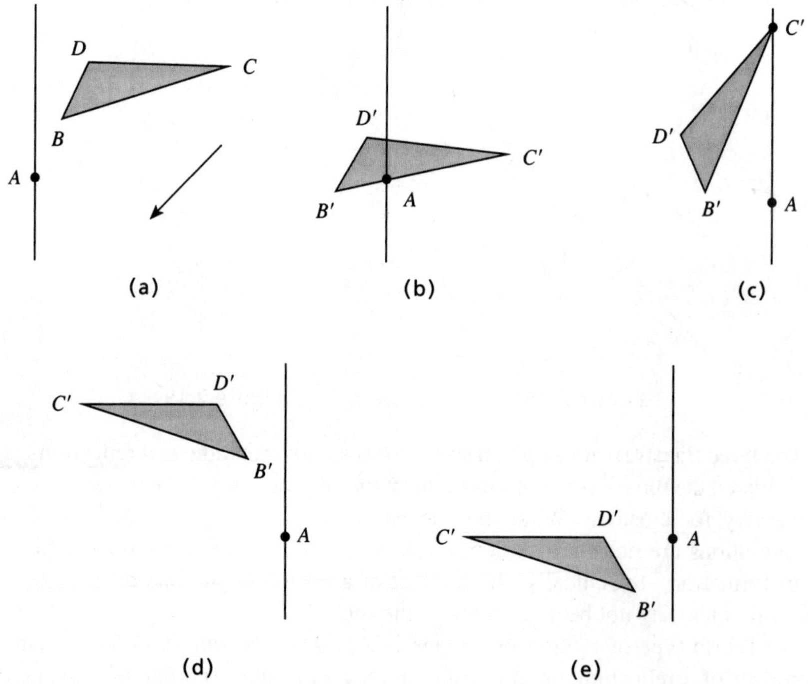

Figure 2.21

The language of Euclidean motions can often be used effectively as an alternate way of explaining elementary concepts in geometry.

Example ➤ **a)** See Figure 2.22a. Traditional: Point A is on the perpendicular bisector of segment AB if and only if it is equidistant from A and B. Using Motions: Point A is on the perpendicular bisector of segment AB if and only if A and B are corresponding points in a reflection about the line through P that is perpendicular to segment AB. **b)** See Figure 2.22b. Traditional: The diagonals of a parallelogram bisect each other. Using Motions: Since B is the image of D and C is the image of A under a rotation of 180 degrees about E, $EB = ED$ and $EC = EA$.

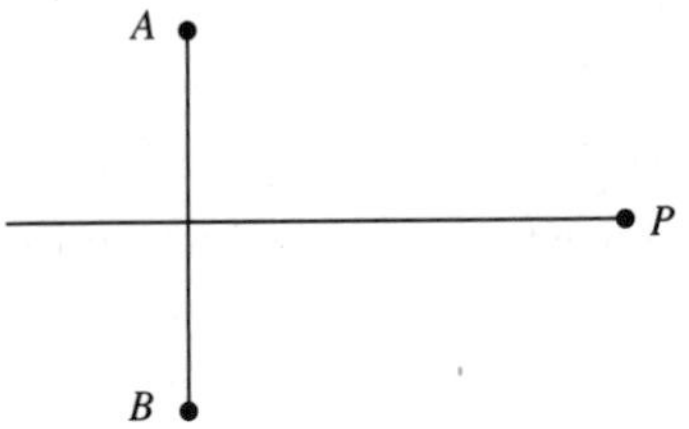

(a)

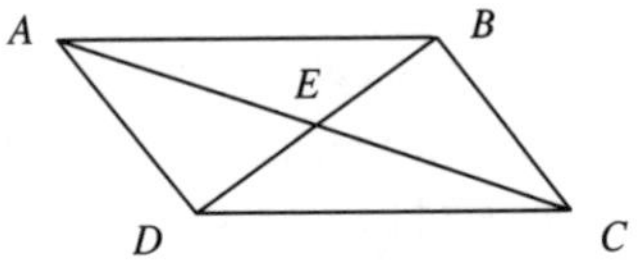

(b)

Figure 2.22

One geometric application of various motions of the plane is in simple geometric designs that form plane–filling repeated patterns. The Dutch artist M. C. Escher extended this idea to very intricate plane–filling patterns involving such figures as fish, birds, horsemen, and reptiles. One Escher drawing is shown in Figure 2.23. (© M.C. Escher Heirs, c/o Cordon Art–Baarn–Holland). The interested reader will enjoy studying these works of art in *The Graphic Work of M.C. Escher*.

Figure 2.23

The somewhat intuitive introduction to plane motions in this section is followed by three sections developing the concepts in greater detail. Section 2.4 develops the material using analytic geometry, Section 2.5 uses matrices to define transformations, and Section 2.6 discusses the set of all motions of the plane from a more general point of view.

▲ EXERCISES 2.3 ▲

1. For a translation, is the measure of the angle between two rays an invariant?

2. Could a translation other than one represented by a zero vector be its own inverse?

3. Does the set of all reflections about lines in a plane include an identity?

4. What is meant by the identity rotation?

5. Explain how a segment and its image might be parallel for a rotation other than the identity.

6. Find the product $R(A, \alpha)R(A, \beta)$.

7. Show that a reflection is an involutory transformation.

8. Describe the inverse of a rotation using a positive angle instead of a negative one.

9. Give an example in which a segment and its image are parallel under reflection.

10. For a glide reflection, show that the image point will be the same if the translation is followed by the reflection.

11. Describe the inverse of a glide reflection.

12. When are a segment and its image parallel under a glide reflection?

13. Can a model for a triangular region be made to coincide with its image under a glide reflection by sliding?

For Exercises 14–17, use Figure 2.24a and sketch the image of $\triangle ABC$ under each.

14. A translation represented by the vector shown

15. A rotation $R(D, 120°)$

16. A reflection about the line shown

17. A glide reflection about the line shown, with a translation vector upward the same length, but not the same direction, as the one shown

go straight up

For Exercises 18–21, use Figure 2.24b and sketch the image of $\triangle ABC$ under each.

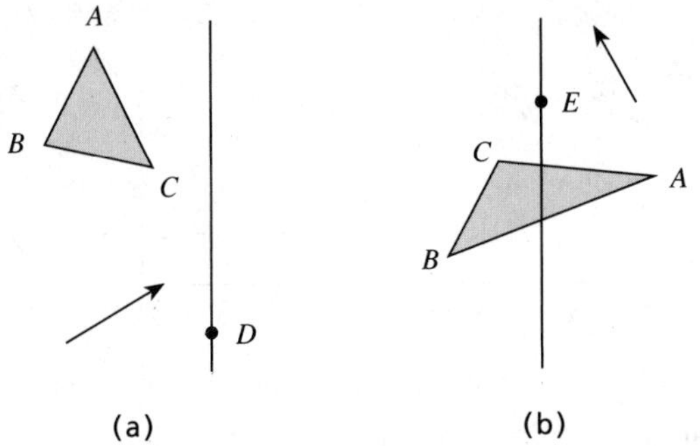

Figure 2.24

18. A translation represented by the vector shown

19. A rotation $R(E, 90°)$

20. A reflection about the line shown

21. A glide reflection about the line shown, with a translation vector the same length, pointing upward, as the one shown

22. Using a set of 6 students, 3 to represent the original vertices and 3 to represent the images, demonstrate by physical movements the result of each Euclidean motion on a triangle.

23. Investigate various Escher drawings using motions, and analyze the types of motions used in each.

24. Study the article, "How to Draw Tessellations of the Escher Type" by Joseph L. Teeters, *Mathematics Teacher* 67 (1974); then draw a simple Escher-type plane-filling pattern.

25. Read and report on one of the articles in *M. C. Escher: Art and Science*, Proceedings of the Interdisciplinary Congress on M. C. Escher, Rome, Italy, 26–28 March 1986.

▲ 2.4 Sets of Equations for Motions of the Plane

Motions of the plane previously explained in Section 2.3 can also be expressed as sets of linear equations. In Figure 2.25a for example, if $P(1, 2)$ has the image $P'(3, 3)$, then the translation may be represented by the equations

$$x' = x + 2$$
$$y' = y + 1$$

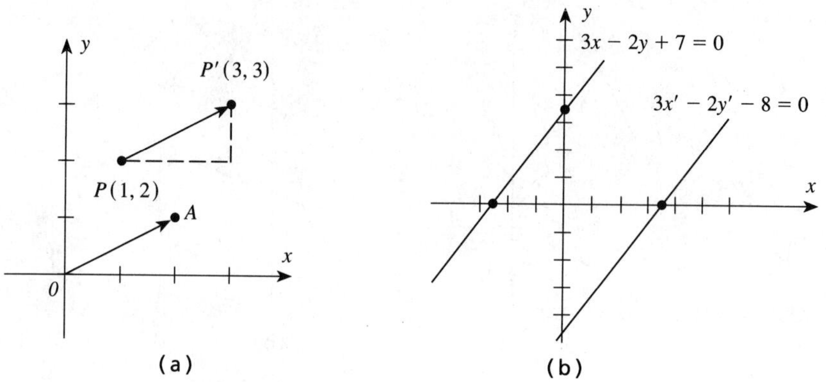

Figure 2.25

A translation can be defined as a transformation with equations of the form

$$x' = x + a$$

$$y' = y + b$$

for a and b any real numbers. The inverse of this transformation has the equations

$$x = x' - a$$

$$y = y' - b$$

Example ➤ Find the image of $(5, 2)$ under the translation with equations $x' = x + 7, y' = y + 3$. The solution is $(12, 5)$.

Example ➤ Find the image of the line $3x - 2y + 7 = 0$ under the translation with equations $x' = x + 7, y' = y + 3$. See Figure 2.25b.

Since values for x and y are needed to substitute for x and y in the equation, first solve the equations for the translation for x and y:

$$x = x' - 7$$

$$y = y' - 3$$

Then $3x - 2y + 7 = 0$ has the image $3(x' - 7) - 2(y' - 3) + 7 = 0$, which can be written as $3x' - 2y' - 8 = 0$.

In analytic geometry textbooks, it is shown that the equations for a rotation about the origin are

$$x' = x \cos \alpha - y \sin \alpha$$

$$y' = x \sin \alpha + y \cos \alpha$$

Example ➤ Find the image of $P(2, 3)$ under a rotation of $60°$ about the origin.

The equations are

$$x' = x \cos 60^\circ - y \sin 60^\circ$$

$$y' = x \sin 60^\circ + y \cos 60^\circ$$

Then

$$x' = (2)\left(\frac{1}{2}\right) - (3)\left(\frac{\sqrt{3}}{2}\right)$$

$$y' = (2)\left(\frac{\sqrt{3}}{2}\right) - (3)\left(\frac{1}{2}\right)$$

The coordinates of P' are

$$\left(1 - \frac{3}{2}\sqrt{3}, \sqrt{3} + \frac{3}{2}\right)$$

The equations of rotation about a point (h, k) are

$$x' - h = (x - h) \cos \alpha - (y - k) \sin \alpha$$

$$y' - k = (x - h) \sin \alpha + (y - k) \cos \alpha$$

To derive these equations, start with the translation that relates the coordinates of point $P(x, y)$, Figure 2.26, to the xy and $x''y''$ axes:

$$x'' = x - h$$

$$y'' = y - k$$

Then the rotation with respect to the origin of the $x''y''$ axes is

$$x'' = (x - h) \cos \alpha - (y - k) \sin \alpha$$

$$y'' = (x - h) \sin \alpha + (y - k) \cos \alpha$$

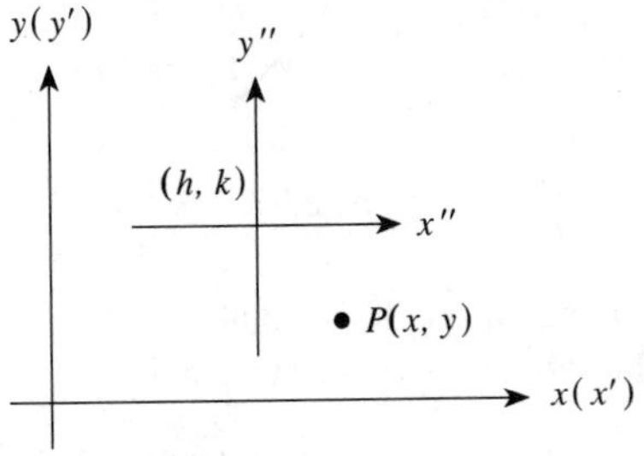

Figure 2.26

Now translate back to the original axes, relabeled $x'y'$, so that

$$x'' = x' - h$$

$$y'' = y' - k$$

The sets of equations for reflections are not as familiar as those for the transformations of translation and rotation. For the special case of reflection about the x-axis, the equations are

$$x' = x$$
$$y' = -y$$

The forms are also simple when the reflection is about the y-axis.

The production of equations for other reflections will also be developed, beginning with simpler cases. A reflection about the line $y = mx$ is shown in Figure 2.27, where $m = \tan\theta$. It is possible to derive the equations by looking at the congruent triangles OPR and $OP'R$, but an alternative method allows us to use ideas from earlier in the section. The reflection about $y = mx$ is a reflection about the x-axis, combined with a rotation through an angle of 2θ about the origin. This can be seen from Figure 2.27 because of the congruences $\angle POB \cong \angle P'OB$ and $\angle POA \cong \angle P''OA$. The equations for this reflection

$$x' = x\cos 2\theta + y\sin 2\theta$$
$$y' = x\sin 2\theta - y\cos 2\theta$$

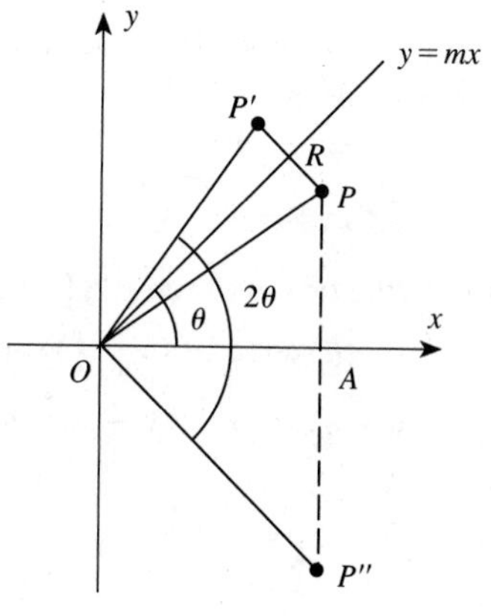

Figure 2.27

are obtained by writing the equations for a rotation of 2θ about the origin, then substituting $-y$ for y from the reflection about the x-axis.

Example ➤ Find the image of (2, 3) under a reflection about the line $y = x$. Since $\cos 2\theta = 0$ and $\sin 2\theta = 1$,

$$x' = y$$
$$y' = x$$

hence,

$$x' = 3$$
$$y' = 2$$

and the point is (3, 2), as should be expected.

The equations for reflection about a line $y = mx + b$ can be obtained from the special case of reflection about a line through the origin by a translation through a distance twice the perpendicular distance d from $y = mx$ to $y = mx + b$. See Figure 2.32 in Section 2.6 for an explanation of why the distance for the translation is twice the distance between the parallel lines.

Since the translation is in the direction indicated by ϕ in Figure 2.28, with $\phi + (90° - \theta) = 360°$, the equations of the translation are

$$x' = x + 2d\cos\phi$$
$$y' = y + 2d\sin\phi$$

and the desired equations for the general reflection are

$$x' = x\cos 2\theta + y\sin 2\theta + 2d\cos\phi$$
$$y' = xsin2\theta - y\cos 2\theta + 2d\sin\phi$$

See Figure 2.28b for additional details of the trigonometry involved.

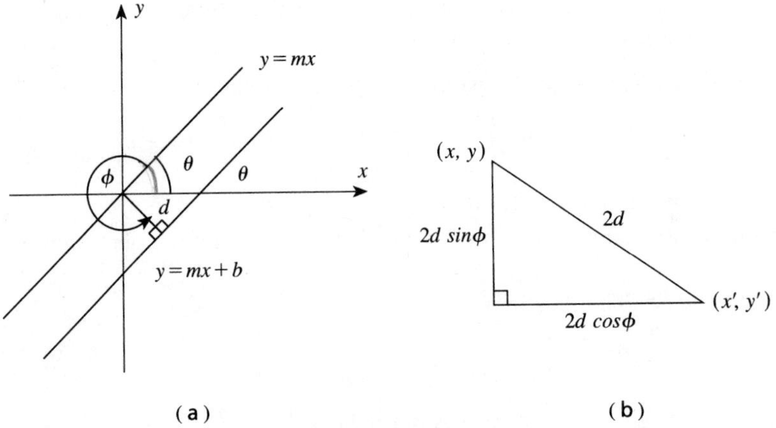

Figure 2.28

Example ➤ Find the image of point (2, 6) under a reflection about the line such that $\theta = 30°$, $\phi = 300°$, and $d = 2$.

$$x' = x\cos 60° + y\sin 60° + 4\cos 300°$$
$$y' = x\sin 60° - y\cos 60° + 4\sin 300°$$
$$x' = (1/2)x + (\sqrt{3}/2)y + 2$$
$$y' = (\sqrt{3}/2)x - (1/2)y - 2\sqrt{3}$$

For $x = 2$, $y = 6$,

$$x' = 1 + 3\sqrt{3} + 2 = 3 + 3\sqrt{3}$$
$$y' = \sqrt{3} - 3 - 2\sqrt{3} = -3 - \sqrt{3}$$

The equations for a glide reflection are derived from those for a reflection and those for a translation in a direction parallel to the line of reflection. The images can be found using the two sets of equations given previously for these transformations.

A study of the form of the sets of equations for the four types of motions of the plane shows that all these sets are special cases of the sets of equations in the following theorem.

THEOREM 2.2 ➤ If a transformation is a plane motion, then it has equations of the form

$$\begin{aligned} x' &= ax + by + c \\ y' &= +(-bx + ay) + d \end{aligned} \qquad \text{or} \qquad \begin{aligned} x' &= ax + by + c \\ y' &= -(-bx + ay) + d \end{aligned}$$

for a, b, c, d real numbers satisfying $a^2 + b^2 = 1$.

The *converse* of Theorem 2.2 is also true, although the proof is not given here. The converse of a theorem stated as a conditional is obtained by interchanging the *if* and *then* parts of the statement. The converse is:

THEOREM 2.3 ➤ If a transformation has equations of the form

$$\begin{aligned} x' &= ax + by + c \\ y' &= +(-bx + ay) + d \end{aligned} \qquad \text{or} \qquad \begin{aligned} x' &= ax + by + c \\ y' &= -(-bx + ay) + d \end{aligned}$$

for a, b, c, d real numbers satisfying $a^2 + b^2 = 1$, then it is a plane motion.

Example ➤ The equations

$$x' = (1/\sqrt{2})x + (1/\sqrt{2})y + 3$$
$$y' = (-1/\sqrt{2})x + (1/\sqrt{2})y + 7$$

having $a^2 + b^2 = \frac{1}{2} + \frac{1}{2} = 1$, represent a plane motion.

The equations for the inverse of a transformation can be found by solving the set of equations for x and y. This was illustrated previously for a translation. The following example is for a reflection.

Example ➤ Find the inverse of the transformation with equations

$$x' = x\cos 30^\circ + y\sin 30^\circ$$
$$y' = x\sin 30^\circ - y\cos 30^\circ$$

Multiply the members of the first equation by $\sin 30^\circ$ and the members of the second by $\cos 30^\circ$:

$$
\begin{aligned}
x' \sin 30^\circ &= x \sin 30^\circ \cos 30^\circ + y \sin^2 30^\circ \\
\underline{y' \cos 30^\circ} &= \underline{x \sin 30^\circ \cos 30^\circ - y \cos^2 30^\circ} \\
x' \sin 30^\circ - y' \cos 30^\circ &= y
\end{aligned}
$$

Now multiply both members of the original first equation by $\cos 30^\circ$ and the members of the second by $\sin 30^\circ$:

$$
\begin{aligned}
x' \cos 30^\circ &= x \cos^2 30^\circ + y \sin 30^\circ \cos 30^\circ \\
\underline{y' \sin 30^\circ} &= \underline{x \sin^2 30^\circ - y \sin 30^\circ \cos 30^\circ} \\
x' \cos 30^\circ + y' \sin 30^\circ &= x
\end{aligned}
$$

As should be expected, a change of variables x, x' and y, y' will make obvious that the inverse of a reflection is the original transformation itself.

Products of motions can be represented by sets of equations derived by applying the second transformation to the equations for the first.

Example ➤ Find the set of equations that represents the product of two translations, if the first has the vector [7, 3] and the second has the vector [2, 4].

The set of equations for the first motion is $x' = x + 7$, $y' = y + 3$. The set of equations for the second motion is $x'' = x' + 2$, $y'' = y' + 4$. When these values are substituted in the first set, the result is

$$
\begin{aligned}
x'' - 2 &= x + 7 &\qquad& x'' = x + 9 \\
&& \text{or} && \\
y'' - 4 &= y + 3 && y'' = y + 7
\end{aligned}
$$

The concept of a *linear transformation* from algebra can be used to classify some Euclidean motions as special examples of this important kind of transformation. A linear transformation is defined for vector spaces with the same scalar fields. The transformation preserves vector sums and scalar products. Examples of motions that are linear transformations include a rotation about the origin and a reflection about a line through the origin. See Exercise 30I.

In the next section, the introduction of matrix multiplication to represent the product of transformations will make the work easier than deriving various sets of equations. Before studying Section 2.5, you may be interested in discovering for yourself how matrices can be used for transformations of sets of points.

▲ EXERCISES 2.4 ▲

For Exercises 1–4, let the vector for a translation be [3, −5] *and give the image for each point under the translation.*

1. (0, 0) **2.** (5, 7)

3. $(-1, -8)$

4. $(-2, -\frac{3}{2})$

For Exercises 5 and 6, let the vector for a translation be $[5, 9]$. *Each point is the image of which original point?*

5. $(0, 0)$

6. $(5, 9)$

7. Find the image of the line $y = 2x$ under the translation with vector $[-2, -3]$.

8. Find the image of the line $y = 7x$ under the translation with vector $[3, \frac{4}{5}]$.

For Exercises 9 and 10, find the image of the line under the translation with vector $[4, -3]$.

9. $2x - 5y + 7 = 0$

10. $3x + 4y + 2 = 0$

11. Derive the general formula for the image of the line $ax + by + c = 0$ under the translation with vector $[d, e]$.

12. What is the image of the point $(3, 4)$ under a rotation of $45°$ about the origin?

13. Find the image of $(2, 0)$ under a rotation of $30°$ about the origin.

14. What point has the image $(-1, 2)$ under a rotation of $45°$ about the origin?

15. Derive the equations for the inverse of a transformation of rotation about a point (h, k), with angle of rotation θ.

16. Find the image of $(3, 7)$ under a reflection about the line $y = (\sqrt{3}/3)x$.

17. Find the image of $(3, 7)$ under a reflection about the line with $\theta = 60°$, $\phi = 330°$, and $d = 5$.

18. What is the image of $(5, 7)$ under the glide reflection consisting of a reflection about the y-axis followed by a translation of three units in a positive direction parallel to the y-axis?

For Exercises 19–24, find the inverse transformation for the transformation whose equations are given:

19. $x' = x - 5$
$y' = y + 2$

20. $x' = x + 4$
$y' = y - 3$

21. $x' = -x$
$y' = -y$

22. $x' = x$
$y' = -y$

23. $x' = x \cos 30° - y \sin 30°$
$y' = x \sin 30° + y \cos 30°$

24. $x' = x \cos 60° + y \sin 60°$
$y' = x \sin 60° - y \cos 60°$

For Exercises 25 and 26, derive the sets of equations for the product of two translations if the vectors are those given.

25. $[6, 4], [-2, 3]$ add x's and add y's

26. $[-4, 1], [3, -6]$

27. Derive the general formula for the image of a point under the product of the two translations with vectors $[a, b]$ and $[c, d]$.

▷ **28.** Write a set of equations for a glide reflection and use it to find the image of points.

▷ **29.** Prove Theorem 2.3.

▷ **30.** Find the definition of a linear transformation in a textbook on linear algebra. Which motions of the plane are examples of linear transformations? Prove that your answer is correct. Investigate other examples of linear transformations.

▲ 2.5 Applications of Transformations in Computer Graphics

One of the most recent and important applications of transformations has been in the new field of computer graphics. Computer graphics refers to the use of the computer to store, manipulate, and show pictorial information. For a typical computer program, the output may be a set of numbers. But for a typical computer graphics program, the output will be a picture composed of a set of points.

In this section, and in later sections using matrices, computer graphics conventions are followed for rows or columns and the order of operations. If you have previously used other conventions in other courses, you will need to be flexible in following a different pattern.

In computer graphics, a point in two dimensions can be represented as a row matrix $(x \quad y)$. Recall that a matrix is a rectangular array of numbers. For example, the point $(2, 3)$ is represented as the matrix $(2 \quad 3)$. Many transformations of points in two dimensions can be represented as 2×2 matrices, with two rows and two columns. Then matrix multiplication can be used to show the effect of the transformation on the point. Recall that matrix multiplication is a row-by-column process. In general,

$$(x \quad y)\begin{pmatrix} a & b \\ c & d \end{pmatrix} = (ax + cy \quad bx + dy)$$

This is interpreted to mean that the point (x, y) is transformed into the point $(ax + cy, bx + dy)$. In considering the application of computer graphics, we can see the importance of transformations as providing the means, through matrix multiplication, of performing the geometric manipulations that will change the picture from one set of points to a modified set.

Example ➤ Find the image of the point $(3, 4)$ under the transformation with matrix $\begin{pmatrix} 1 & 2 \\ -1 & 3 \end{pmatrix}$:

$$(3 \quad 4)\begin{pmatrix} 1 & 2 \\ -1 & 3 \end{pmatrix} = (-1 \quad 18)$$

The image is $(-1, 18)$.

Here are some of the special 2×2 matrices representing particular geometric manipulations that are useful in computer graphics.

$\begin{pmatrix} 1 & 0 \\ 0 & 1 \end{pmatrix}$	Identity
$\begin{pmatrix} a & 0 \\ 0 & 1 \end{pmatrix}$	A scale change in the x direction
$\begin{pmatrix} -1 & 0 \\ 0 & 1 \end{pmatrix}$	Reflection about the y-axis
$\begin{pmatrix} 1 & 0 \\ 0 & -1 \end{pmatrix}$	Reflection about the x-axis
$\begin{pmatrix} 1 & b \\ 0 & 1 \end{pmatrix}$	Shear in y direction
$\begin{pmatrix} 0 & 1 \\ -1 & 0 \end{pmatrix}$	Rotation of 90°
$\begin{pmatrix} -1 & 0 \\ 0 & -1 \end{pmatrix}$	Rotation of 180°
$\begin{pmatrix} 0 & -1 \\ 1 & 0 \end{pmatrix}$	Rotation of 270°
$\begin{pmatrix} 0 & 1 \\ 1 & 0 \end{pmatrix}$	Reflection about line $y = x$
$\begin{pmatrix} a & 0 \\ 0 & a \end{pmatrix}$	Scaling change in both x and y direction
$\begin{pmatrix} a & 0 \\ 0 & b \end{pmatrix}$	Unequal scaling change
$\begin{pmatrix} \cos\theta & \sin\theta \\ -\sin\theta & \cos\theta \end{pmatrix}$	General rotation about origin

Many of these transformations are already familiar, but a few are new. A scale change of 4 units in the x direction is shown in the following equation:

$$(2 \quad 3)\begin{pmatrix} 4 & 0 \\ 0 & 1 \end{pmatrix} = (8 \quad 3)$$

The result is to multiply the x coordinate by 4. A shear, such as $\left(\begin{smallmatrix} 1 & 2 \\ 0 & 1 \end{smallmatrix}\right)$ has the geometric effect of changing (x, y) to $(x, 2x + y)$, so that the y coordinate is changed by adding a multiple of the x coordinate. For $(2, 3)$, the image is $(2, 7)$. This is a shear in the y direction. Similarly, a shear such as $\left(\begin{smallmatrix} 1 & 0 \\ 2 & 1 \end{smallmatrix}\right)$ is a shear in the x direction.

Changing one point at a time would be a very tedious way of changing the appearance of a picture in computer graphics. Fortunately, however, more than one point can be changed at the same time. Just by changing the two endpoints, an entire line segment can be changed. For example, the transformation $\left(\begin{smallmatrix} 1 & 2 \\ 3 & 1 \end{smallmatrix}\right)$ can be applied to the endpoints (0, 1) and (2, 3) of a segment by means of a single matrix multiplication, in which each row of the matrix on the left represents one point:

$$\begin{pmatrix} 0 & 1 \\ 2 & 3 \end{pmatrix} \begin{pmatrix} 1 & 2 \\ 3 & 1 \end{pmatrix} = \begin{pmatrix} 3 & 1 \\ 11 & 7 \end{pmatrix}$$

The image of (0, 1) is (3, 1) and the image of (2, 3) is (11, 7). Both the length and the slope of the segment were changed in this case. See Figure 2.29.

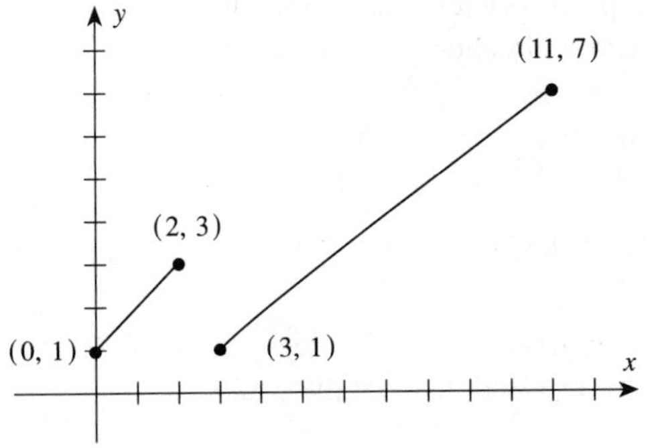

Figure 2.29

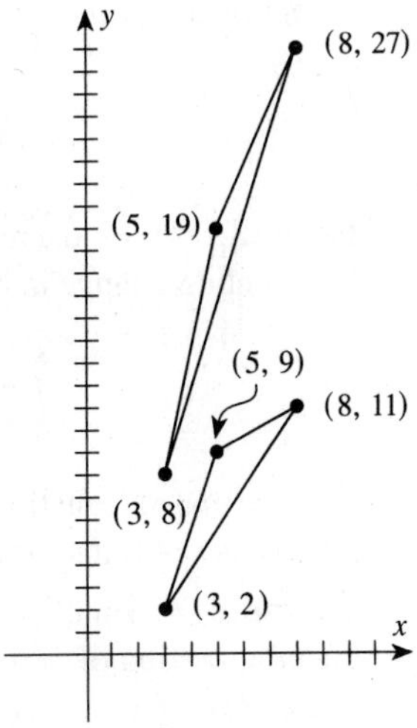

Figure 2.30

By a generalization of this technique, entire polygons can be transformed by transforming the vertices.

Example ➤ A scale change in the x direction for a triangle:

$$\begin{pmatrix} 3 & 2 \\ 5 & 9 \\ 8 & 11 \end{pmatrix} \begin{pmatrix} 3 & 0 \\ 0 & 1 \end{pmatrix} = \begin{pmatrix} 9 & 2 \\ 15 & 9 \\ 24 & 11 \end{pmatrix}$$

Example ➤ A shear in the y direction for a triangle. See Figure 2.30.

$$\begin{pmatrix} 3 & 2 \\ 5 & 9 \\ 8 & 11 \end{pmatrix} \begin{pmatrix} 1 & 2 \\ 0 & 1 \end{pmatrix} = \begin{pmatrix} 3 & 8 \\ 5 & 19 \\ 8 & 27 \end{pmatrix}$$

Example ➤ A quadrilateral is reflected about the y-axis:

$$\begin{pmatrix} 3 & 6 \\ 4 & 3 \\ 6 & 2 \\ 7 & 5 \end{pmatrix} \begin{pmatrix} -1 & 0 \\ 0 & 1 \end{pmatrix} = \begin{pmatrix} -3 & 6 \\ -4 & 3 \\ -6 & 2 \\ -7 & 5 \end{pmatrix}$$

Products of transformations can be represented by a matrix that is the product of the matrices for the transformations.

Example ➤ Find a matrix that represents a rotation of 90° followed by a reflection about the line $y = x$:

$$\begin{pmatrix} 0 & 1 \\ -1 & 0 \end{pmatrix} \begin{pmatrix} 0 & 1 \\ 1 & 0 \end{pmatrix} = \begin{pmatrix} 1 & 0 \\ -1 & -1 \end{pmatrix}$$

Example ➤ Find a matrix that represents a reflection about the y-axis followed by a scaling change in both the x and y directions:

$$\begin{pmatrix} -1 & 0 \\ 0 & 1 \end{pmatrix} \begin{pmatrix} a & 0 \\ 0 & a \end{pmatrix} = \begin{pmatrix} -a & 0 \\ 0 & a \end{pmatrix}$$

A single matrix equation can be used to find the image of the vertices of a polygon under a product of transformations.

Example ➤ Find the image of the triangle with vertices $(2, 5), (7, 3), (6, 4)$ under a rotation of 180° followed by a reflection about the y-axis.

$$\begin{pmatrix} 2 & 5 \\ 7 & 3 \\ 6 & 4 \end{pmatrix} \begin{pmatrix} -1 & 0 \\ 0 & -1 \end{pmatrix} \begin{pmatrix} -1 & 0 \\ 0 & 1 \end{pmatrix} = \begin{pmatrix} 2 & -5 \\ 7 & -3 \\ 6 & -4 \end{pmatrix}$$

Example ➤ Find the image of the triangle with vertices $(2, 5), (7, 3), (6, 4)$ under a scale change of 3 units in the x direction followed by a y-shear of 2 units.

$$\begin{pmatrix} 2 & 5 \\ 7 & 3 \\ 6 & 4 \end{pmatrix} \begin{pmatrix} 3 & 0 \\ 0 & 1 \end{pmatrix} \begin{pmatrix} 1 & 2 \\ 0 & 1 \end{pmatrix} = \begin{pmatrix} 6 & 17 \\ 21 & 45 \\ 18 & 40 \end{pmatrix}$$

Although many transformations in computer graphics can be shown using a 2×2 matrix, a very common one—translation—cannot. A translation in two dimensions can be shown using a 3×3 matrix with a new type of coordinate system that will be introduced in Chapter 7. You may want to try to discover such a 3×3 matrix for yourself, before studying Chapter 7. For this reason, and others to be explained in Chapter 7, projective geometry is ordinarily employed in computer graphics.

The subject of computer animation is a further extension of the application of transformations in computer graphics. Computer animation consists of creating a figure, erasing the figure, using a transformation to recreate the figure (or a modified figure) in a new location, and rapidly continuing this process to simulate motion. Because of the possibility of showing many repetitions each second, the computer is able to simulate movement as in a motion picture film. Using a relatively short computer program with loops, computer science students can create short "movies."

A different application of geometry in the broad area of computer graphics is the expanding field of cellular automata. We can think of using the computer screen to create a stylized universe. Space is represented by a grid, with each individual cell containing only a few bits of data. The laws of this universe are such that any change in the state of a cell is computed by considering that of neighboring cells. Cellular automata can be used to model biological processes as well as combinatorial mathematics. Perhaps the best known early use of cellular automata is in the game of "Life" developed by John Conway in 1970.

This section has considered applications of geometric transformations in computer graphics. There are many other applications of geometry in computer science, particularly in computer-aided design, which are included in computer science texts.

▲ EXERCISES 2.5 ▲

For Exercises 1–4, find the image of the given point under the transformation with the given matrix.

1. $(2, 3)\begin{pmatrix} 1 & -2 \\ 4 & 6 \end{pmatrix}$

2. $(0, 5)\begin{pmatrix} 4 & 0 \\ 2 & -5 \end{pmatrix}$

3. $(3, -2)\begin{pmatrix} -2 & 1 \\ 6 & 2 \end{pmatrix}$

4. $(-4, 1)\begin{pmatrix} 5 & 2 \\ -1 & -3 \end{pmatrix}$

For Exercises 5–14, find the image of the given points under the transformation described.

5. $(3, 4), (2, 7)$ Scale change of 3 units in x direction

6. $(4, 5), (-1, 3)$ Rotation of 180°

7. $(4, 1), (-2, 3)$ Reflection about line $y = x$

8. $(0, 2), (4, -5)$ Scale change of 4 units in both x and y directions

9. $(0, 3), (7, 0), (5, 5)$ Reflection about the x-axis

10. $(2, -1), (0, 4), (3, 3)$ Rotation of 90°

11. $(2, 2), (3, 3), (4, 4), (5, 5)$ Rotation of 180°

12. (3, 4), (1, 6), (2, 3), (9, 5) Scale change of 5 units in x direction
13. (−5, 4), (−2, −3), (−5, −2), (−1, 3), (−6, 5) Scale change of 5 units in both x and y directions
14. The same five points as in Exercise 13, with reflection about line $y = x$

For Exercises 15–20, sketch the original figure, apply the transformation, then sketch the image figure on the same set of axes.

15. A triangle with vertices (6, 5), (7, 3), (8, 2) under a shear with matrix $\begin{pmatrix} 1 & 4 \\ 0 & 1 \end{pmatrix}$
16. A triangle with vertices as in Exercise 15 under an unequal scaling change with matrix $\begin{pmatrix} 3 & 0 \\ 0 & 4 \end{pmatrix}$
17. A quadrilateral with vertices (3, 3), (9, 3), (9, 5), (3, 5) under a rotation of 90°
18. The same quadrilateral as in Exercise 17 under a reflection about the line $y = x$
19. A hexagon with vertices (3, 2), (7, 3), (8, 7), (6, 10), (2, 6), (−2, 4) under a rotation of 270°
20. The same hexagon as in Exercise 19 under a rotation of 180°

For Exercises 21 and 22, find the matrix representing the product of the two transformations.

21. Rotation of 270° followed by reflection about x-axis
22. Reflection about y-axis followed by scale change of −5 units in the x direction

For Exercises 23 and 24, find the image of the triangle whose vertices are given, with respect to the product of the transformations.

23. (3, 9), (5, 2), (7, 4) Reflection about the x-axis followed by rotation of 270°
24. (−3, 6), (1, 5), (4, 2) y-shear of 3 units, followed by scaling change of 4 units in x and 5 units in y
25. Show that a translation cannot be done by multiplication by a 2×2 matrix. (*Hint:* Consider the point (0, 0).)

▷ 26. List nine specific transformations that can be represented by matrices; then find one matrix that represents the product of the transformations in some specific order.

▷ 27. Investigate the game of "Life" as an example of cellular automata.

▲ 2.6 Properties of the Group of Euclidean Motions

The Euclidean motions have been discussed from various points of view in each of the last three sections. This section provides more information about the structure of the set of motions and investigates some invariant properties of sets of points under these transformations.

THEOREM 2.4 ➤ The four types of Euclidean motions of the plane constitute a group of transformations.

The proof of Theorem 2.4 consists of showing that there exists an inverse for each element and that the product of any two elements is another element of the set.

It is easy enough to verify that an element of each of the four types of motions has an inverse element. For example, the inverse of a translation is another translation in the opposite direction. The check for closure with respect to multiplication is somewhat more complex. A synthetic proof is normally facilitated by showing that motions of the plane may be discussed as products of reflections.

THEOREM 2.5 ➤ If a transformation is a plane motion, then it is the product of three or fewer reflections, and conversely.

The theorem is proved in two parts. The first part is for translations and rotations, while the second part is for reflections and glide reflections.

1. Every translation or rotation is the product of two reflections, and conversely.

There are two cases, depending on whether the two lines of reflection intersect or are parallel.

a. If l_1 and l_2 are any two lines of the plane intersecting at 0, and if the angle from l_1 to l_2 is θ, then $R(O, 2\theta) = R_{l_2}R_{l_1}$.

In Figure 2.31, the rotation that transforms $\overline{PQ}$ into $\overline{P''Q''}$ can be considered the product of two reflections about l_1 and l_2.

b. If l_1, and l_2 are parallel lines, and if they are perpendicular to a translation vector **OP** and at a distance $|OP|/2$ apart, then $T_p = R_{l_2}R_{l_1}$.

In Figure 2.32, the motion that transforms points A and B onto A'' and B'' may be considered the product of two reflections about l_1 and l_2.

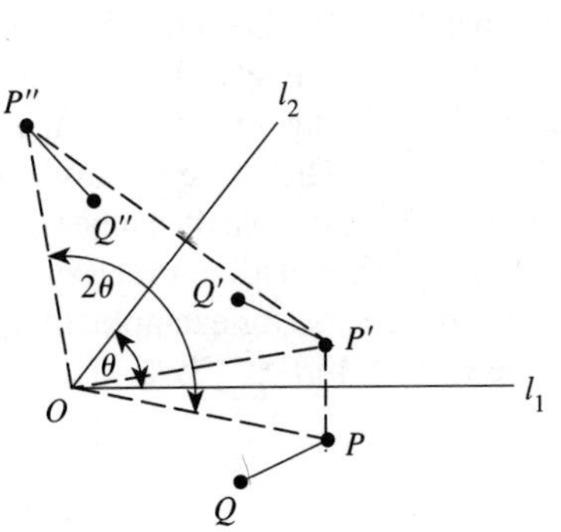

Figure 2.31

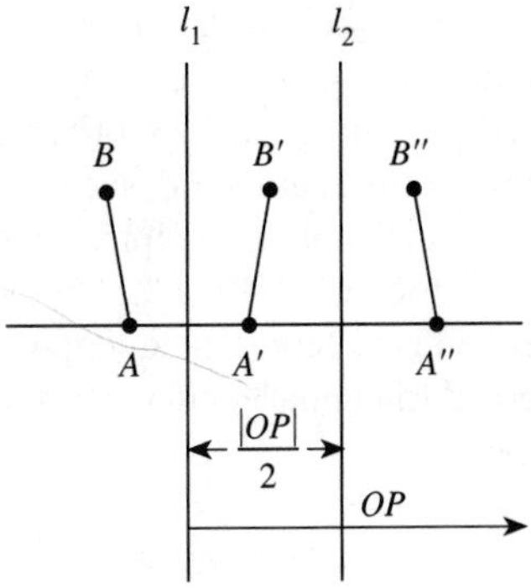

Figure 2.32

2. Every reflection or glide reflection is the product of three or fewer reflections. This statement is trivial for a reflection itself. Since a glide reflection is the product of a reflection and a translation, and since a translation is the product of two reflections, a glide reflection is the product of three reflections.

The proof of the converse of Theorem 2.5 is not completed. The converse states that if a transformation is the product of three or fewer reflections, then it is a plane motion. The type of motion depends on the relative position of the lines of reflection. The possibilities are summarized in Table 2.2.

Table 2.2

Product of Two Reflections	A. If the two lines of reflection are parallel, then the motion is a translation. B. If the two lines of reflection are nonparallel, then the motion is a rotation.
Product of Three Reflections	A. If two of the lines of reflection coincide, then the motion is a reflection. B. If the three lines of reflection are parallel, then the motion is a reflection. C. If two lines of reflection intersect at a point on the third, then the motion is a reflection. D. If two lines of reflection intersect at a point not on the third, then the motion is a glide reflection.

In Section 2.3, experimentation led to the conclusion that some motions could be represented by sliding a cardboard model in the plane, while others could not. This matter is now reconsidered in the light of Theorem 2.5.

DEFINITION ➤ A plane motion is a *direct* motion if it is the product of an even number of reflections. It is an *opposite* motion if it is the product of an odd number of reflections.

According to this definition, rotations and translations are direct motions, whereas reflections and glide reflections are opposite motions. Intuitively, the difference between direct and opposite motions can be seen by considering a model of a triangular region cut out of cardboard, such as $\triangle ABC$ in Figure 2.33. This piece of cardboard can be moved in the plane to represent any direct motion. For example, it can be translated to the position of $\triangle A'B'C'$. On the other hand, the paper must be turned over if the movement is to represent an opposite motion. It is necessary, for example, to turn the cardboard triangular region over for it to coincide with $\triangle A''B''C''$.

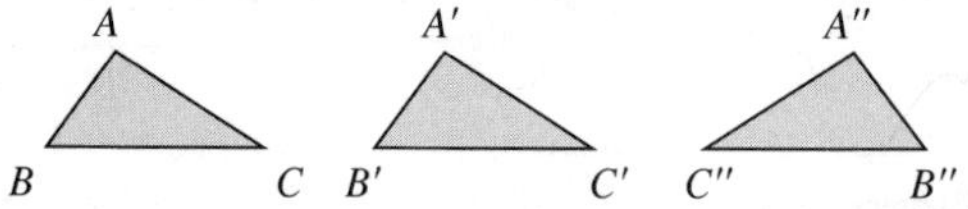

Figure 2.33

An additional intuitive idea concerning the distinction between direct and opposite motions is that direct motions preserve orientation, while opposite motions change it. For example, in Figure 2.33, in the original figure you can think about moving around the boundary from A to B to C, with the interior always on your left. This same orientation holds for $\triangle A'B'C'$. But in $\triangle A''B''C''$, if you think about moving around the border from A'' to B'' to C'', the interior will be on your right, so the orientation has changed.

The two sets of equations for a motion in Theorem 2.2 differ by a sign in the second equation. The plus sign indicates a direct motion, and the negative sign indicates an opposite motion. This may be verified by a check of each individual type of equation for a motion.

The products of direct and opposite motions are summarized in Table 2.3.

Table 2.3

	Direct +	**Opposite** −
Direct	Direct	Opposite
Opposite	Opposite	Direct

As one with sufficient algebraic background might suspect, the system of direct and opposite motions under multiplication is isomorphic to the system of even and odd numbers under addition (or the integers modulo 2). This can be seen by comparing Table 2.3 with the addition tables derived for the other two systems mentioned. The concepts of direct and opposite motions finally make it possible to show closure for Theorem 2.4.

Recall that a geometry consists of the study of properties invariant under a group of transformations. The group of plane motions, for example, are the transformations of ordinary Euclidean geometry. In addition to studying other invariant properties, we can also study invariant points. An invariant point is its own image under a transformation. Before reading ahead, you may want to see if you can discover the set of invariant points for each type of motion.

For each type of motion, the possibilities for numbers of invariant points without being the identity transformation are summarized in Table 2.4.

Table 2.4

Motion	**Possibilities for Invariant Points**
Translation	None
Rotation	One point
Reflection	One line
Glide reflection	None

At most, one line of the plane may remain invariant in a motion that is not the identity. In the light of this discovery, it is significant to ask about the minimum number of points and their images needed to uniquely determine a plane motion other than the identity. Proof of the following theorem will establish that the answer is three noncollinear points.

THEOREM 2.6 ➤ A motion of the plane is uniquely determined by an isometry of one triangle onto another.

PROOF: Let $\triangle ABC \cong \triangle A'B'C'$ in Figure 2.34. This notation means the two triangles are congruent and that one is the image of the other under a motion. Furthermore, the notation indicates that A, A'; B, B'; and C, C' are pairs of corresponding points. It is necessary to show there is exactly one motion of the plane agreeing with this isometry.

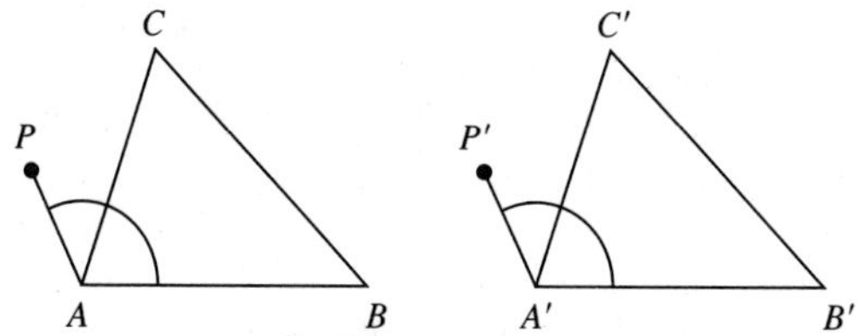

Figure 2.34

Let P be any point of the plane other than A, B, or C. If the isometry is direct, then P' is uniquely determined by constructing $\angle B'A'P' \cong \angle BAP$ and $A'P' = AP$. (The symbol $A'P'$ means the distance from A' to P', rather than the segment itself.) If it is opposite, the angle is constructed in the opposite sense.

Let P_1 and P_2 be two points of the plane in Figure 2.35. $\triangle AP_1P_2 \cong \triangle A'P'_1P'_2$, since $AP_1 = A'P'_1$, $AP_2 = A'P'_2$, and $\angle P_1AP_2 \cong \angle P'_1A'P'_2$. Thus, $P_1P_2 = P'_1P'_2$, so that a motion is determined because distance between two points is preserved.

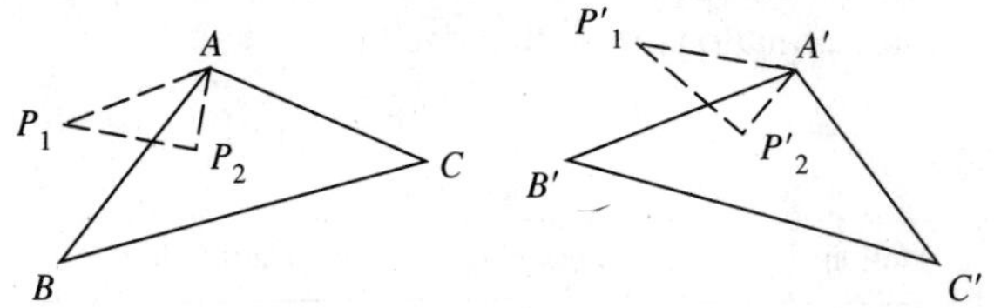

Figure 2.35

Since, for any given point P, there is only one point of the plane P' such that $PA = P'A'$, $PB = P'B'$, and $PC = P'C'$, P' is the image of P, and the motion is the only one for which $\triangle ABC$ has as its image $\triangle A'B'C'$.

One implication of Theorem 2.6 is that the coefficients in the equations of Theorem 2.3 can all be determined uniquely if three noncollinear points and their images are known. The motion, although unique, could actually be carried out as the product of other motions in numerous ways.

The properties studied in elementary geometry, such as congruence of segments, triangles, and angles, area of regions, and intersection of lines, are all properties of sets of points that are invariant under the group of motions.

The proof of the invariance of a property under a group of transformations can be either synthetic or analytic. A synthetic proof is one that does not use coordinates, and an analytic proof is one that does use coordinates. Both types of proofs are illustrated in examples.

Example ➤ Prove that a segment and its image are parallel under a translation.

In Figure 2.36 let $\overline{AB}$ be the segment and $\overline{A'B'}$ its image. $\triangle AA'B' \cong \triangle B'BA$, because pairs of corresponding sides are congruent. Thus, $\angle AB'A' \cong \angle B'AB$, and $\overline{AB}$ and $\overline{A'B'}$ are parallel.

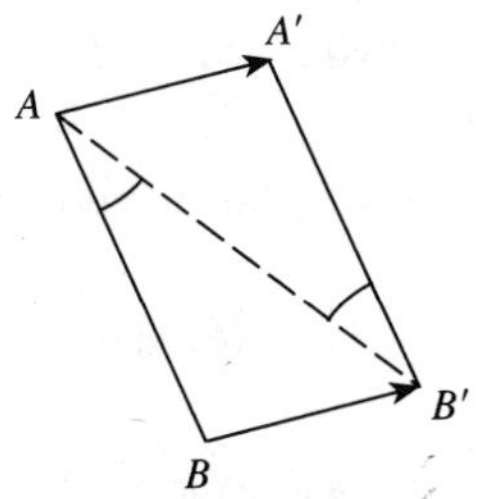

Figure 2.36

Example ➤ Prove analytically that a line and its image are parallel under a translation.

Let the equation of the line be $ax + by + c = 0$ and the equations of the translation be $x' = x + d$, $y' = y + e$. The equations of the inverse are $x = x' - d$, $y = y' - e$, so that the image of the line is $a(x' - d) + b(y' - e) + c = 0$, or $ax' + by' + (-ad - be + c) = 0$. This equation and the original equation have the same slope, so that the lines are parallel.

The power of analytic proofs is perhaps better illustrated by the next example, which uses a made-up property called "color," for which the invariance or lack of invariance may not be so obvious.

Example ➤ Is the property of color, defined as $c - (a + b)$ of a set of points $ax + by^2 + cy + d = 0$, invariant under a translation?

Let the equations of the translation be $x' = x + e$, $y' = y + f$. The image of the set of points is $a(x' - e) + b(y' - f)^2 + c(y' - f) + d = 0$, which can be simplified to the form $ax' + by'^2 + (c - 2bf)y' + k = 0$, where k is some constant. Color is invariant if $(c - 2bf) - (a + b) = c - (a + b)$. If these expressions are equal, then $c - 2bf$ must equal c, so that $2bf$ must equal 0, which is not generally true. Thus, the property is not invariant.

The realization that the group of motions allows very few changes in significant properties leads to the need to investigate more general types of transformations that do not have distance as an invariant property. One such group, the similarities, is introduced in Section 2.8.

▲ EXERCISES 2.6 ▲

1. Draw a figure for each case in Table 2.2.
2. Verify, as suggested, that the equations for direct motions have a + symbol and the equations for opposite motions have a − symbol.

For Exercises 3–6, use congruent triangles.

3. Draw a specific example showing that the product of a reflection and a rotation could be a glide reflection.
4. Draw a specific example showing that the product of a reflection and a rotation could be a reflection.
5. Draw a specific example showing that the product of two glide reflections could be a translation.
6. Draw a specific example to show that the product of two rotations about different points could be a rotation about a third point.
7. Does the set of all glide reflections form a subgroup of all the motions of a plane?
8. Does the set of all glide reflections and all translations form a subgroup of the group of all motions of the plane?
9. Does the set of all translations and all reflections form a subgroup of the group of all motions of the plane?
10. Except for special cases, you must know how many pairs of corresponding points in order to determine a unique motion of the plane?
11. Show analytically that a translation has no invariant points.
12. Investigate analytically the results of setting $x = x'$ and $y = y'$ in the equations of a rotation.
13. Use Theorem 2.6 to prove that, if three noncollinear points of a plane remain invariant under a motion, the motion is the identity.
14. Prove analytically that, for a translation, the image of two intersecting lines is two intersecting lines and the image of the original point of intersection is the new point of intersection.
15. Prove analytically that the angle between two intersecting lines is an invariant under the set of all translations.

16. Prove analytically that, for a reflection about the y-axis, the image of two intersecting lines is two intersecting lines and the image of the original point of intersection is the new point of intersection.

17. Prove analytically that the angle between two intersecting lines is an invariant for a reflection about the y-axis.

18. Draw a triangle and several other triangles congruent to it. In each case, describe how the first triangle could be transformed into the second (a) by one motion, and (b) by products of motions in various ways.

19. Use an analytic proof to show whether or not "color" is invariant under reflection with respect to the x-axis or the y-axis.

20. Make up a property (analogous to "color") for a set of points, and use an analytic proof to show whether or not it is invariant under translation.

21. Investigate, using analytic proofs, invariance of various properties with respect to rotation about the origin.

22. Investigate, using analytic proofs, invariance of various properties with respect to reflection about the line $y = x$.

23. Before reading the next section, see if you can discover some facts about the motions of three-space. For example, how many types exist? Are some direct and some opposite?

▲ 2.7 Motions of Three-Space

The motions of three-space can be applied to the set of points in Euclidean three-space without changing the distance between any two points. The motions of three-space are six in number. The three simplest motions are translation, rotation, and reflection, but these concepts must be clarified for three-space.

The definition of translation in space is analogous to the definition of translation for the plane.

DEFINITION ➤ The transformation T_A such that $T_A(X) = X + A$ is a translation of the space in the direction OA.

For example, in Figure 2.37, if A is the point $(2, 4, 8)$ and X is the point $(1, 2, 3)$, then $T_A(X) = (3, 6, 11)$. The image point can easily be found by addition of coordinates: Verify the fact that a translation in three-space sends all the points of a plane into the points of a parallel plane.

DEFINITION ➤ The rotation $R(l, \alpha)$ in three-space through an angle α about a fixed line is a counterclockwise rotation in a plane perpendicular to l. See Figure 2.38. The assumption is made here that a right-handed coordinate system has been established.

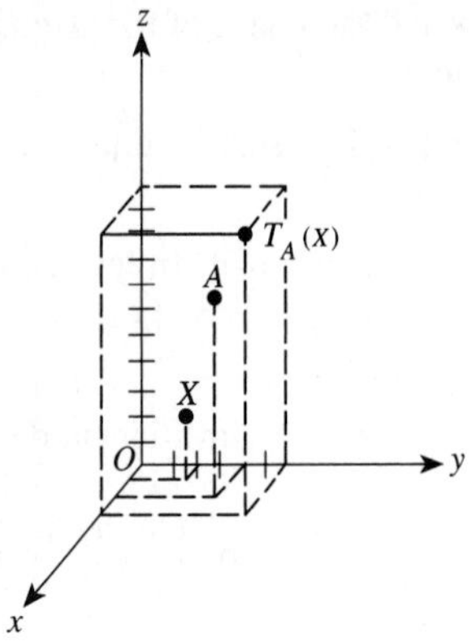

Figure 2.37

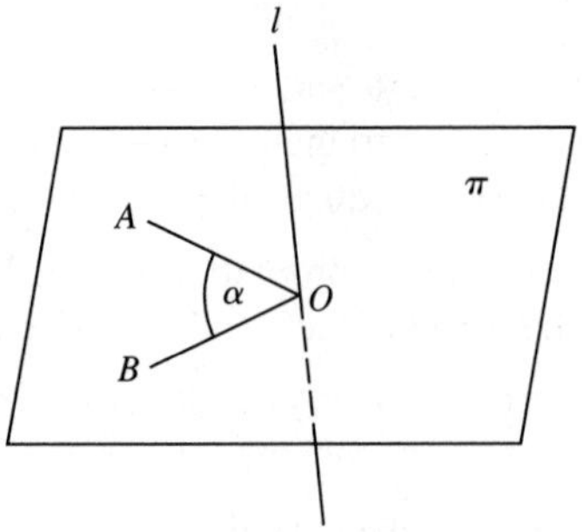

Figure 2.38

The image of point A not on l is a point B such that $\overleftrightarrow{AO}$ and $\overleftrightarrow{OB}$ determine a plane π perpendicular to l.

DEFINITION ➤ A reflection R_π in three-space is a reflection about a plane such that, for P and its image P', π is the perpendicular bisector of $\overline{PP'}$. See Figure 2.39. A set of points and its image under reflection are symmetric with respect to the plane of reflection.

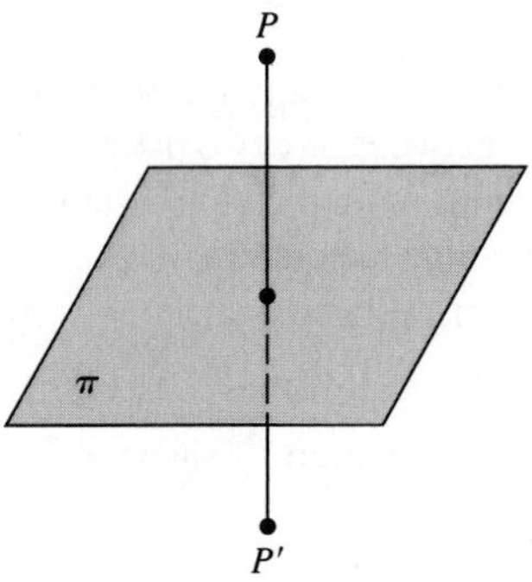

Figure 2.39

Example ➤ Find the image of the point (2, 3, 7) under a reflection about the plane $y = 1$.

The x and z coordinates are unchanged. Since the y coordinate is 2 units above 1, the y coordinate of the image must be 2 units below 1, so that the image is $(2, -1, 7)$.

The Platonic solids and other polyhedra are discussed in Section 8.5, and provide many examples of rotational symmetry and symmetry with respect to a plane in three-space. See Exercises 25I and 26I, Exercises 2.7.

The complete analytic presentation of motions for three dimensions is not given, but special cases are considered. The following equations for a translation in three-space are analogous to those for translations in a plane:

$$x' = x + a$$
$$y' = y + b$$
$$z' = z + c$$

For a rotation in space, if the fixed line is the z-axis and the plane is the xy-plane, then the image of a point in that plane can be found using the rotation equations given previously for rotation in a plane about the origin. For a point in a plane parallel to the xy-plane, the z coordinate will remain constant.

For a reflection about the xy-plane, the image of (a, b, c) is $(a, b, -c)$, hence

$$x' = x$$
$$y' = y$$
$$z' = -z$$

Corresponding sets of equations can be written for reflections about the xz- and yz-planes.

The matrices to accomplish a translation in three-space, like those for two-space, involve the use of homogeneous coordinates and will be discussed in Chapter 7. The matrix for the reflection about the xy-plane ($z = 0$) is

$$\begin{pmatrix} 1 & 0 & 0 \\ 0 & 1 & 0 \\ 0 & 0 & -1 \end{pmatrix}$$

In addition to the three simple motions, it is necessary to define three additional types of motions that are products of the basic ones so that the set of motions of three-space will be closed. Table 2.5 shows the definitions of the additional motions as products of two simpler motions.

Table 2.5

Motion	**Explanation**
a. Screw displacement	Rotation and translation along axis of rotation
b. Glide reflection	Reflection and translation parallel to a line in the plane of reflection
c. Rotatory reflection	Reflection and rotation with axis perpendicular to plane of reflection

A complete analysis of the motions of three-space is not given here, but the theorems are analogous to the theorems in two dimensions.

THEOREM 2.7 ➤ The motions of three-space constitute a group of transformations.

While the complete proof is not given here, it is analogous to the proof of Theorem 2.4 in that the analysis depends on classifying motions in three-space as direct or opposite and showing that each space motion can be expressed as the product of reflections about a plane. Table 2.6 shows which motions are direct and which are opposite. For three dimensions, the opposite motions are those that include a reflection. The table also gives a description of the invariant points for each.

Table 2.6

Motion in Three-Space	Direct or Opposite	Invariant Points
Rotation	Direct	A line
Translation	Direct	None
Reflection	Opposite	Plane of reflection
Rotatory reflection	Opposite	One point
Screw displacement	Direct	None
Glide reflection	Opposite	None

The table makes it possible to indicate the product of any two motions, but the results cannot be determined uniquely without more information. For example, the product of a screw displacement and a glide reflection is an opposite transformation; hence it is a reflection, a rotatory reflection, or a glide reflection.

The study of Table 2.6 shows that a reflection has an invariant plane and that no other motion has as extensive a set of invariant points. The following theorem is analogous to Theorem 2.6.

THEOREM 2.8 ➤ A motion in space is uniquely determined by a tetrahedron and its image.

The proof is left as an exercise.

One interesting application of geometry in three dimensional computer graphics is based on the need to change the data base consisting of sets of ordered triples, representing coordinates of points in 3 dimensions to a data base with 2 coordinates. This change makes it possible to choose the pixels to illuminate on the two dimensional screen so that the picture looks three dimensional.

Historically, this problem was solved (particularly in engineering courses) by applying trigonometry. For example, Figure 2.40 shows how the point with coordinates (3, 4, 5) can be made to correspond to the point with x, y coordinates $x = 4 - 3 \sin 30°$ and $y = 5 - 3 \cos 30°$, assuming that the angle between the x axis and the y and z axes is made 120 degrees.

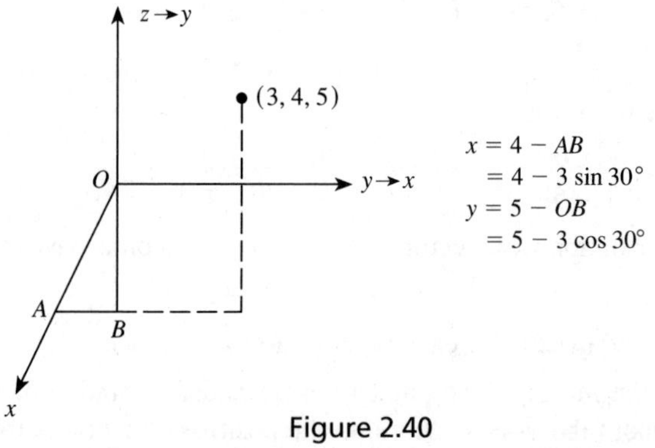

Figure 2.40

A second solution often used in practice is simply to drop the z coordinate. Geometrically, that would mean projecting the point onto the $z = 0$ plane. The disadvantage of this method is that the result doesn't look very three dimensional, although the results can be improved by applying rotations and/or perspectivities (discussed in Chapter 7) first.

A third solution is for the user simply to superimpose a grid over the three dimensional axes system and then establish a new system of x and y axes from which the new coordinates can be read. Although this method puts more work on the user instead of on the computer, it can be quick and simple to use, to find coordinates of the vertices of a polyhedron, for example. The method is illustrated in Figure 2.41, in which the points A, B, C with x, y, z coordinates given can be represented on the computer screen by the corresponding points A', B', C' with approximate x, y coordinates. One big advantage of superimposing a two dimensional grid over the three dimensional axes is that the scale can be changed by rescaling the grid in any way desired. A further advantage is that the figure can be rotated just by rotating the grid when it is superimposed.

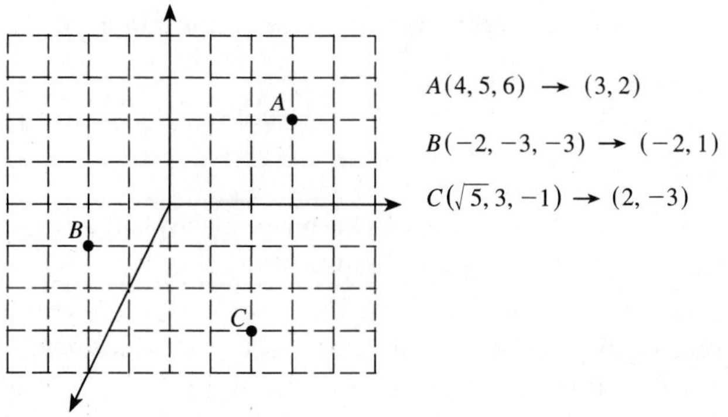

Figure 2.41

▲ EXERCISES 2.7 ▲

1. Let the vector for a translation be $[-2, 1, 7]$, and give the image for each point under the translation.

a. $(0, 0, 0)$ **b.** $(2, 5, 3)$
c. $(-2, -3, -5)$ **d.** $(0, -4, 1)$

2. In the translation with vector $[-1, -3, 4]$, what points have these images?

a. $(0, 0, 0)$ **b.** $(2, -1, 3)$

3. Describe the inverse for each type of motion of space.

4. What is the image of the point $(2, 4, 5)$ under a rotation of $45°$ in the plane $y = 4$ about the y-axis? Assume the positive direction is from the x- to the z-axis.

For Exercises 5–8, give the image of the point under a reflection about the given plane.

5. $(3, 8, 1)$, xy-plane **6.** $(-2, 4, 3)$, xz-plane
7. $(2, 3, 9)$, yz-plane **8.** $(-1, 4, -2)$, $x = 3$

9. Write 3×3 matrices for reflection about the coordinate planes and use matrix multiplication to verify the answers to Exercises 5–7.

10. Sketch an example of each of the last three types of motions of space.

11. Prove that, if four noncoplanar points of space remain invariant under a motion, the motion is the identity.

12. Name the possibilities for the type of motion that is the product of:

a. A rotatory reflection and a translation
b. A rotation and a glide reflection

13. Derive the equations for a reflection about the plane $z = 2$.

14. Prove Theorem 2.8.

Make a sketch showing a tetrahedron and its image for the following:

15. Translation

16. Rotation

17. Reflection

18. Derive a general formula for finding the image of the plane $ax+by+cz+d = 0$ for a translation in three-space with equations $x' = x+e$, $y' = y+f$, $z' = z+g$.

19. Derive a general formula for finding the image of a sphere $x^2 + y^2 + z^2 = d^2$ with respect to a translation in three-space with equations $x' = x + a$, $y' = y + b$, $z' = z + c$.

20. State and prove a space theorem analogous to the plane theorem that states that each plane motion is the product of three or fewer reflections.

21. Thinking of the classroom as part of the first octant, describe in words the result of students changing positions under various rotations, translations, and reflections in three-space. (For example, use a reflection about the plane of the floor.)

22. Develop sets of equations for additional motions of three-space besides the sets of equations given here for translation and a special case of reflection.

23. Investigate more about the use of trigonometric formulas for finding x, y coordinates on the two-dimensional computer screen corresponding to x, y, z coordinates for points in three dimensions.

24. Construct several grids and use them to convert from three-dimensional to two-dimensional coordinates for a computer screen.

25. Investigate rotational symmetry for the polyhedra in Section 8.5.

26. Investigate symmetry under reflection with respect to a plane for the polyhedra in Section 8.5.

27. Investigate the application of finite groups of symmetries in three-space in crystallography.

▲ 2.8 Similarity Transformations

The purpose of this section is to present similarities as examples of a somewhat more general type of transformation than the motions, including them as special cases. The transformations are applied to the points of the Euclidean plane or three-dimensional space. You are probably familiar with the study of similar triangles from secondary school geometry, and probably realize that ideas from the geometry of similarity are used in applications such as photographic enlarging, mapmaking, and the production of scale models. Distance between points is not always preserved but is modified in a consistent way throughout the plane, or space.

DEFINITION ➤ A plane similarity is a transformation of the plane onto itself such that if A, A' and B, B' are corresponding points, the length of $\overline{A'B'}$ is r times the length of $\overline{AB}$, where r is some nonzero real number. See Figure 2.42.

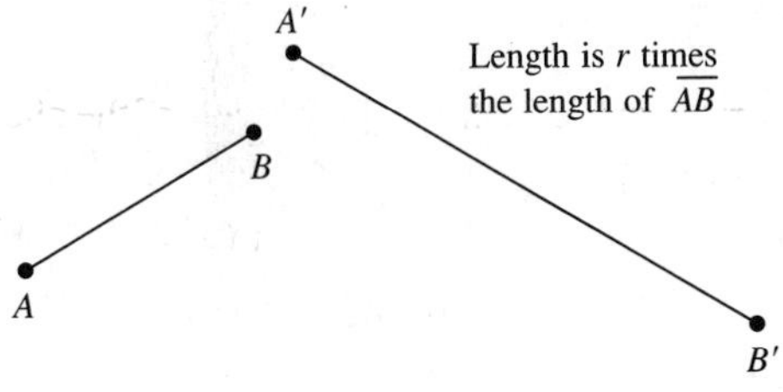

Figure 2.42

For this example, $r = 2$, so that $\overline{AB}$ is stretched by the similarity transformation into a segment twice its length. For $r = 1/3$, it would have been shrunk into a segment one-third of its length. For $r = 1$, the similarity transformation is a motion. The number r is called the *ratio of similarity*.

The geometry of similarity is not mutually exclusive from Euclidean geometry, but instead is a more general geometry that includes Euclidean motions as a special case in which the ratio of similarity is one. High school geometry texts ordinarily present the geometry of similarity.

THEOREM 2.9 ➤ The set of all similarities of the plane is a transformation group.

The inverse of a similarity with ratio r has the ratio $1/r$. The ratio of a product of two similarities is the product of the ratios of the two similarities. For example, if the first ratio of similarity is 2 and the second ratio of similarity is 3, then the ratio of similarity for the product transformation is 6. Theorem 2.9 makes it possible to think of the motions of a plane as a subgroup, with $r = 1$, of the group of similarities of the plane.

In Section 2.6, it was proved that a unique motion is determined if three noncollinear points and their images are known. The same proof can be modified to apply to the more general group of similarities. A brief outline of the proof follows the statement of the theorem.

THEOREM 2.10 ➤ A plane similarity is uniquely determined when a triangle and its image are given.

In Figure 2.43, $\triangle A'B'C'$ is the image of $\triangle ABC$. The distances $P'A'$, $P'B'$, and $P'C'$ and the point P' can all be found.

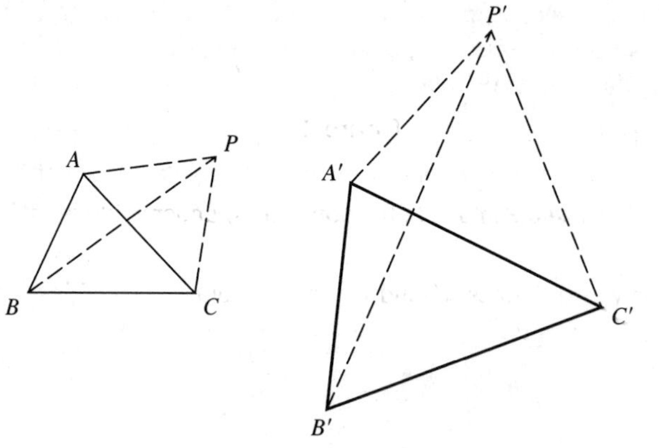

Figure 2.43

An intuitive exploration of how to explain similarities as a series of simple steps leading from a figure to its image indicates that there can be both a change in position and a uniform change in size (but not a change in shape). Think of using physical models in the demonstration. For example, in Figure 2.43, changes of position can result in triangle ABC being situated so that A and A' coincide, so that $\overline{AB}$ lies along $\overline{A'B'}$, and so that $\overline{AC}$ lies along $\overline{A'C'}$. A change in size is then necessary if B and B' are to coincide. Since motions of the plane can accomplish the change in position, the new transformation needed is one to accomplish the uniform change in size. The new type of transformation must be a special similarity in which the image of a set of points is a similar figure with corresponding sides parallel.

DEFINITION ➤ A dilation $H(O, r)$ is a plane similarity that maps a point P to a point P', where $OP' = rOP$, the points P, O, P' are collinear, and $r \neq O$. If $r < O$, then O is between P and P', while if $r > O$, the points P and P' are on the same side of O. The point O is called the center, and the number r the ratio of the dilation. A dilation is sometimes called a dilatation or a homothety.

The center of dilation is an invariant point, according to this definition. The ratio of the distances from the center to the image point and to the original point is a constant, the ratio for the dilation. For example, Figure 2.44a shows a dilation with $OP'/OP = 3/2$, and Figure 2.44b shows a dilation with $OP'/OP = -3/2$. Remember that the ratio of similarity always compares the image with the original.

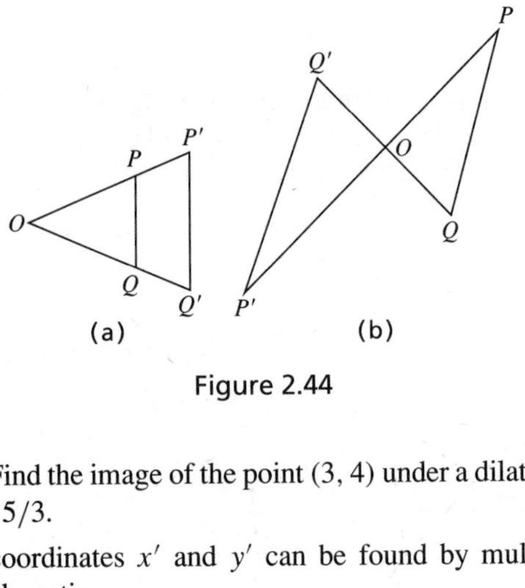

Figure 2.44

Example ➤ Find the image of the point (3, 4) under a dilation with center at the origin and ratio 5/3.

The new coordinates x' and y' can be found by multiplying the original coordinates by the ratio.

$$x' = 5/3(3) = 5$$
$$y' = 5/3(4) = 20/3$$

The image is (5, 20/3).

The exploration of the need for a dilation showed that motions could result in an image in the correct relative position; thus corresponding segments under a dilation should be parallel. That the property of parallelism is an invariant under a dilation is proved as the next theorem.

THEOREM 2.11 ➤ The image of a segment under a dilation is a parallel segment.

Proof: In Figure 2.44, $OP'/OQ' = OP/OQ$, and $\triangle OP'Q' \sim \triangle OPQ$; hence $\overline{PQ} \parallel \overline{P'Q'}$. The proof must be modified for the special case in which O, P, Q are collinear. See Exercise 15.

A figure and its image under a dilation are sometimes called homothetic figures. It is sometimes said that homothetic figures are both similar and similarly placed, since their corresponding sides are parallel. For example, in Figure 2.45, $\triangle ABC$ and $\triangle A'B'C'$ are similar but not homothetic, whereas $\triangle ABC$ and $\triangle A''B''C''$ are homothetic. Also, $\triangle A'B'C'$ and $\triangle A''B''C''$ are similar but not homothetic. The property of parallelism is preserved under a dilation.

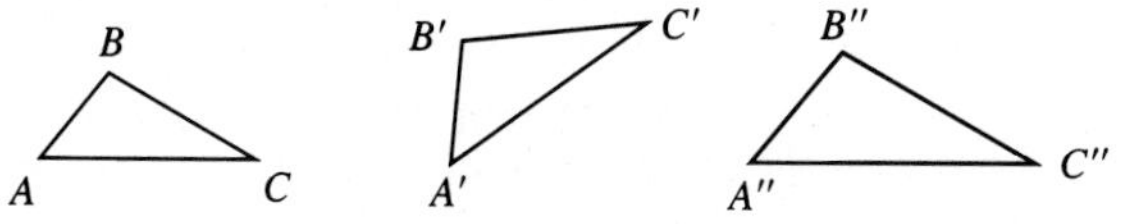

Figure 2.45

Another invariant property is the property of being a circle. There are several ways to prove that a set of points is a circle. The way used in the following proof is to show that all the points are a fixed distance from the center.

THEOREM 2.12 ➤ The image of a circle under a dilation is another circle. See Figure 2.46.

Figure 2.46

Proof: Let O be the center of dilation and P' and Q' the images of a point P on the given circle and its center, respectively. Assume P is not collinear with O, Q. Since $\overline{PQ} \parallel \overline{P'Q'}$ by Theorem 2.11, $P'Q'/PQ = OQ'/OQ$, or

$$P'Q' = \frac{PQ \cdot OQ'}{OQ}$$

But each segment in the expression on the right has a fixed length; hence $P'Q'$ is a constant. For any position of point P, P' lies on a circle with Q' as center.

So far it has been determined that a similarity can be considered the product of a dilation and a motion. Because a dilation also can accomplish a translation, the following stronger statement can be established.

THEOREM 2.13 ➤ A plane similarity with r a positive number not equal to 1 is the product of a dilation and a rotation or a reflection.

Proof: It is assumed here, and can be proved using theorems from topology, that every plane similarity not a motion has exactly one fixed point. In Theorem 2.13, the rotation is about the fixed point and the reflection is about a line through the fixed point.

In Figure 2.47, let O be the fixed point and $\overline{AB}$ any segment. A' and B' are the images of A and B under a dilation with O as center. Let $f(A)$ and $f(B)$ be the images of A and B under any similarity that is not a motion. $\triangle OA'B' \sim \triangle Of(A)f(B)$. $\triangle OA'B'$ can be mapped into $Of(A)f(B)$ either by a rotation through angle $AOf(A)$ or by a reflection about the bisector of this angle. In the first case, the similarity is direct, and in the second case the similarity is opposite.

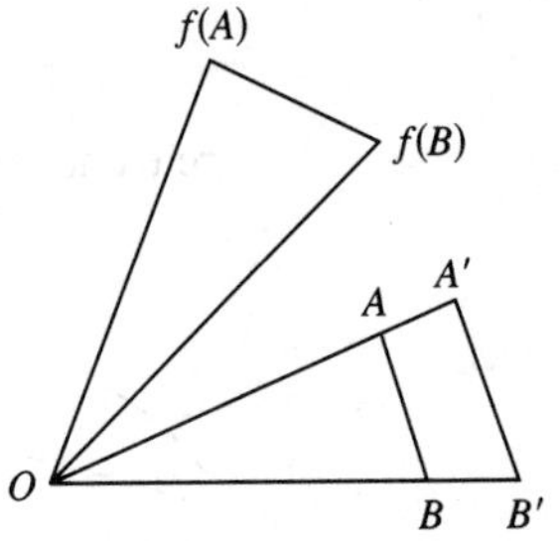

Figure 2.47

The analytic development of the equations for similarity transformations begins with the observation that the equations for a dilation with center at the origin are

$$x' = rx$$
$$y' = ry$$

Since a similarity is the product of a motion and a dilation, the equations for a similarity are of the form

$$x' = r(ax + by + c)$$
$$y' = \pm r[(-bx + ay) + d]$$

for $a, b, c, d \in R$ and $a^2 + b^2 = 1$.

Similarities in three-space also form a group having the motions of three-space as a subgroup. The theory and definitions are analogous to those for two dimensions.

THEOREM 2.14 ➤ A space similarity is uniquely determined when a tetrahedron and its image are given.

THEOREM 2.15 ➤ The set of all similarities of space is a transformation group.

THEOREM 2.16 ➤ A space similarity that is not a motion of space is the product of a dilation about its fixed point and a rotation about a line passing through the fixed point.

PROOF: Consider, because of Theorem 2.14, any tetrahedron $OABC$ with the fixed point O as one vertex. It is assumed here, as for two dimensions, that a fixed point exists for any similarity. Let r be the ratio of similarity. Then the image of $OABC$ under the dilation $H(O, r)$, $OA'B'C'$, is congruent to the image $OA''B''C''$ under the similarity. If the motion connecting $OA''B''C''$ and $OA'B'C'$ is direct, then the similarity is the product of $H(O, r)$, for r positive, and a rotation about some line through O. If the similarity is opposite, then r is negative. Figure 2.48 shows a tetrahedron and its image under a direct dilation.

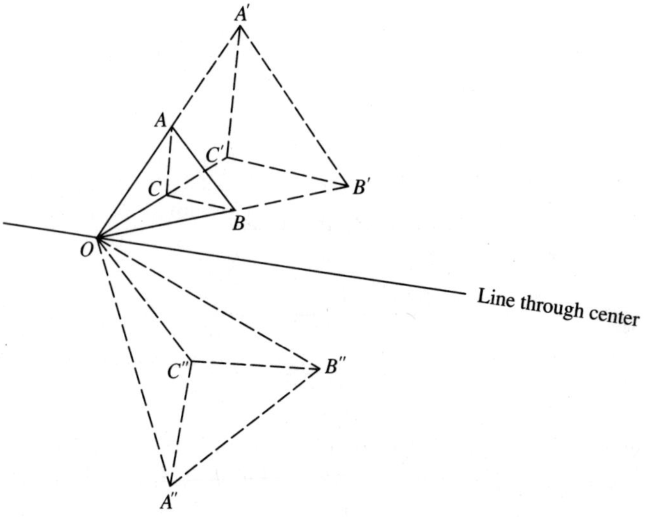

Figure 2.48

▲ EXERCISES 2.8 ▲

1. Find the length of the image of a 3-cm segment under a similarity with ratio 4/3.
2. Find the ratio of similarity if a 5-cm segment has a 6-cm image.
3. Name some properties preserved under all motions that are not preserved under all similarities.
4. Sketch two segments such that the ratio of similarity is 3/2.
5. Sketch two similar triangles such that the ratio of area is 3/2.
6. Sketch two similar tetrahedrons such that the ratio of volume is 3/2.
7. What is the effect of a similarity on the area of a triangular region?
8. Find the ratio of similarity if a square region with an area of 12 square units has an image with an area of 17 square units.
9. In Figure 2.46, compare the ratio of the radii of the two circles with the ratio of similarity.
10. Find the image of (3, 5) under a dilation with center at (0, 0) and with ratio 3/4.
11. Find the image of (4, 3) under the similarity with equations

$$x' = 2[(3/\sqrt{10})x - (1/\sqrt{10})y + 5]$$

$$y' = 2[(1/\sqrt{10})x + (3/\sqrt{10})y + 2]$$

12. Prove analytically that if two lines intersect, their images under a similarity also intersect.
13. Is the set of dilations with a given invariant point a group of transformations?
14. Prove analytically that the image of the parabola $y = ax^2$ under a dilation with (0, 0) the invariant point is a parabola.
15. Modify the proof of Theorem 2.11 for the special case in which O, P, Q are collinear.
16. Complete the detail of the proof of Theorem 2.12 by considering the possibility of P being collinear with O and Q.
17. How many different similarities carry a specific segment $\overline{AB}$ to its specific image $\overline{A'B'}$? Use Theorem 2.10 to prove that your answer is correct.
18. ▷ According to Theorem 2.10, a plane similarity is uniquely determined if you know the vertices of a triangle and the vertices of its image triangle. Take a specific example, using coordinates for the vertices of two similar triangles, to see if you can obtain the equations for the similarity.

2.9 Introduction to the Geometry of Fractals

The mathematical topic of *fractals* has quickly become popular, but the impression sometimes given is that they are simply pretty pictures produced by a computer. Actually, fractals are a good example of a very modern idea that is a geometric application. This application is made practical by the ability of the modern computer to show graphic output resulting from recursive procedures, and to show it in intricate detail. Although fractals might have been inserted in Chapter 4 or Chapter 8, they are included in Chapter 2 because, as shapes containing self-similarity under repeated dilations, they are an application of the geometry of similarity transformations.

The geometry of fractals was first introduced by Benoit B. Mandelbrot in 1977. One of the simplest examples of a fractal, and the one that is probably most helpful in beginning the study of the geometric background of the subject, is the "snowflake curve," or more technically, the triadic von Koch curve. The initiator of this curve is an equilateral triangle with unit sides. See Figure 2.49.

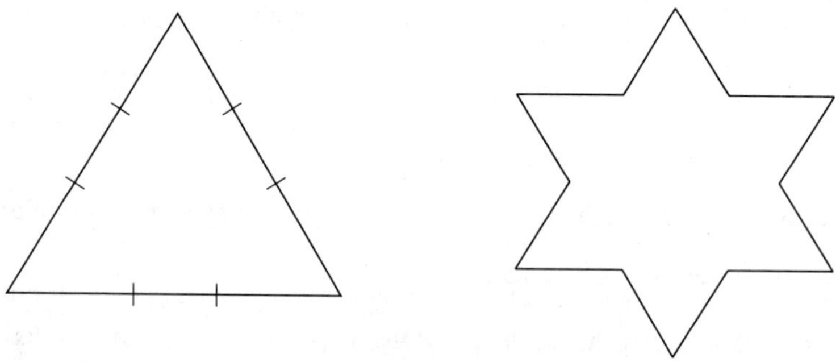

Figure 2.49

On each side of this triangle, the next step is to divide the side into three parts. On the middle part, two segments with sides $1/3$ unit long are used as shown. Thus the original side, broken into thirds, is reconstructed using four of these pieces, and the new length is $4/3$. The snowflake curve is produced by repeating this process many times. Using computer graphics, the curve is reproduced in great detail by a recursive procedure. A snowflake curve after several recursions is shown in Figure 2.50.

One interesting feature of the snowflake curve involves the length of the curve and the area of the region within the curve. The length of one part of the curve, if the process is continued indefinitely, takes on the values $1, 4/3, (4/3)(4/3), \ldots$, which diverge. Strangely enough, the area within the curve does not become infinite, but approaches a limit which is $8/5$ the original area. See Exercise 1, Exercises 2.9.

Figure 2.50

Probably an even more significant feature of the snowflake curve is its fractal dimension. One of the unusual features of fractals is the definition of fractal dimension that allows for real numbers instead of just positive integers. Since the snowflake curve is constructed by dividing the sides into three parts and then reconstructed by using four parts, the fractal dimension is defined to be Log 4/ Log 3 $\approx$ 1.2618.

More generally, if an object is scaled by S ($S = 3$ in the snowflake example) and if N of the pieces ($N = 4$ in the snowflake example) are required to reconstruct it, then the fractal dimension is D, where $N = S^D$, or $D = \text{Log } N / \text{Log } S$.

Example ➤ Suppose that a fractal was constructed by dividing each side of an equilateral triangle into five parts, and on the middle part of each side, two parts were used, making a total of six parts used for the reconstruction as shown in Figure 2.51. Then the fractal dimension is defined to be Log 6/Log 5 $\approx$ 1.1133.

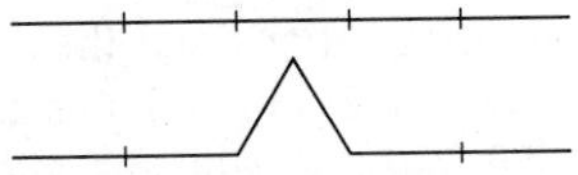

Figure 2.51

Using the idea of fractal dimension, a fractal can be described as a set of points, curves, or surfaces with the fractal dimension larger than the ordinary (topological) dimension. For example, any curve is thought of as a one-dimensional object, with length only, but we have seen that the fractal dimension of the snowflake curve is a real number greater than one.

It is fascinating that the concept of fractal dimension actually makes it possible for a curve to have a fractal dimension of 2. This is true of Hilbert's plane-filling curve developed in Figure 2.52. The dotted line is the first step and the reconstructed curve wraps around it. In this example, $S = 2$ and $N = 4$, so that $4 = 2^D$ and $D = 2$. Other examples of fractals are given in the exercises.

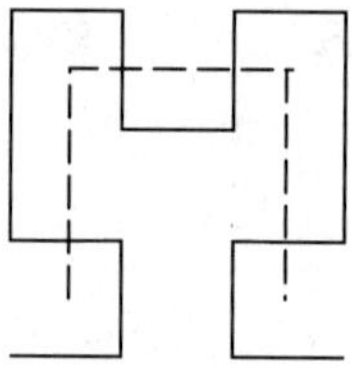

Figure 2.52

Recall the feature of fractals as shapes containing self-similarity under repeated dilations. It is this feature that leads to one of the practical applications of fractals in connection with recursive procedures that allow a computer to generate realistic images of such natural features as coastlines, mountain ranges, star-clusters, lightning and river systems, as well as snowflakes. Mandelbrot's book title, *The Fractal Geometry of Nature,* seems justified by the wealth of these applications. Examples of complex computer-generated pictures of fractals are shown on the endpapers.

This chapter has enlarged the idea of Euclidean geometry by showing how it can be considered the study of those properties that are invariant under the group of motions. Euclidean geometry has also been considered as a special case of the geometry of similarities. Through this new approach, additional significance is attached to the study of congruent and similar triangles. Even more important, however, is the anticipation that more general transformations exist (such as those given in Chapter 7, for example), with similarities and Euclidean geometry as special cases.

One type of high school geometry course in limited use is a revised version of Euclidean geometry with the concepts of motions included in the postulates. Although this approach leads to the simplification of some theorems, it requires a more complex set of postulates. Anyone interested in teaching at the secondary school level should realize that intuitive ideas of transformations and their applications are common in elementary and middle schools, as well as high schools and community colleges.

Probably the greatest value of transformations is that just a few weeks of study of motions prepares the way for the introduction of more general transformations used in projective geometry, topology, and inversive geometry. If you wish to study the more general transformations immediately, you should begin with Chapter 7.

The study of Euclidean geometry has been revitalized during the present century in many ways besides the approach through transformations. Chapters 3, 4, and 5 contain material that is a part of modern Euclidean geometry. A classification and detailed study of sets of points in Euclidean geometry utilizing the modern concept of convexity is the subject of Chapter 3. Chapter 4 considers the advanced Euclidean geometry of the polygon and circle, along with new applications, while Chapter 5 contains the modern theory of constructions.

▲ EXERCISES 2.9 ▲

1. Verify the fact that the area within the von Koch snowflake curve approaches a limit that is 8/5 of the original triangle.

For Exercises 2–5, suppose a fractal is created by the process shown in the appropriate figure, in which each side is divided into a specified number of parts, then the side is reconstructed as shown. In each case, find the fractal dimension for the curve.

2. The curve represented in Figure 2.53a.

3. The curve represented in Figure 2.53b.

4. The curve represented in Figure 2.53c.

5. The curve represented in Figure 2.53d.

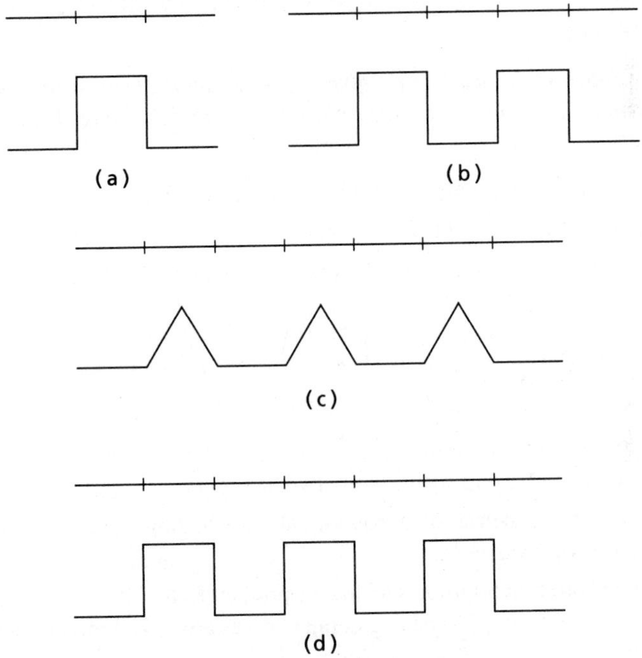

Figure 2.53

For Exercises 6–9, sketch the result of the next step in the recursive process for each of the sides shown.

6. The curve in Exercise 2.

7. The curve in Exercise 3.

8. The curve in Exercise 4.

9. The curve in Exercise 5.

For Exercises 101–121, investigate the particular fractal named.

▷ **10.** The Sierpinski Carpet

▷ **11.** The Dragon

▷ **12.** The Mandelbrot Set

▷ **13.** Study more about the theory of fractals and specific examples and applications of fractals in *The Fractal Geometry of Nature* or another reference.

▲ CHAPTER 2 REVIEW EXERCISES ▲

1. Let $f = \{(2, 3), (1, 7), (5, 6)\}$ and $g = \{(3, 1), (7, 5), (6, 2)\}$. Find fg and gf.

2. Find the point that has $(3, 1, 6)$ as its image under the translation with vector $[4, -2, 3]$.

For Exercises 3–5, find the image of the given points under the transformation described, using matrix multiplication. Sketch the original figure and the image figure on the same axes.

3. Segment with endpoints $(3, 7)$, $(5, 13)$ under a reflection about the line $y = x$

4. Triangle with vertices $(1, 2)$, $(5, 6)$, $(3, 9)$ under a rotation of 180°

5. Quadrilateral with vertices $(1, 1)$, $(3, 2)$, $(4, 7)$, $(2, 5)$ under a shear with matrix

$$\begin{pmatrix} 1 & 0 \\ 3 & 1 \end{pmatrix}$$

6. A transformation group has what for its members?

7. Give an example from the finite geometry of the symmetries of an equilateral triangle of a subgroup with exactly two elements.

8. In the set of symmetries of an equilateral triangle, how many of the six elements are their own inverses?

9. Use the permutation symbols to find the product $R_1 R(240)$ for the symmetries of an equilateral triangle. The beginning vertices are 1, 2, 3 in a counterclockwise direction.

10. Let F be the transformation such that x has the image x^2, and let G be the transformation such that x has the image $x + 2$. Find the product FG.

11. Find the image of the line $x + y = 6$ under the translation with vector $[2, 4]$.

12. Find the image of the line $2x - y = 5$ under a reflection about the x-axis.

13. Which two types of the four types of motions of the plane have no invariant points?

14. What is the minimum number of points, no three collinear, of a plane that must remain invariant under a motion in order for that motion to be the identity?

15. Find the image of (2, 0) under a reflection about the line $y = (\sqrt{3}/3)x$.

16. Find the image of the point (3, 4, 5) with respect to a reflection about the plane $y = -4$.

17. For a plane similarity with ratio 3/2, find the area of the image of a triangle whose original area is one unit.

18. Sketch one more complete step in the development of a fractal initiated by dividing the sides of a square and reconstructing them as shown in Figure 2.54.

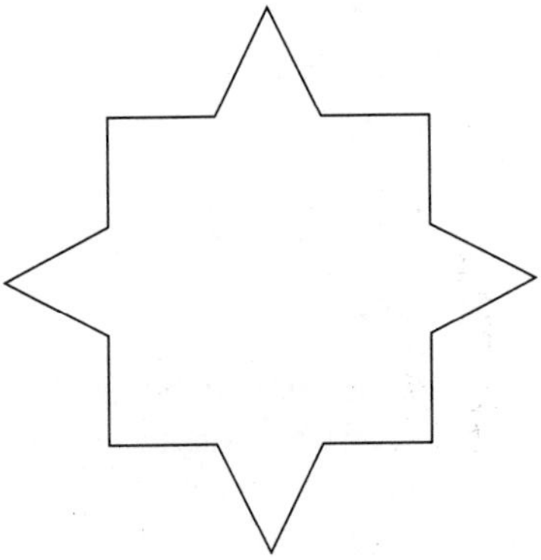

Figure 2.54

19. If a fractal is constructed from an equilateral triangle by dividing each side into nine parts and then using 13 of these parts to reconstruct it, find the dimension.

For Exercises 20–27, is the set of transformations a group of transformations?

20. Rigid motions of the plane

21. Opposite plane motions

22. All plane translations represented by a vector 1 cm long or longer

23. All plane translations represented by a vector 1 cm long or shorter

24. All similarities in a plane

25. All similarities that are not motions

26. All plane similarities with a ratio equal to or less than 1

27. All dilations about a given point

Additional chapter review exercises may be selected from any exercises not already assigned from previous exercise sets in this chapter.

3 Convexity

▲ 3.1 Basic Concepts

The study of convexity is an important part of current research in modern Euclidean geometry. In this chapter, the common properties that some familiar sets of points share with all convex sets will enrich some basic concepts of Euclidean geometry and lead to applications you should find significant.

The words *convex* and *convexity* are common outside of mathematics. *Convex* means curving outward. For example, a convex lens bulges outward. Fortunately, the use of the word *convex* to describe particular sets of points in mathematics is only a careful restatement of this common meaning. Study Figure 3.1, parts a and b, to discover the basic difference between convex sets of points and those that are nonconvex. The example on the far right in Figure 3.1a is in the shape of a convex-convex lens in optics, while the last example on the far right in Figure 3.1b is in the shape of a convex-concave lens.

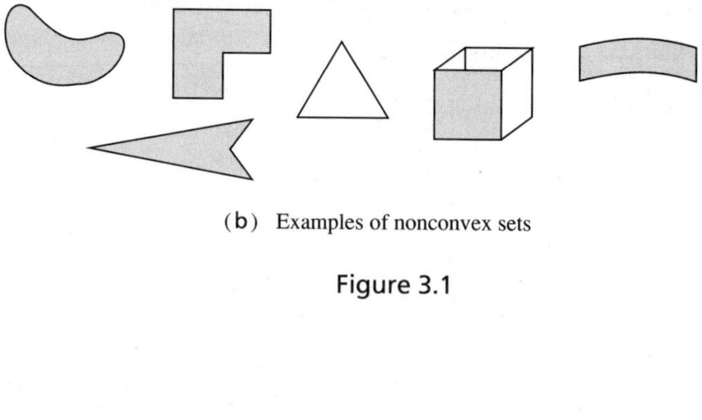

(a) Examples of convex sets

(b) Examples of nonconvex sets

Figure 3.1

For a convex set of points, such as K in Figure 3.2a, any segment with endpoints A and B in the set lies wholly in the set. This is not true for all pairs of points in the nonconvex set in Figure 3.2b. For example, $\overline{CD}$ includes points outside of set K'.

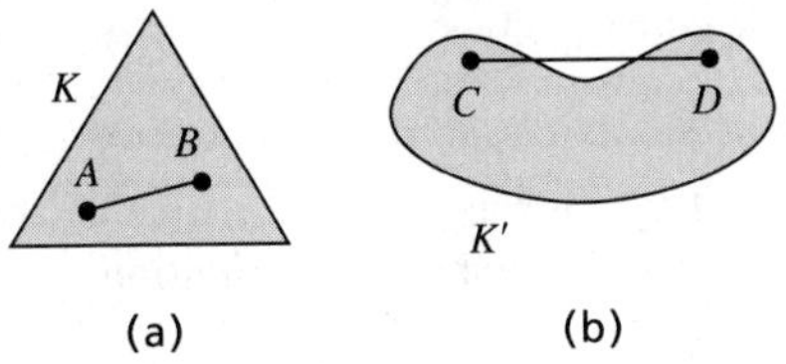

Figure 3.2

The symbol $\subseteq$ means "is a subset of." The symbol $\subset$ means "is a proper subset of."

DEFINITION ➤ A *convex* set is a set of points K such that if A and B are any points in K, then the segment $\overline{AB} \subseteq K$.

This definition of convex set includes many of the common sets of points studied in Euclidean geometry. Examples are a circular region, some polygonal regions, segments, an angle (with measure less than π) and its interior, and a spherical region. The empty set and a set consisting of a single point are both convex by agreement.

To show that a particular set is convex by using the definition is not always easy. In addition to Theorem 3.1, which follows, analytic geometry is often used. You should assume here and elsewhere in this chapter, unless stated to the contrary, that the variables for coordinates of points are elements of the set of real numbers.

THEOREM 3.1 ➤ The intersection of two convex sets is a convex set.

PROOF: In Figure 3.3, let K_1 and K_2 be any two convex sets and let S be their intersection. For $A, B \in S$, all points of $\overline{AB}$ are elements of K_1 because K_1 is convex. All points of $\overline{AB}$ are also elements of K_2, because it is also convex. Hence, $\overline{AB} \subseteq (K_1 \cap K_2)$; thus, $K_1 \cap K_2$ is convex.

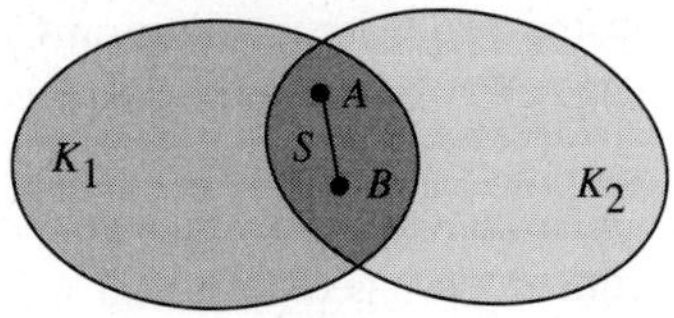

Figure 3.3

Theorem 3.1 may be generalized to more than two sets, as in Exercise 3 of Exercises 3.1. The theorem may be used to show the convexity of some common sets of points. For example, an angle (measure less than π) and its interior can be defined as the intersection of two half-planes, which are easily shown to be convex, so that Theorem 3.1 applies.

An alternative approach is to use analytic geometry to prove that sets of points are convex. This approach can even be used for a half-plane.

Example ➤ Let S be the set of points with coordinates (x, y) such that $x > 0$. Show that S is a plane convex set.

Let $A(x_1, y_1)$ and $B(x_2, y_2)$ be any two points of S, as in Figure 3.4, such that $x_1 > 0$ and $x_2 \geq x_1$. Then, for any point $P(x, y)$ of $\overline{AB}$, $x_2 \geq x \geq x_1$, hence $x > 0$ and $\overline{AB} \subset S$, so that S is convex by definition.

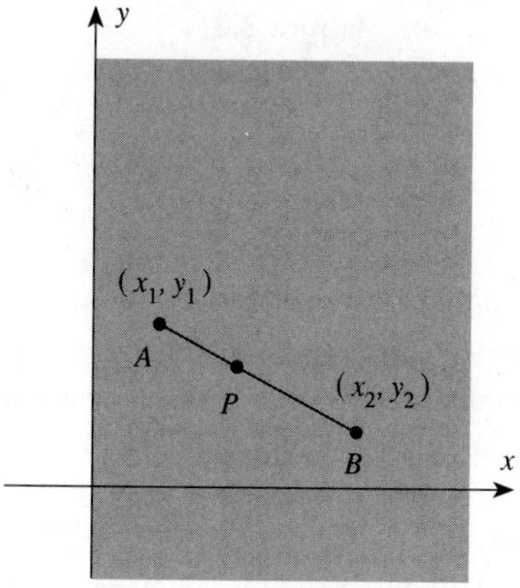

Figure 3.4

Credit for beginning the systematic study of convex sets in the early twentieth century is given to H. Brunn and H. Minkowski. The 1934 work, *Theorie der konvexen Körper*, by T. Bonnesen and W. Fenchel showed much progress. During the 1940s and 1950s many applications were studied, especially to the field of optimization. The *Proceedings of Symposia in Pure Mathematics*, Volume VII, was the publication resulting from the 1963 Symposium on Convexity sponsored by the American Mathematical Society. Convexity continues to be an area of current research in geometry.

Before exploring more complex properties of convex sets, it is necessary to introduce several basic concepts from modern geometry that will be useful both here and in later chapters.

Recall that a function $y = f(x)$ is continuous at $x = a$ if and only if, given $\varepsilon > 0$, there exists a δ such that

$$|f(x) - f(a)| < \varepsilon \qquad \text{if} \qquad |x - a| < \delta$$

A function is continuous if it is continuous at every point of its domain.

For example, the function in Figure 3.5a is continuous, whereas the function in Figure 3.5b is not.

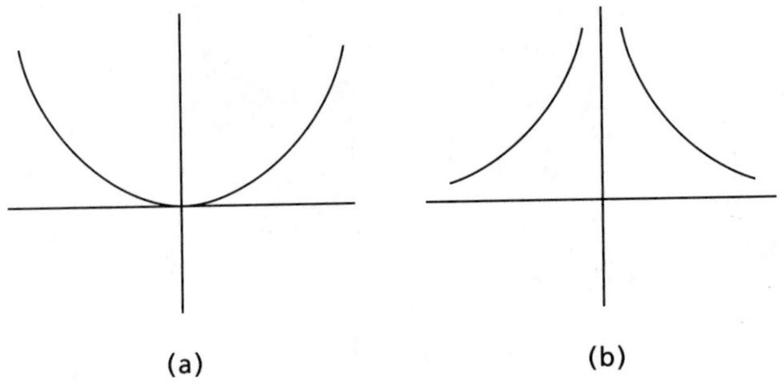

Figure 3.5

A plane *curve* is the graph of a set of equations of the form $x = f(t)$, $y = g(t)$, for f and g continuous functions on an interval, I, of the set R of real numbers. The curve is the image of a mapping from I to the plane. If the mapping is one-to-one, the curve is a *simple curve*. If the points, $(f(a), g(a))$ and $(f(b), g(b))$, corresponding to both endpoints of the defining interval $[a, b]$ are identical, the curve is a *closed curve*. If the curve is both simple and closed, then it is called a *simple closed curve*. Intuitively, a simple closed curve is thought of as a curve that begins and ends at the same point but does not cross itself; thus, there is only one interior. (See Figure 3.6.) The set of points on and inside a simple closed curve or an angle in a plane is called a *plane region*.

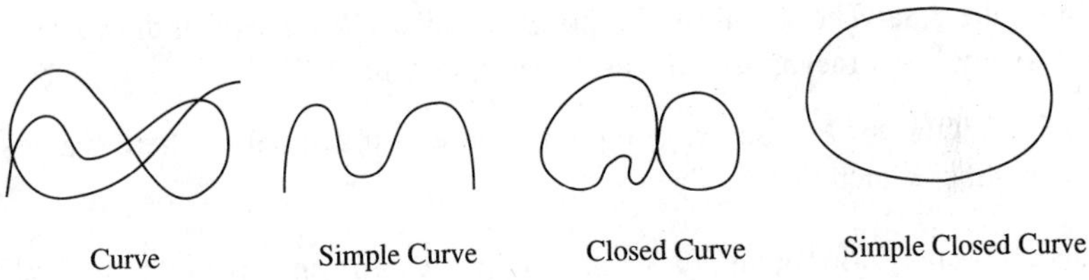

Figure 3.6

Precise definitions of interior, exterior, and boundary points of a set of points such as a plane region depend on the concept of neighborhood of a point.

DEFINITION ➤ *For Two Dimensions:* The *open circular neighborhood* with radius r of a point P is the set of points inside a circle of radius r with P as center. (See Figure 3.7a.) This definition can be written as

$$N(P, r) = \{A : PA < r\}$$

For Three Dimensions: The *open spherical neighborhood* with radius r of a point P is the set of points in space inside a sphere of radius r with P as center. (See Figure 3.7b.) This definition can be written as

$$N(P, r) = \{A : PA < r\}$$

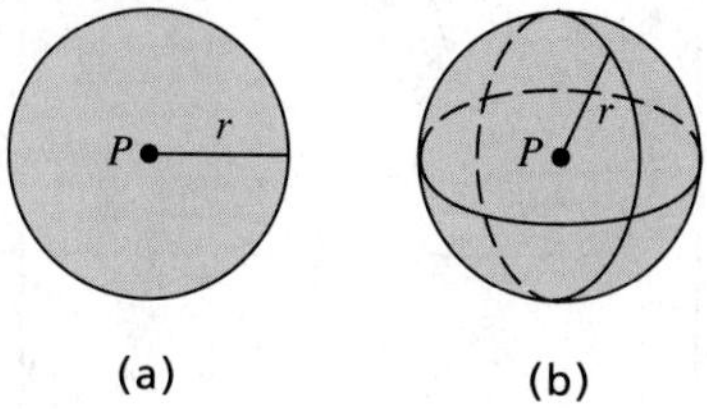

Figure 3.7

For Two or Three Dimensions: A *closed neighborhood* includes the points on the circle or sphere as well as those inside. The definition of closed neighborhood, using set symbolism, is

$$N[P, r] = \{A : PA \leq r\}$$

Both open and closed neighborhoods are convex sets of points.

DEFINITION ➤ P is an *interior point* of a set S if and only if there is some positive real number r such that $N(P, r) \subseteq S$.

This definition means intuitively that a point is an interior point of a set if every point sufficiently near it is an element of the set. For example, point A is an interior point in Figure 3.8a. The set of interior points is called the *interior* of the set.

The symbol $\sim S$ means "not S," or the complement of S.

DEFINITION ➤ P is an exterior point of a set S if and only if there is a positive real number r such that $N(P, r) \subseteq \sim S$.

A point is an exterior point of a set if every point sufficiently near it is a member of the complement of the set. For the set of points $x^2 + y^2 < 1$, the exterior is $x^2 + y^2 > 1$. Point B is an exterior point in Figure 3.8b. The set of exterior points of a set is called the *exterior* of the set.

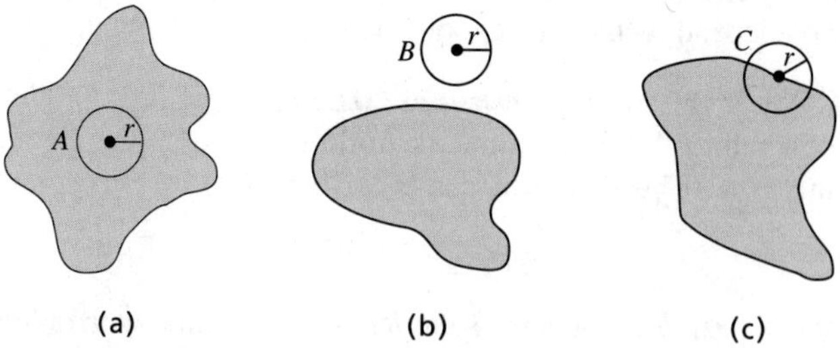

Figure 3.8

A point that is neither an interior nor an exterior point of a set is called a *boundary point*. The set of boundary points for a set is called the *boundary* of the set. Note that if P is a boundary point of S, then it is also a boundary point of the complement of S.

For the set $x^2 + y^2 < 1$, the boundary is $x^2 + y^2 = 1$, Point C is a boundary point in Figure 3.8c. The boundary of a half-plane is the line determining the half-plane. A simple closed curve is the boundary of the set of points in its interior.

Example ➤ Describe the interior, boundary, and exterior for the plane set $\frac{x^2}{4} + \frac{y^2}{9} \leq 1$.

The interior is the set of points inside the ellipse, so that $\frac{x^2}{4} + \frac{y^2}{9} < 1$. The boundary is the ellipse $\frac{x^2}{4} + \frac{y^2}{9} = 1$, and the exterior is the set of points outside the ellipse, satisfying the inequality $\frac{x^2}{4} + \frac{y^2}{9} > 1$.

You should realize that the boundary points of a set may or may not be elements of the set. An important idea is that, to decide whether a point is an interior, an exterior, or a boundary point, it is necessary to consider the number of dimensions. For example, in Figure 3.9, $\{(x, y, 0) : x^2 + y^2 \leq 1\}$ has no interior points in three-space. Each point of the set is a boundary point, since every spherical neighborhood for a point of the set contains both points of the set and points of its complement.

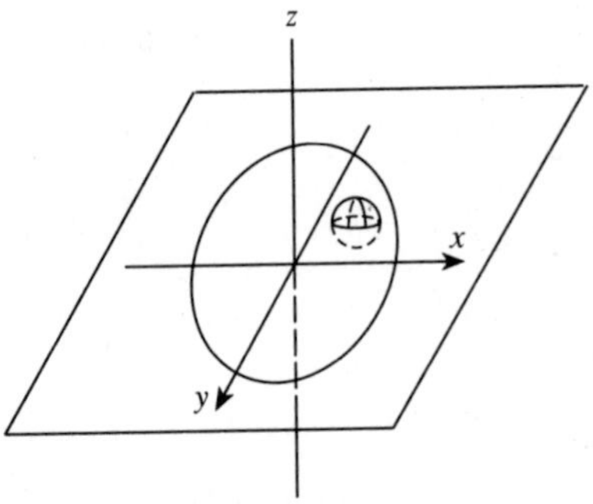

Figure 3.9

The classification of points as interior, exterior, or boundary points for a set leads to useful classifications of the sets themselves.

DEFINITION ➤ An *open* set has only interior points.

An open neighborhood is an open set, as is the interior of any simple closed curve.

DEFINITION ➤ A *closed* set contains all its boundary points.

Examples of closed sets are $x^2 + y^2 \leq 1$, polygonal regions (polygons and their interiors), and a segment $\overline{AB}$. It is important to observe that these definitions of open and closed sets are not mutually exclusive, nor do they include all possible sets of points. A set may be both open and closed or neither open nor closed. An example of a set both open and closed is the entire plane. Since this set has no boundary, it includes all its boundary points. At the same time, all of its points are interior points. Examples of sets that are neither open nor closed are given in Figure 3.10. The dotted portion of the boundary is not considered to be part of the set.

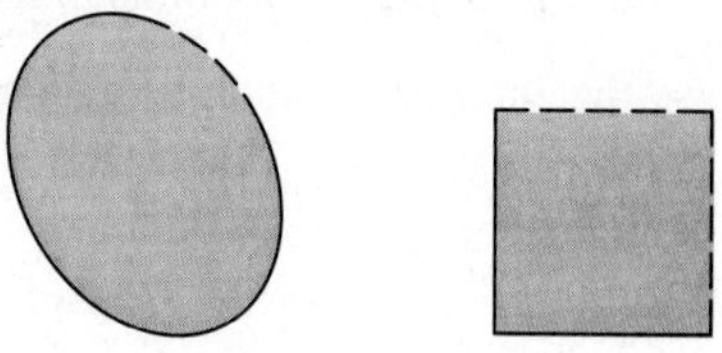

Figure 3.10

DEFINITION ➤ A *bounded set* in two dimensions is one that is a subset of some circular region. In three dimensions, a bounded set is a subset of some spherical region.

Some bounded sets are shown in Figure 3.11. Bounded sets cannot extend indefinitely. For example, the parabola $y = x^2$ is not a bounded set because no circle can enclose the set of points. The concept of a bounded set should not be confused with the boundary of a set.

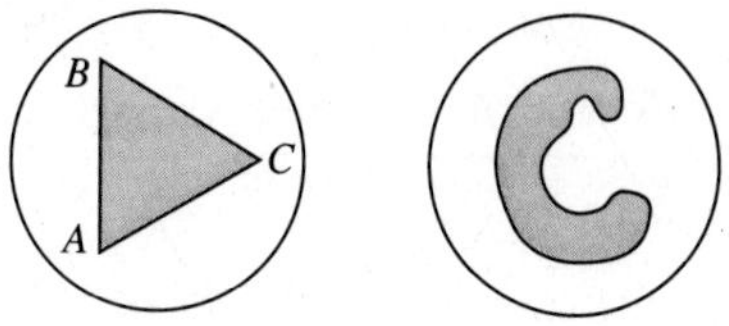

Figure 3.11

▲ EXERCISES 3.1 ▲

1. Which of these sets are convex?

   a. An angle
   b. A rectangular region
   c. An interior of an ellipsoid
   d. A straight line
   e. A single point
   f. A ray
   g. A triangular region
   h. A triangle

2. Which of these sets are convex sets?
   a. A circular region with one point on the boundary removed
   b. A rectangular region with one vertex removed
   c. A rectangular region with one point, not a vertex, on the boundary removed
   d. A circular region with one interior point removed
3. Prove that the intersection of any collection of convex sets is a convex set.
4. Is the union of two convex sets ever a convex set? Is it always a convex set?
5. Prove that a triangular region is a convex set.
6. Show analytically that $S = \{(x, y) : x > 3\}$ is a convex set.
7. Show analytically that $T = \{(x, y) : x > 3 \text{ and } y > 4\}$ is a convex set.
8. Give a definition for a one-dimensional neighborhood.
9. Describe the interior, boundary, and exterior for these plane sets.
   a. $\{(x, y) : x^2 + y^2 \leq 1\}$
   b. $\{(x, y) : y > x^2\}$
   c. $\{(x, y) : y < |x|\}$
   d. $\left\{(x, y) : x^2 + \frac{y^2}{4} \leq 1\right\}$
   e. $\{(x, y) : x \text{ and } y \text{ are integers}\}$
   f. $\{(x, y) : x \text{ and } y \text{ are rational numbers}\}$
10. Describe the interior, boundary, and exterior for these sets of points in space.
   a. $\{(x, y, z) : x^2 + y^2 + z^2 \leq 1\}$
   b. $\{(x, y, 0) : x^2 + y^2 \leq 9\}$
   c. $\{(x, y, 0) : x > y\}$
   d. $\{(x, y, 0) : x = 2\}$
11. Classify these sets as open, closed, neither, or both.
   a. The set of all points with rational coordinates on a number line
   b. The set of all points on a number line in the interval [0, 1]
   c. The sets in Exercise 9
   d. The sets in Exercise 10
   e. A rectangular region with one vertex removed
12. Which of the sets in each exercise are bounded?
   a. The sets in Exercise 9
   b. The sets in Exercise 10
13. Prove that the boundary of a set of points is also the boundary of the complement of the set.
14. Prove that the complement of a closed set is open.
15. What can you say about the complement of a set that is neither closed nor open?
16. Give examples of (a) a bounded set that does not contain its boundary and (b) a set that contains its boundary but is not bounded.
17. Is the intersection of two closed sets always a closed set? Prove that your answer is correct.
18. Prove that every one-dimensional convex set is a segment, a ray, or a line.
19. Prove that if P and Q are boundary points of a convex set, then no point of $\overline{PQ}$ can be an exterior point of the set.

▷ **20.** Take specific equations and numerical values as examples to illustrate the concepts of open circular neighborhood, open spherical neighborhood, closed neighborhood, interior point, and exterior point.

▲ 3.2 Convex Sets and Supporting Lines

Using the basic ideas of convex sets introduced in Section 3.1, it is now possible to explore some additional concepts associated with convex sets. In this section, it is assumed that the geometry being considered is the geometry of two-space.

Some properties known to be true for a specific convex curve such as a circle can also be shown to be true for all convex simple closed curves. An example is the following theorem attributed to Auguste Miquel ("Mémoire De Géométrie," *Journal De Mathématiques Pure et Appliquées*, January, 1844.)

THEOREM 3.2 ➤ For any polygon inscribed in any convex curve, the sum of all the angles exterior to the polygon and interior to the curve is equal to four right angles. A representative figure is shown in Figure 3.12.

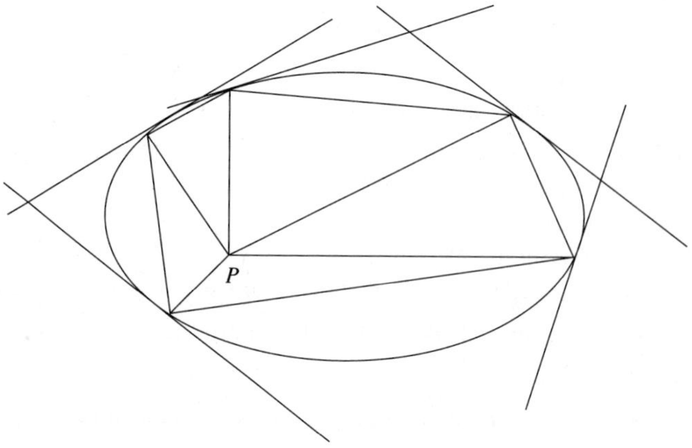

Figure 3.12

The property of convexity is used to establish the existence of an interior point such as P, that partitions the interior of the polygon into triangles. At each vertex, the sum of the interior and exterior angles on one side of the tangent is 180 degrees, so the total for the n vertices of the polygon is $n(180)$. The sum of the exterior angles is this total $n(180)$ minus the sum of the angles in the triangles, except for those angles at the interior point P, so the exterior angle sum is

$$n(180) - ([n(180)] - 360) = 360$$

or four right angles as claimed.

A concept especially important in discussing convex sets, but applicable to other sets of points also, is the concept of supporting line.

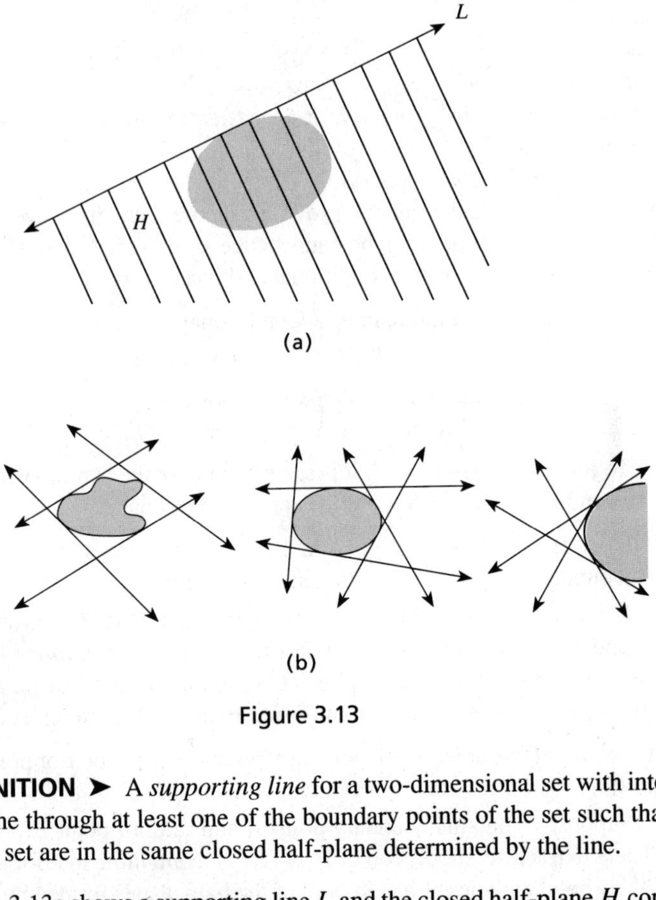

Figure 3.13

DEFINITION ➤ A *supporting line* for a two-dimensional set with interior points is a line through at least one of the boundary points of the set such that all points of the set are in the same closed half-plane determined by the line.

Figure 3.13a shows a supporting line L and the closed half-plane H containing the set. Figure 3.13b shows three examples of sets, with several supporting lines for each.

The proof of the following theorem results in an alternative description for supporting lines of convex two-dimensional sets with interior points.

THEOREM 3.3 ➤ A line is a supporting line for a convex set of points if it goes through at least one boundary point of the set but no interior points, and conversely.

To prove a theorem and its converse, you must prove two separate statements. Theorem 3.3 includes two conditionals:

a. If a line goes through at least one boundary point of the set but contains no interior points, then it is a supporting line for the convex set of points.

b. If a line is a supporting line for a convex set of points, then it contains at least one boundary point of the set but no interior points.

Associated with a conditional are three other conditionals: the converse, inverse, and contrapositive. These are shown symbolically as follows:

Original conditional	$p \to q$
Converse	$q \to p$
Inverse	$\sim p \to \sim q$ (not p implies not q)
Contrapositive	$\sim q \to \sim p$

Sometimes, especially in geometry, it is easier to prove the inverse and/or contrapositive than some other form of the conditional. A conditional and its contrapositive are logically equivalent. So are the converse and the inverse. Proving an "if and only if" theorem is the same as proving a theorem and its converse. Proving $p \to q$ and $q \to p$ results in proving the equivalence $p \longleftrightarrow q$. Any of these pairs of statements may be proved to prove a conditional and its converse:

$$\begin{pmatrix}\text{Conditional}\\ \text{converse}\end{pmatrix}\begin{pmatrix}\text{Conditional}\\ \text{inverse}\end{pmatrix}$$

$$\begin{pmatrix}\text{Contrapositive}\\ \text{inverse}\end{pmatrix}\begin{pmatrix}\text{Contrapositive}\\ \text{converse}\end{pmatrix}$$

Proof: To establish Theorem 3.3, it is easiest to prove the contrapositive and the inverse of conditional a, which is equivalent to conditional b.

Contrapositive: If a line is not a supporting line for a convex set in two dimensions but contains a boundary point, then it contains an interior point.

Proof: If l is not a supporting line, as it is not in Figure 3.14a, then interior points A and B of set K can be found in different half-planes formed by l. The intersection of $\overline{AB}$ and l is not empty. But C is an interior point of set K, since K is convex and A and B are interior points. See Exercise 18, Exercises 3.2.

Inverse: If a line contains at least one boundary point and also interior points, then it is not a supporting line for the convex set in two dimensions.

Proof: Suppose l contains boundary point A and interior point B, as in Figure 3.14b. There exists a neighborhood of B, $N(B, e)$, containing interior points. But this neighborhood includes points in both of the half-planes formed by l; hence, l is not a supporting line.

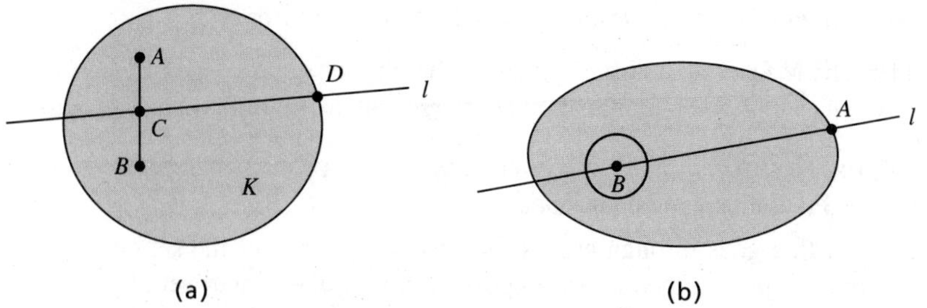

Figure 3.14

The concept of supporting line is closely related to the more familiar idea of a *tangent*. From calculus, the idea of a tangent to a curve is that of a line intersecting the curve at a point and having the same slope as the curve at that point. As Figure 3.15 shows, supporting lines and tangents are not necessarily the same thing.

As used with convex sets, the concept of tangent is somewhat different from the idea used in calculus.

(a) A tangent that is not a supporting line

(b) A supporting line that is not a tangent

Figure 3.15

DEFINITION ➤ The *tangent cone* for a boundary point of a convex set is the set of all rays that: (1) have the boundary point as endpoint and (2) also contain other points of the convex set or its boundary.

Figure 3.16 shows three tangent cones. For Figure 3.16a, $\overrightarrow{AC}$ and $\overrightarrow{AD}$ are the boundaries of the tangent cone. (The notation $\overrightarrow{AC}$ means ray AC.) The boundaries are not themselves rays of the tangent cone. The boundaries of the tangent cone are called *semitangents*. In Figure 3.16a, $\overrightarrow{AB}$ and $\overrightarrow{AD}$ are semitangents. In Figure 3.16b, $\overrightarrow{BE}$ and $\overrightarrow{BF}$ are semitangents and are collinear.

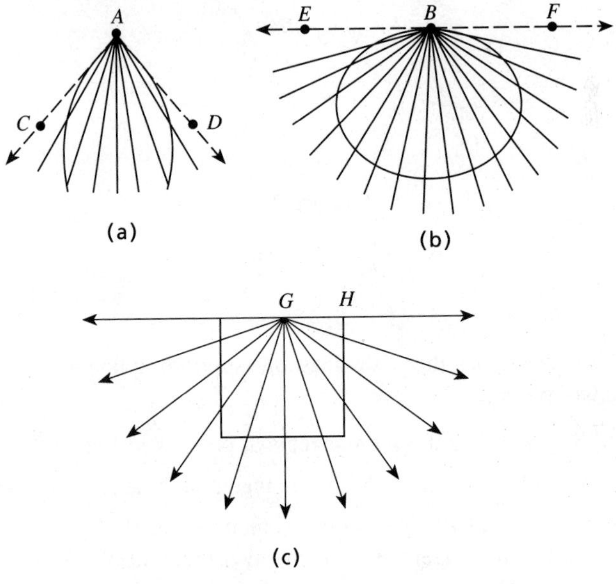

(a)

(b)

(c)

Figure 3.16

DEFINITION ➤ A line is the *tangent* to a convex set at a point if it is the union of two collinear semitangents.

From Figure 3.16, or from a consideration of the definition of tangent, it should be clear that there are tangents at some boundary points of a convex set and not at others. It is possible to classify boundary points of a convex set as *regular* or *corner* points on the basis of whether there is or is not a tangent to the curve determined at that boundary point. There is a tangent at every regular point. In Figure 3.16, A and H are corner points, whereas B and G are regular points. Other examples of corner points are the vertices of a convex polygonal region. On the other hand, all of the points on a circle are regular points on the boundary of the circular region.

One of the main reasons for introducing the concept of supporting line is that it can be used to distinguish a convex set from one that is nonconvex.

THEOREM 3.4 ➤ The interior of a simple closed curve is a convex set if and only if through each point of the curve there passes at least one supporting line for the interior.

Figure 3.17 helps to explain intuitively the significance of Theorem 3.4. If a set is not convex, then points (such as A) exist on the boundary such that any line through A contains interior points. Through any point A' on the boundary of a convex set, at least one line can be drawn containing no interior points.

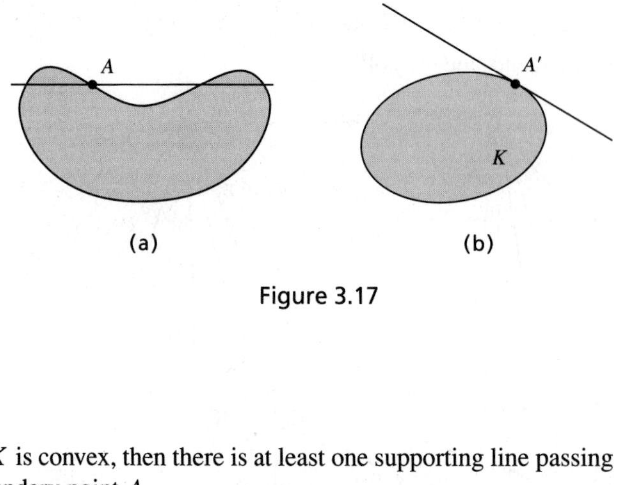

Figure 3.17

PROOF:

a. If K is convex, then there is at least one supporting line passing through each boundary point A.

Let $\overrightarrow{AB}$ and $\overrightarrow{AC}$ be the semitangents at point A in Figure 3.18. If the measurement of $\angle BAC$ is π, then $\overleftrightarrow{BC}$ is a tangent and, hence, a supporting line. If the measurement of $\angle BAC$ is less than π, then $\overleftrightarrow{AB}$ and $\overleftrightarrow{AC}$ are both supporting lines, and there is no tangent at A. If the measurement of $\angle BAC$ is greater than π, then the original set is not convex, contrary to the beginning assumption. See Exercise 16, Exercises 3.2.

b. If S' is not a convex set, then there is not a supporting line at every boundary point.

This statement is the inverse of the conditional in (a), so proving it will complete the proof of Theorem 3.4. (See Figure 3.19.) Since S' is not convex, an interior point C and a boundary point B can be found such that a boundary point A lies on $\overline{BC}$. $\overleftrightarrow{BC}$ is not a support line, since it contains interior points. Any other line through A has B and C in opposite halfplanes; hence, it cannot be a supporting line. There is no supporting line to S' through A.

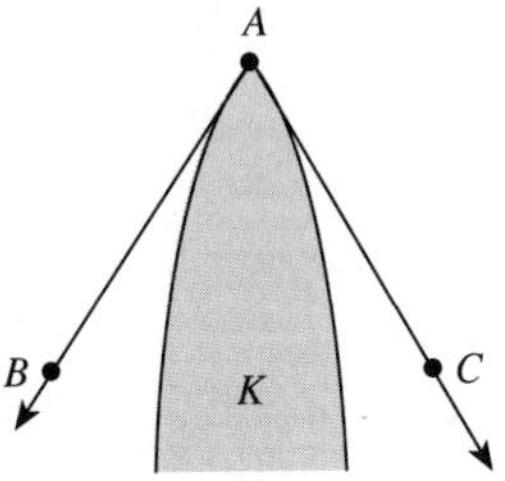

Figure 3.18

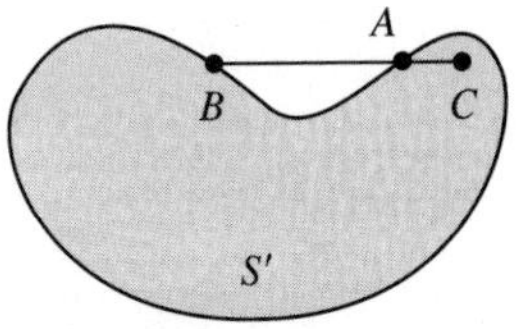

Figure 3.19

If a line is a supporting line for a set in two dimensions, the closed half-plane formed by the line and containing the set is called a *supporting half-plane*. This concept can also be used to give an additional property of convex sets.

THEOREM 3.5 ➤ A plane closed convex set that is a proper subset of a plane is the intersection of all its supporting half-planes.

An example of this theorem for a quadrilateral and its interior, showing four of the supporting planes formed by the sides, is shown in Figure 3.20.

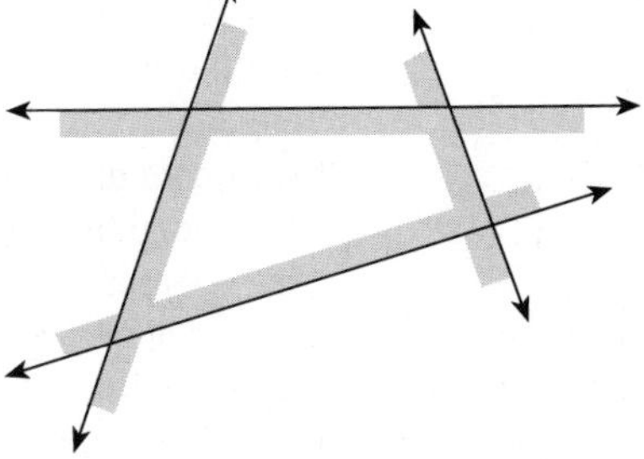
Figure 3.20

PROOF: Theorem 3.5 states that the set of points in the convex set is identical to the set of points in the intersection of the supporting half-planes. A convenient way to prove that two sets are identical is to prove that each is a subset of the other, since the sets must then have exactly the same elements. Let the convex set be K and the intersection of the supporting half-planes be K'. By definition of supporting half-planes, K is a subset of K'.

Now let A be a point in the complement of K, as in Figure 3.21. It can be proved that some point of K, say B, is the closest point of K to A. Let l be a supporting line to K at B that is perpendicular to $\overleftrightarrow{AB}$. Then A is not in the supporting half-plane of K determined by l; therefore, $A \notin K'$. Since any point in the complement of K is also in the complement of K', $K' \subseteq K$; hence, $K = K'$. In this proof, the existence of supporting line l is not proved. See Exercise 20 in Exercises 3.2.

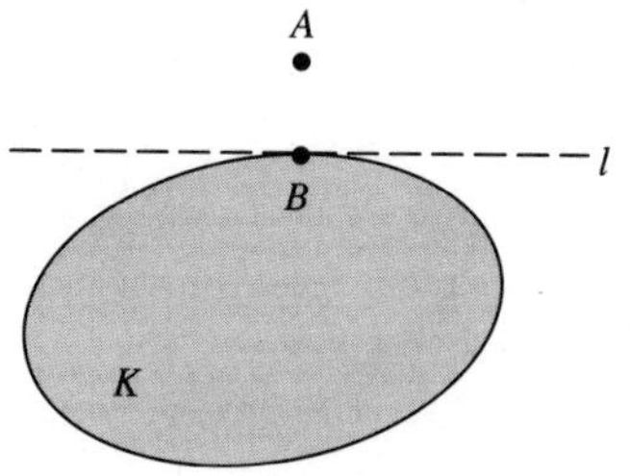

Figure 3.21

▲ EXERCISES 3.2 ▲

For Exercises 1–8, tell whether the line is or is not a supporting line for the set.

Set	**Line**
1. Circular region	Line through center
2. Circular region	Tangent
3. Square region	Line containing side of square
4. Square region	Line containing diagonal of square
5. Convex polygonal region	Line containing one side
6. Convex polygonal region	Tangent at vertex
7. Convex polygonal region	Line containing diagonal
8. Convex polygonal region	Line joining midpoint of two sides

For Exercises 9–12, describe the regular points and corner points for each of these sets.

9. Circular region

10. Square region

11. Angle and its interior—measure less than π

12. Convex polygonal region
13. Use Theorem 3.5 to give a definition of:
    a. A triangular region   b. A convex polygonal region
14. Draw three nonconvex sets and show one point on the boundary of each through which no supporting line passes.
15. State the converse, inverse, and contrapositive of the second conditional in Theorem 3.3.
16. Prove that each interior angle of a convex polygonal region has a measurement less than π.
17. Prove that a tangent cone to a convex set at a boundary point is itself a convex set.
18. Prove that if A is an interior point of convex set K and B is any other point of K, then every point in $\overset{\bullet\!-\!\circ}{AB}$ ($\overline{AB}$ except for point B) is an interior point of K. (*Hint:* Use Theorem 3.4, part a .)
19. Prove that if A and B are boundary points of a convex set K as well as points of K, then either $\overset{\circ\!-\!\circ}{AB}$ ($\overline{AB}$ without its endpoints) is entirely in the interior of K or $\overset{\circ\!-\!\circ}{AB}$ is entirely in the boundary of K.
20. Let A be a point in the complement of K, a plane closed convex set that is a proper subset of a plane, and let B be the point of K closest to A. Prove the existence of a supporting line for K at B that is perpendicular to $\overline{AB}$.
21. ▷ Take specific examples and explain the difference between the concept of tangent and supporting line.

▲ 3.3 Convex Bodies in Two-Space

Many of the common sets of points from Euclidean geometry are convex sets. Many of these are a special type of convex set called a convex body.

DEFINITION ➤ A *convex body* is a convex set of points that is *closed, bounded, nonempty,* and has *interior points.*

Examples of convex bodies include convex polygonal regions and circular regions. On the other hand, the interior of a parabola, a line, and the entire plane are convex sets that are not convex bodies.

The following theorem gives an important characteristic of a two-dimensional convex body.

THEOREM 3.6 ➤ A simple closed curve S and its interior form a convex body K if and only if every line through an interior point of K intersects S in exactly two points.

a. If a simple closed curve S is the boundary of a convex body K, then every line through an interior point of K intersects S in exactly two points.

See Figure 3.22. Let A be an interior point of K. Any ray $\overrightarrow{AP}$ intersects K in a segment $\overline{AB}$. This is true because the intersection of two convex sets is a convex set. $\overline{AB}$ is a one-dimensional convex body. But $\overrightarrow{AB'}$ (B', A, and B collinear) also intersects K in a segment $\overline{AB'}$, and B' is the second of the two boundary points on the line through A. The argument in this paragraph does not include all the details needed for a complete proof. We must also show that the line $\overleftrightarrow{AB}$ intersects S in no more than two points; this part of the proof is not included.

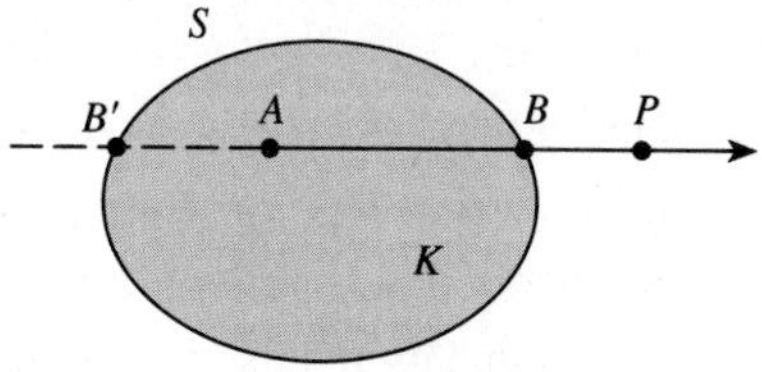

Figure 3.22

b. If every line through each interior point of K intersects the simple closed curve S in exactly two points, then S is the boundary of a convex body K. (It can be shown that the contrapositive of this second conditional, which is logically equivalent to the conditional itself, holds. The contrapositive states that if K is nonconvex, then some line through an interior point of K does not intersect the boundary S in exactly two points.)

In Figure 3.23, if K is nonconvex, interior points A and B can be found such that C on $\overline{AB}$ is an exterior point. Then it can be assumed that boundary points D and E for K exist on $\overline{AB}$. The ray $\overrightarrow{DA}$ also has a third boundary point F, so $\overleftrightarrow{AB}$ intersects S in at least three points.

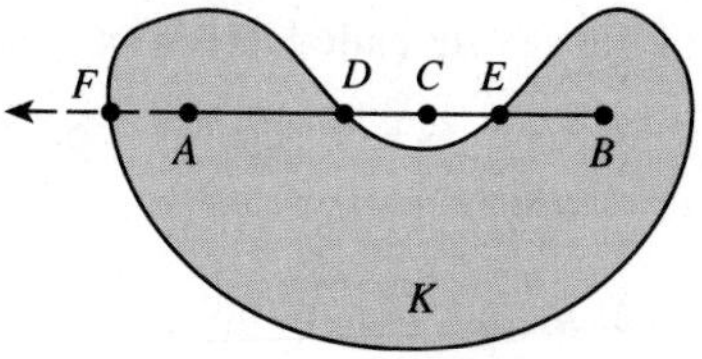

Figure 3.23

Another theorem about a convex body and its boundary is assumed here. A proof appears in Benson, *Euclidean Geometry and Convexity* (see bibliography).

THEOREM 3.7 ➤ The boundary of a convex body in two-space is a simple closed curve.

Of course, the converse of Theorem 3.7 is not true, because simple closed curves include a great variety of shapes in addition to those that are convex. Another condition for determining whether a particular simple closed curve is the boundary of a convex body is provided by the next theorem.

THEOREM 3.8 ➤ A simple closed curve S is the boundary of a two-dimensional convex body K if and only if each closed polygon $T = P_0P_1 \ldots P_nP_0$ determined by successive points on S is the boundary of a convex polygonal region.

The theorem is illustrated in Figure 3.24. Any polygonal region inscribed in the simple closed curve of Figure 3.24a will be convex. For the closed curve in Figure 3.24b, however, some of the inscribed polygonal regions, such as the one shown, are nonconvex.

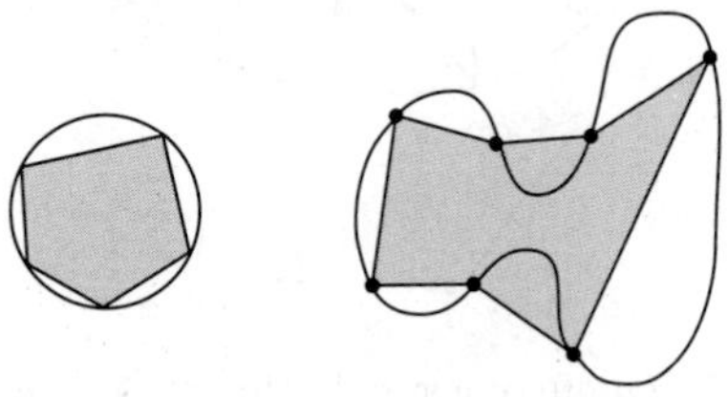

(a) (b)

Figure 3.24

The proof of the only if conditional of Theorem 3.8 involves showing that any inscribed polygon for a convex body has a supporting line at each point.

Let l be the line containing one side $\overline{AB}$ of an inscribed polygon. Since K is convex, $\overline{AB} \subset K$, so that $\overset{\circ\!-\!\circ}{AB}$ contains only points of S or only interior points of K (Exercise 19, Exercises 3.2). If $\overset{\circ\!-\!\circ}{AB}$ contains only points of S, then it can contain no interior points of T and is a supporting line for both K and T. If $\overset{\circ\!-\!\circ}{AB}$ contains interior points of K, as in Figure 3.25, then S is partitioned into two curves lying on opposite sides of l. All the vertices of T lie in one supporting half-plane H determined by l, since otherwise A and B would not be successive vertices. Then l is a supporting line for $H \cap K$, and thus is a supporting line for T and its interior. Every point of T has a supporting line passing through it, so that T is the boundary of a convex polygonal region.

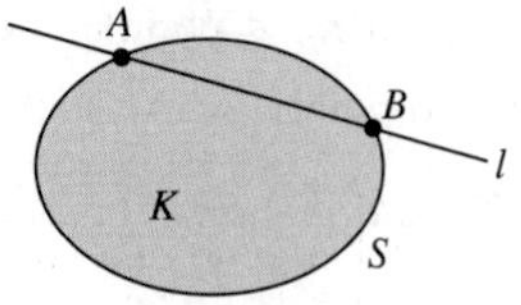

Figure 3.25

The proof of the if statement can be accomplished by showing that, if the set K with boundary S is not convex, there exists an inscribed quadrilateral that is not the boundary of a convex region. Such a quadrilateral is pictured as $ABCD$ in Figure 3.26, but the details of the proof are left as Exercise 15 of Exercises 3.3.

Two other theorems concern a special property of the length of the boundaries of two convex bodies.

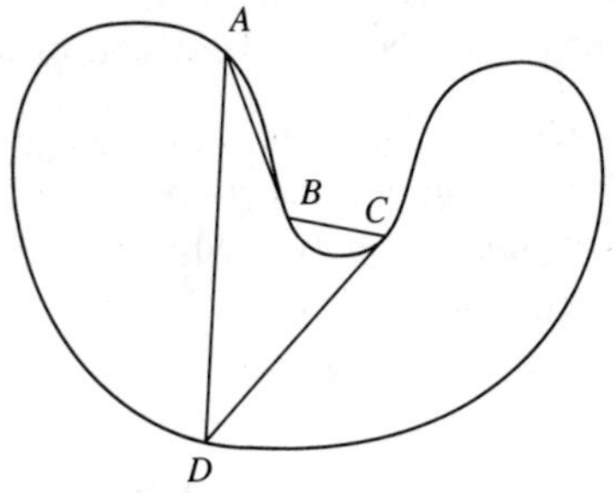

Figure 3.26

DEFINITION ➤ The *length* of a simple closed curve is the *least upper bound* of the length of all inscribed polygonal curves. Recall that the least upper bound of a set of real numbers is the smallest real number that is greater than or equal to each number of the set. The length of the boundary of a region is called the *perimeter* of the region.

THEOREM 3.9 ➤ If K_1 and K_2 are convex polygonal regions with $K_1 \subseteq K_2$, then the perimeter of K_1 is less than or equal to the perimeter of K_2.

PROOF: Let the vertices of K_1 be, in order, $P_0, P_1, \ldots, P_n$, and let l_i, $i = 1, \ldots, n$, be the line through P_i and P_{i-1}. The typical picture is shown in Figure 3.27 to help with this and the following notation. Let S_i be a polygonal region such that $S_i = H_i \cap S_{i-1}$, $i = 1, \ldots, n$, where H_i is the supporting half-plane of K_1 with boundary l_1 and S_0 is the greater region with boundary K_2. In Figure 3.27, $S_1 = H_1 \cap S_0$ is shaded. Since $AB < (AC + CD + DB)$, the perimeter of $S_1 < S_0$. For all i from 1 to n,

$$\text{perimeter of } S_1 \leq \text{perimeter of } S_{i-1}$$

Since the boundary of S_n is the boundary of K_1, then

$$\text{perimeter of } K_1 \leq \text{perimeter of } K_2$$

Theorem 3.9 can be generalized from polygonal regions to convex bodies.

THEOREM 3.10 ➤ If K_1 and K_2 are two-dimensional convex bodies with $K_1 \subseteq K_2$, then

$$\text{perimeter of } K_1 \leq \text{perimeter of } K_2$$

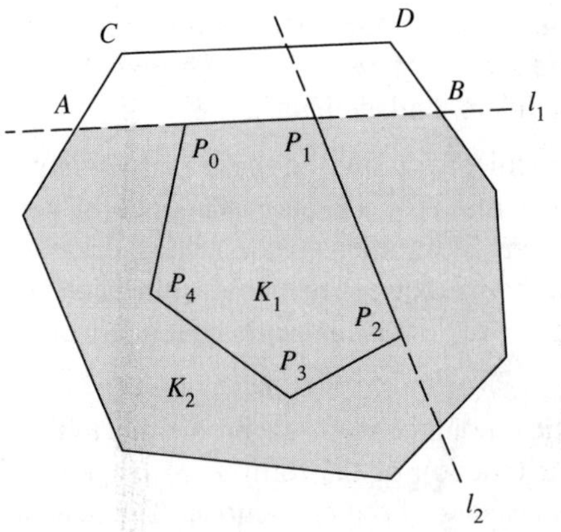

Figure 3.27

In Figure 3.28, let A be an interior point of K_1. Each ray with endpoint A intersects the boundaries of K_1 and K_2 in pairs of corresponding points, such as B and B'. There is a one-to-one correspondence between a polygon inscribed in K_1 and a polygon inscribed in K_2, determined by choosing corresponding vertices collinear with A and on the same side of A. For example, $BCDEF$ is paired with $B'C'D'E'F'$. Since each polygon S_1 of K_1 is contained in the corresponding polygon S_2 of K_2, from Theorem 3.9, the length of S_1 is less than or equal to the length of S_2. The least upper bounds satisfy the inequality

$$\text{least upper bound of } S_1 \leq \text{least upper bound of } S_2$$

so the perimeter of K_1 is less than or equal to the perimeter of K_2.

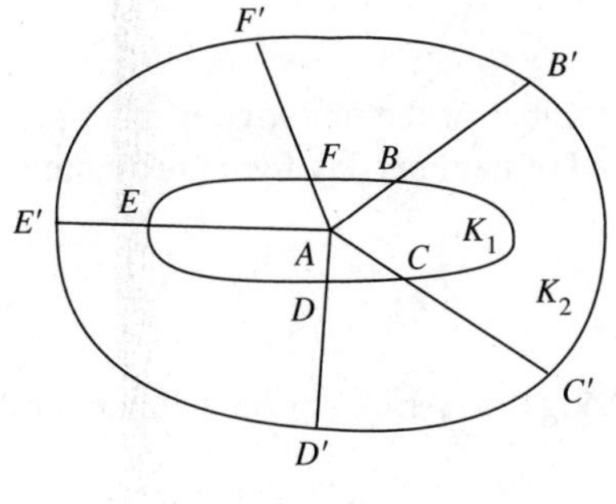

Figure 3.28

One important modern application of the theory of convex bodies (convex polygonal regions in this case) is in *linear programming*. The basic idea on which linear programming depends is that the maximum and minimum values of a linear function defined for all points on a convex polygonal region occur at vertices of the polygon. A brief explanation of the theory, and an example, are necessary here in order to make the application meaningful. For a thorough study of linear programming, see, for example, Leon Cooper and David Steinburg, *Methods and Applications of Linear Programming*, published by W. B. Saunders, Philadelphia, 1974.

To use a very simplified example, suppose that a company plans its production for maximum profit and that they sell just two kinds of items, for different prices. Other limiting restrictions in addition to the selling price must be considered—for example, the number of hours required to manufacture each item and the availability of machines and labor. The problem is to decide how many of each item to manufacture for maximum profit.

From a mathematical point of view, the problem is to find values for x and y that will maximize a linear function of the form $ax + by + c$, where x and y can take on only certain values because of the restrictions. The general theorem on which the application depends (Theorem 3.11) is stated without proof.

THEOREM 3.11 ➤ If a function $ax + by + c$ is defined for each point of a convex polygonal region, the maximum value occurs for the coordinates of one vertex, and the minimum value occurs for the coordinates of another vertex.

Example ➤ In order to produce the first kind of item for a profit of \$50, the first machine must be used for one hour and the second machine for two hours. In order to produce the second kind of item for a profit of \$40, the first machine must be used for one hour and the second machine for one-half hour. Each machine cannot be operated more than 12 hours a day. How many of each item should be manufactured each day for a maximum profit?

If x represents the number of the first kind of item per day and y represents the number of the second kind of item per day for a maximum profit, then

$$x \geq 0$$

$$y \geq 0$$

The inequalities, stating the restrictions on the time for the machines, are:

First machine	$x + y \leq 12$
Second machine	$2x + \frac{1}{2}y \leq 12$

Each of the four inequalities is represented in Figure 3.29 as a closed half-plane, and the intersection of these four convex sets is the convex polygonal region $ABCD$.

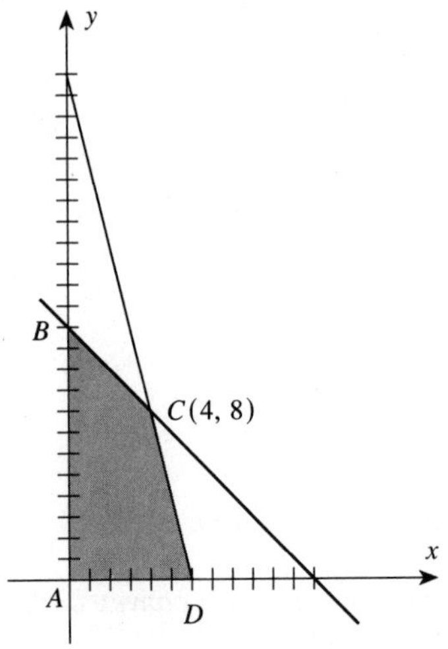

Figure 3.29

For the example, the function giving the profit is $f = 50x + 40y$, and the values at the vertices of the polygon are:

$$\text{At } A(0,0) \qquad f = 0$$

$$\text{At } B(0,12) \qquad f = 480$$

$$\text{At } C(4,8) \qquad f = 520$$

$$\text{At } D(6,0) \qquad f = 300$$

For a maximum profit, the company should make four of the first item and eight of the second. The profit will be \$520 per day.

A recent application of convexity is in computer graphics. Certain polygonal regions are simpler than others to fill with color. In addition to convex and nonconvex, a third category, horizontal-convex, is introduced. For example, the region in Figure 3.30a is called horizontal-convex. In this case, the intersection of each horizontal line with the region is a point or a segment. This is not true for the nonconvex polygon in Figure 3.30b. For convex and horizontal convex polygons, one algorithm for filling the shape with color depends on checking each horizontal line. This process is simplified because there is at most a single segment to consider.

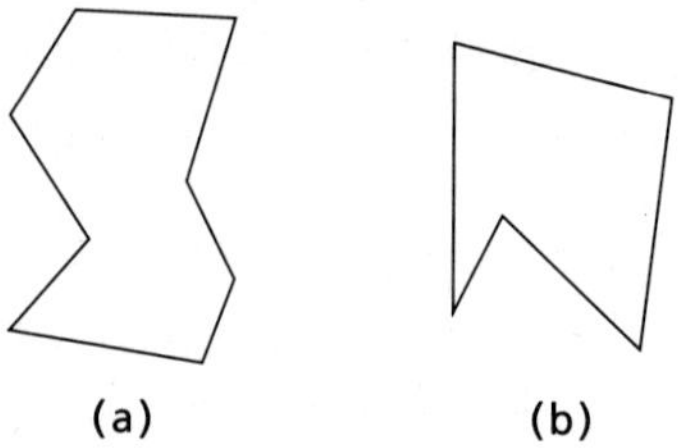

Figure 3.30

▲ EXERCISES 3.3 ▲

For Exercises 1–10, indicate which sets are convex bodies. For each set that is not necessarily a convex body, explain why not.

1. Triangular region
2. Line
3. Angle and its interior
4. Circle
5. Half-plane
6. Nonempty intersection of two convex sets
7. Ray
8. Convex polygonal region
9. Circular region with one boundary point missing
10. Intersection of two convex bodies
11. For a convex body, a ray with an interior point as its endpoint intersects the boundary of the convex body in exactly how many points?
12. If the intersection of two circular regions is such that it contains more than one point, is the intersection always a convex body?
13. Is the boundary of a convex body always a convex set?
14. Sketch two examples to show that Theorem 3.8 does not necessarily hold if the points P_i are not successive.
15. Complete the proof of Theorem 3.8.
16. Sketch two examples to show that Theorem 3.9 does not necessarily hold if the regions are nonconvex.
17. Describe all possible one-dimensional convex bodies.

18. Show that a plane convex body has a tangent at boundary point A if and only if there is exactly one supporting line for the body at A.

19. Rework the example given for linear programming with the following conditions: for the first item, the first machine must be run two hours and the second machine three hours; the profit on the first item is \$90 per item.

20. A rancher has space for 300 cows. He wants no more than 200 of one breed. If he makes a profit of \$40 on a Hereford and \$50 on a Black Angus, how many of each should he raise for a maximum profit?

▷ **21.** Study the proof of Theorem 3.11 in a text on linear programming.

▷ **22.** Write and solve a linear programming example similar to the one given following Theorem 3.11, but with more than four inequalities.

▷ **23.** Investigate the simplex method for solving more complex problems in linear programming.

▷ **24.** Sketch several additional examples of polygonal regions that are horizontal-convex but not convex. Investigate computer algorithms for filling such regions.

▷ **25.** The next section, on convex bodies in three-space, has many ideas analogous to those already developed for two-space. What can you discover about convex bodies in three-space, prior to studying the section?

▲ 3.4 Convex Bodies in Three-Space

The concepts and theorems for convex bodies in three dimensions are closely analogous to those already discussed for two dimensions.

DEFINITION ➤ Plane π is called a *supporting plane* of a three-dimensional set S if and only if π contains at least one boundary point of S and S lies entirely in one of the closed half-spaces determined by π. (See Figure 3.31.) The closed half-space determined by π that contains S is called a supporting half-space of S.

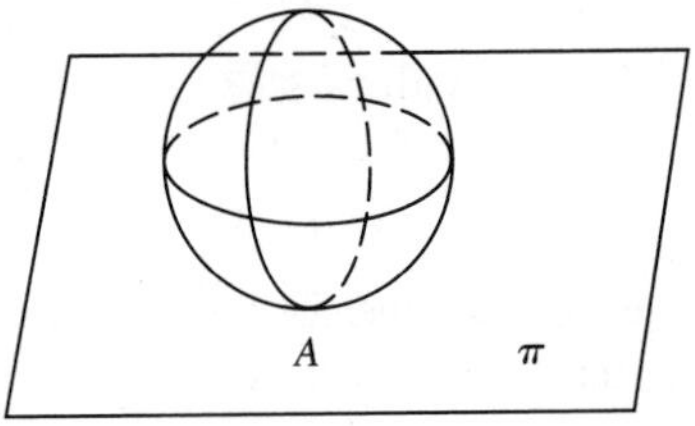

Figure 3.31

Things from the physical world that suggest sets and supporting planes are a basketball on a floor and a polar map, which is made by projecting points onto a plane tangent to the globe at the north pole, as shown in Figure 3.32a. The lines meeting at the pole are lines of longitude and the circles are lines of latitude (Figure 3.32b).

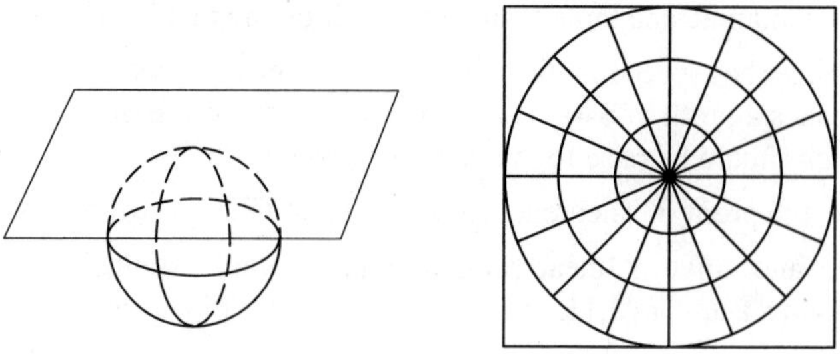

(a) (b)

Figure 3.32

For three-space, a tangent cone is actually the interior of an ordinary cone, as in Figure 3.33. There is no direct analogue in three-space to semitangents, since the boundary of the tangent cone is the ordinary cone itself. If the boundary of the tangent cone for a point A is a plane π, then π is called the *tangent plane* at A.

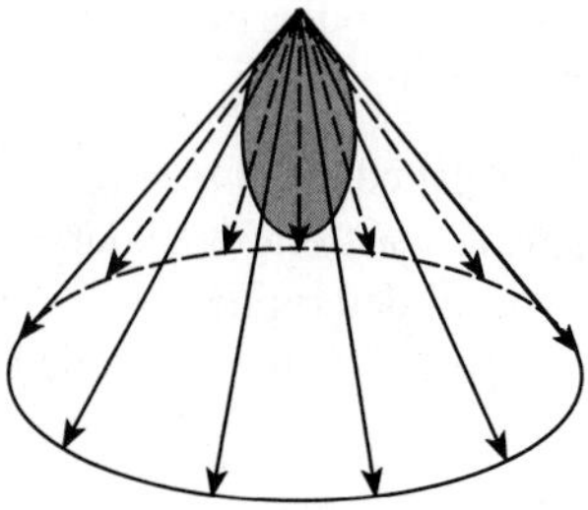

Figure 3.33

Supporting lines may also be defined for three-dimensional convex sets, but the definition of supporting lines used for two dimensions will not suffice, since a line does not determine *unique* half-planes or half-spaces in space. Line l is called a *supporting line* of a three-dimensional convex set K if and only if l contains at least one boundary point but no interior points of K. This definition guarantees that l in Figure 3.34 will not be considered as a supporting line for a two-dimensional set considered in three dimensions, since the definition applies only to three-dimensional sets.

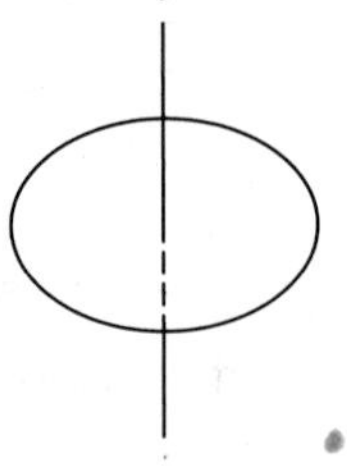

Figure 3.34

Theorem 3.3 from Section 3.2 corresponds to the following theorem in three-space. The proof is quite similar and is left as an exercise.

THEOREM 3.12 ➤ π is a supporting plane of the three-dimensional set K if and only if π contains boundary points but not interior points of K.

The concepts of regular point and corner point can be extended to three-space. A boundary point of a convex set K is regular if and only if K has a tangent plane at the point. The vertices of a convex polyhedral region are examples of corner points, whereas each point on a sphere is a regular point for the spherical region.

The next theorem is the three-space analogue of Theorem 3.4. Before stating the theorem,we must introduce the concept of surface. A *surface* is the graph of a set of equations of the form $x = f(s, t)$, $y = g(s, t)$, $z = h(s, t)$, for f, g, and h continuous functions and s and t an interval of real numbers. A *simple closed surface* has a single interior. It partitions every spherical neighborhood into two disjoint sets if the center is a point of the surface.

THEOREM 3.13 ➤ A simple closed surface S is the boundary of a convex set K if and only if through each point of S there passes at least one supporting plane of K.

PROOF:

a. Proof of the inverse of the only if statement.

Points A and B can be found with A an interior point such that $\overline{AB}$ contains a boundary point C if set K is nonconvex, as in Figure 3.35. No plane through A and B can be a supporting plane, since A is an interior point. Any other plane through C separates A from B and cannot be a supporting plane. Thus, there is no supporting plane of K through C.

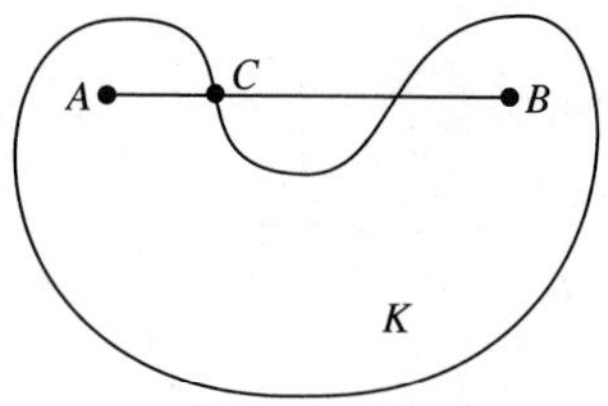

Figure 3.35

b. The proof of the only if statement depends on the concept of projection, to be explained in Chapter 7. It consists of showing that if the curve is the boundary of a convex set, then there is a supporting plane at each boundary point. This proof is not given.

The next theorem relates two-dimensional and three-dimensional convex bodies.

THEOREM 3.14 ➤ Let l be a supporting line of the three-dimensional convex body K at point B, and let π be the plane determined by l and an interior point A of K. Then $\pi \cap K$ is a two-dimensional convex body that has l as a supporting line.

Proof: A is also an interior point of $\pi \cap K$, so $\pi \cap K$ is a two-dimensional convex body. But suppose l is not a supporting line of $K \cap \pi$. Since B lies on l and B is a boundary point of $K \cap \pi$, l contains an interior point C of $K \cap \pi$, as in Figure 3.36. There is a point D of $K \cap \pi$ such that C lies between A and D. Since A is an interior point of K and $D \in K$, it follows that C is an interior point of K. This means that l is not a supporting line of K, contrary to the assumption.

Figure 3.36

The last theorem of this section is a three-space counterpart of Theorem 3.5.

THEOREM 3.15 ➤ A three-dimensional closed convex set (a proper subset of space) is identical to the intersection of all its supporting half-spaces.

This theorem is proved in a way analogous to the proof of Theorem 3.5, but the proof is left as an exercise.

▲ EXERCISES 3.4 ▲

1. Give other examples from the physical world representing supporting planes and three-dimensional sets.
2. Describe the tangent cone for a vertex of a cubical solid.
3. Is every line in a supporting plane for a convex set a supporting line for the set? Why?

Which of the sets of points in Exercises 4–11 are three-dimensional convex bodies?

4. Tetrahedral solid
5. Sphere
6. Half-space
7. Convex polyhedral solid
8. Cubical solid
9. The nonempty intersection of three spherical solids
10. Nonempty intersection of two tetrahedral solids
11. Simple closed surface
12. Prove Theorem 3.12.
13. Use Theorem 3.15 to give a definition for a tetrahedral solid.
14. Prove Theorem 3.15.
15. Prove that if π is a plane through an interior point of a three-dimensional convex set K and if l is a supporting line of $\pi \cap K$, then l is a supporting line of K.
16. Prove that every boundary point of a three-dimensional convex set lies on a supporting line of the set.
17. ▷ Give specific examples to distinguish between the concepts of supporting plane and tangent plane for a three-dimensional set.

▲ 3.5 Convex Hulls

In this section, a major new idea is introduced—that of a convex hull.

DEFINITION ➤ The *convex hull* (sometimes called the *convex cover*) of a set of points S is the smallest convex set containing S.

For the convex hull to be the smallest convex set containing S means that the convex hull is a subset of every other convex set containing S.

The definition does not specify that a convex hull must be a convex body.

Example ➤ Examples of convex hulls are shown in Figure 3.37.

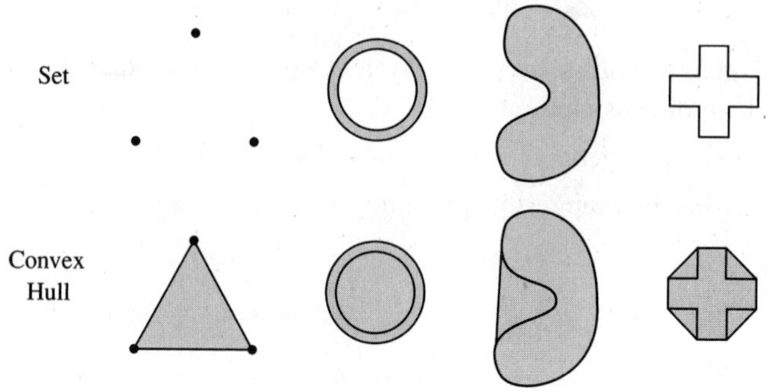

Figure 3.37

The concepts of convex set and convex hull are further related by the following theorem, whose proof is left as an exercise.

THEOREM 3.16 ➤ A set is convex if and only if it is its own convex hull.

In many problems that involve packing objects in convex containers, the convex hull must fit within the container, so the convex hull, rather than the object itself, must be considered. For example, a shoe box is rectangular, rather than the shape of a pair of shoes; a typewriter cover is not usually the shape of a typewriter.

The points on the boundary of a convex set have previously been classified as corner points or regular points. In order to investigate additional relationships between sets and their convex hulls, it is necessary to introduce a second classification based on whether a boundary point is or is not an extreme point.

DEFINITION ➤ A point A of convex set K is called an *extreme point* of K if and only if there are no two points P_1 and P_2 of K such that A is a point of $\overset{\circ\!-\!\circ}{P_1P_2}$.

In Figure 3.38, A, B, C, D, E, H, J are extreme points, whereas F, G, L are not.

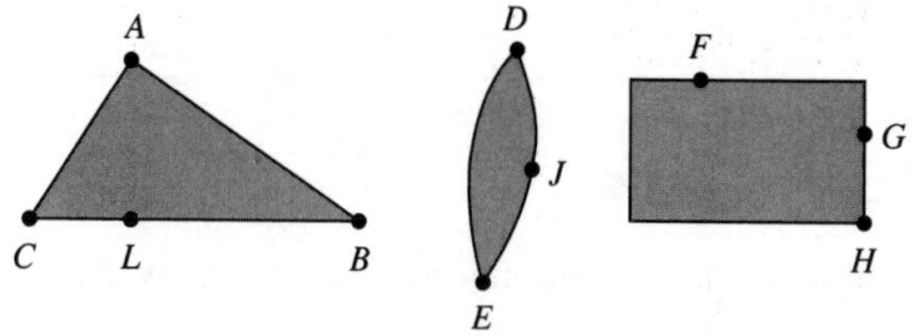

Figure 3.38

Every point on the boundary of a circular region is an extreme point. The connection between extreme points and convex hulls for some sets is made clear in the following two theorems.

THEOREM 3.17 ➤ A convex polygonal region S is the convex hull K of its extreme points.

PROOF: The extreme points are the vertices. Since each pair of extreme points belongs to K, each edge of S joining extreme points belongs to K. Any line through an interior point of S intersects the boundary in two points belonging to K. Then $S \subseteq K$. Since K is the smallest convex set containing the vertices, $K \subseteq S$ and therefore $S = K$.

THEOREM 3.18 ➤ The convex hull of a finite number of points in a plane is a convex polygonal region. Examples of convex hulls for a finite number of points in a plane are shown in Figure 3.39.

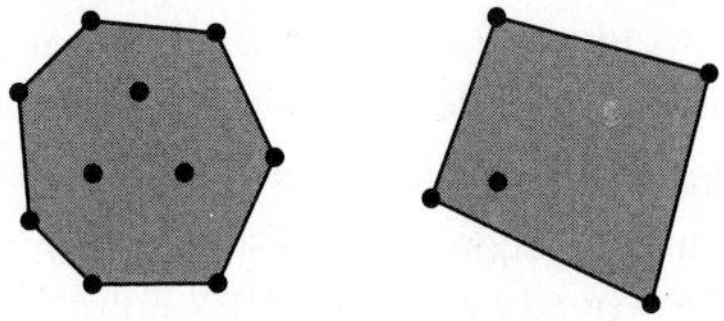

Figure 3.39

PROOF: Let K be the convex hull of the set of points $S = \{P_1, \ldots, P_n\}$ lying in a plane. If K is one-dimensional, then K is a line segment that may be considered a special case of a convex polygonal region.

If K is two-dimensional, it is bounded and the boundary is a simple closed curve T. Use a typical figure (Figure 3.40) to help with the notation.

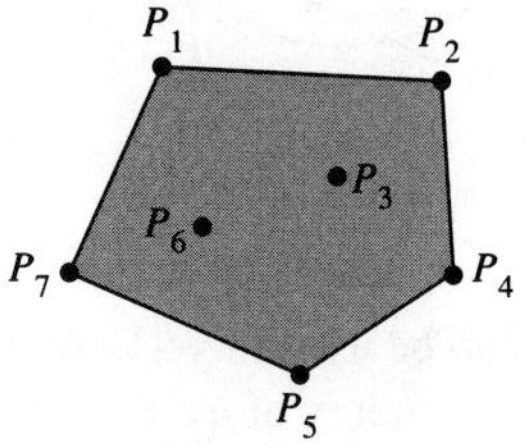

Figure 3.40

Let U be the intersection of the finite number of supporting half-planes of S whose boundaries contain at least two points of S. For example, $\overleftrightarrow{P_1P_2}$ and $\overleftrightarrow{P_2P_4}$ determine supporting half-planes, whereas $\overleftrightarrow{P_1P_3}$ and $\overleftrightarrow{P_6P_7}$ do not.

Since U is convex and $S \subseteq U$, then $K \subseteq U$. The rest of the proof involves showing that $U \subseteq K$.

Suppose, as in Figure 3.41, that $A \notin K$ and that B is the point of K at a minimum distance from A. The line l perpendicular to $\overline{AB}$ is a supporting line of K at B.

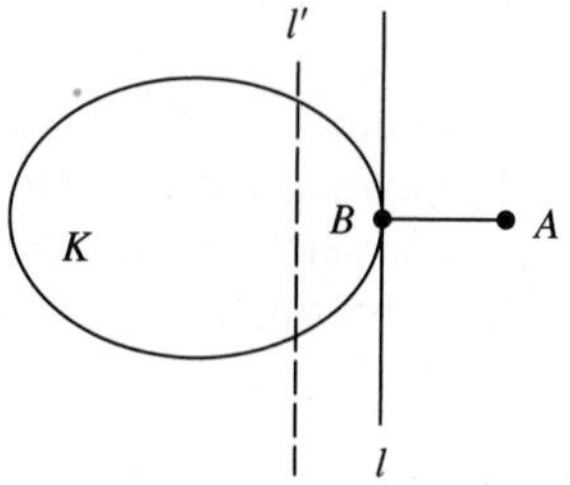

Figure 3.41

At least one point, say P_1, of S must lie on l; otherwise, l' could be located parallel to l, and there would be a convex proper subset of K on one side of l' that would contain S, contrary to the definition of K as the convex hull.

Now think of line l being rotated about P_1, as in Figure 3.42, keeping point A in the same half-plane until a second point P_2 of S is contained in l. The half-plane opposite point A can be proved to be one of the supporting half-planes of U. This means that A, which is not in K, is not in U. The conclusion is that U is a subset of K and, finally, that $K = U$. The boundary of K consists of the union of line segments, so that K is a convex polygonal region.

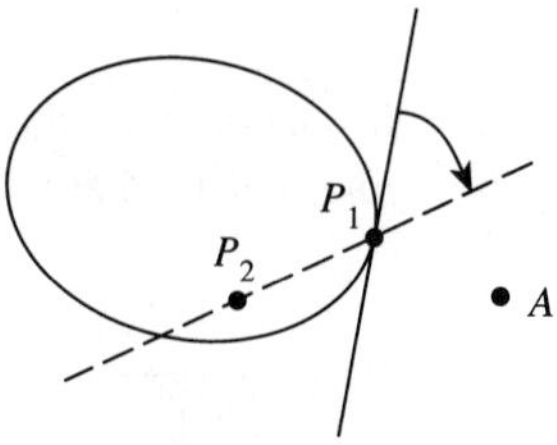

Figure 3.42

The idea of a convex hull can be used in yet another test to determine whether or not a simple closed curve is the boundary of a convex set.

THEOREM 3.19 ➤ Let S be a finite set of points on a simple closed curve T. T is the boundary of a convex set K if and only if, for all such sets S, no point of S is an interior point of the convex hull K' of S.

Figure 3.43 helps to make clear the meaning of this theorem. For the convex set in Figure 3.43a, the convex hull of any finite number of points, such as B, C, D, does not contain any other points on the boundary. For Figure 3.43b, however, the convex hull for $EGHI$ contains boundary point F.

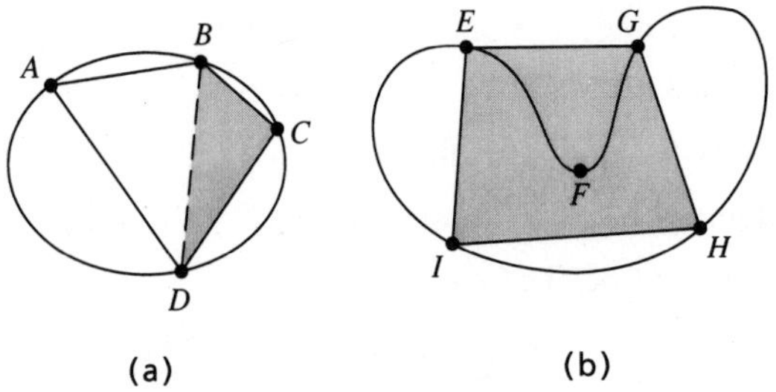

Figure 3.43

PROOF:

a. If T bounds a convex set K, then, by definition of convex hull, $K' \subseteq K$. Therefore, all interior points of K' are also interior points of K.

b. The proof of the converse consists of showing that if K is not convex, then there is a set of points S on T such that some member of S is an interior point of the convex hull of S (Exercise 15, Exercises 3.5).

The concept of convex hull is easily extended to sets in three dimensions. A convex polyhedral solid is the convex hull of its extreme points, the vertices. Similarly, the convex hull of a finite number of points in space is a convex polyhedral solid.

The final theorem in this section is also a three-space generalization of Theorem 3.19.

THEOREM 3.20 ➤ Let S be a finite set of points on a simple closed surface T. T is the boundary of a convex set if and only if no point of S is an interior point of the convex hull U of S for all such sets S.

▲ EXERCISES 3.5 ▲

For Exercises 1–10, describe or sketch in two dimensions the convex hull for each set listed below:

1. Circle

2. Triangle

3. Convex polygonal region

4. Five collinear points

5. Two intersecting lines

6. Two nonintersecting circles

7. Parabola

8. Hyperbola

9. Angle

10. Four distinct points, no three of which are collinear

11. Give examples involving practical applications of packing objects in convex containers.

12. Prove Theorem 3.16.

13. Can an extreme point be a regular point?

14. Can an extreme point be a corner point?

15. Complete the proof of Theorem 3.19.

16. Give an example of a set of points that is closed, but whose convex hull is not closed.

17. Prove that a convex body is the convex hull of its extreme points.

For Exercises 18–21, describe the convex hull of:

18. A dihedral angle

19. Two skew lines

20. A sphere

21. Four distinct noncoplanar points

22. Can the convex hull of any two-dimensional set be three-dimensional?

23. Prove Theorem 3.20.

24. Prove that every supporting plane of the convex hull of a closed bounded set contains at least one point of the set.

▷ **25.** Read and report on a recent article containing research involving convex hulls.

▷▷ **26.** Make an extended list of sets of points (in addition to those in Ex.18–21) and describe the convex hull of each.

▲ 3.6 Width of a Set

Some of the most useful applications of convexity depend on a property of a set of points called the *width* of the set.

DEFINITION ➤ The perpendicular distance along a line l between two parallel supporting lines of a bounded two-dimensional set or two parallel supporting planes of a bounded three-dimensional set is the width of the set in the direction indicated by the line l.

In Figure 3.44, w is the width of the set in the direction of l. A set may have different widths in different directions. Figure 3.45 shows two examples of width in particular directions for a two-dimensional set.

The diameter of any set can be defined as the least upper bound of the distances between any two points of the set. Then, for a closed bounded set, it can be proved that the diameter is equal to the maximum width of the set. The following two theorems depend on this result.

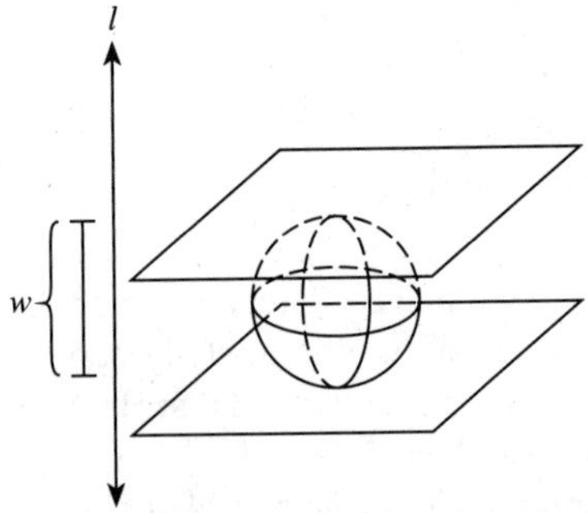

Figure 3.44

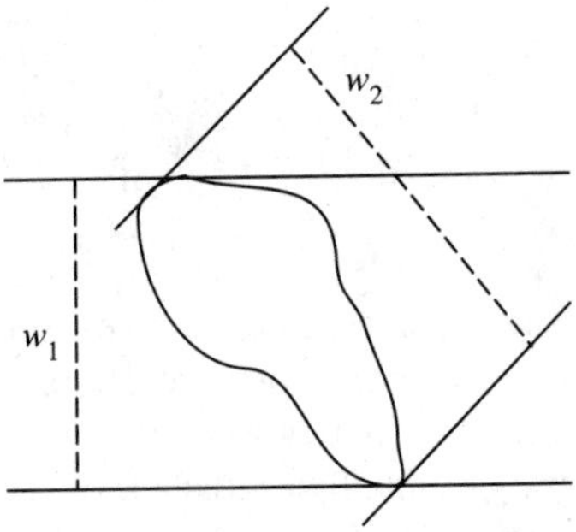

Figure 3.45

THEOREM 3.21 ➤ Let π and π' be parallel supporting planes for a set S in a direction of maximum width. If A is any point of $\pi \cap S$, then the line l through A perpendicular to π and π' intersects π' in a point B that is a point of S.

Proof: Assume the contrary, that B is not a point of S, as in Figure 3.46. Then there is a point C of $\pi' \cap S$; but $AC > AB$, so that line l is not in a direction of maximum width, contrary to assumption.

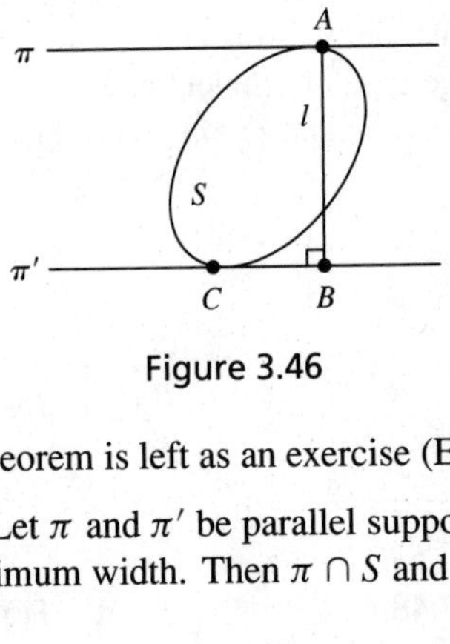

Figure 3.46

The proof of the next theorem is left as an exercise (Exercise 24, Exercises 3.6)

THEOREM 3.22 ➤ Let π and π' be parallel supporting planes for a closed set S in a direction of maximum width. Then $\pi \cap S$ and $\pi' \cap S$ each contain exactly one point.

Example ➤ Find the minimum width and the diameter of a rectangular region 3 cm by 4 cm.

The minimum width is 3 cm and the diameter is 5 cm, the length of the diagonals.

For some cases, the width of a set is the same for all directions. Sets of this kind are called sets of *constant width.* Before reading ahead, see how many different types of sets of constant width you can discover.

A set of constant width does not need to be convex, as is shown by the examples in Figure 3.47.

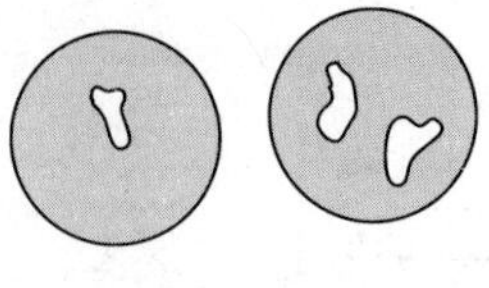

Figure 3.47

The most common convex set of constant width is a circular region in two dimensions. Euler, prior to 1800, seems to have been the first to study convex sets of constant width that were not circular regions. It was not until 1953 that Hammer and Sobyzk gave the first satisfactory general account of all such sets.

The simplest example of a convex set of constant width that is not a circular region is shown in Figure 3.48. Let $\triangle ABC$ be an equilateral triangle with the length of each side w. With each vertex as center, construct the shorter arc (of radius w) joining the other two vertices. The union of these three arcs is the boundary of a set of constant width w called a *Reuleaux triangle*, after an early contributor to the study of sets of constant width. A Reuleaux triangle is actually a region, rather than a curve.

An interesting difference between the Reuleaux triangle and a circular region is that a Reuleaux triangle has corner points at A, B, and C. In recent years, the mathematical properties of the Reuleaux triangle have led to some extremely important applications. One such application is in the cylinder of a Wankel engine. The Reuleaux triangle inside the double elliptical chamber, as shown in Figure 3.49, allows more than one phase of operation of the engine to take place at the same time. This type of engine is used in some automobiles. The same type of efficient engine has been used in snowmobiles. For further information, consult *Scientific American*, February 1969.

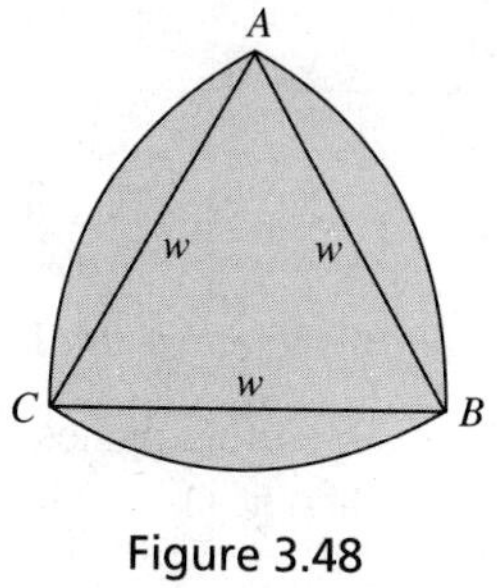

Figure 3.48

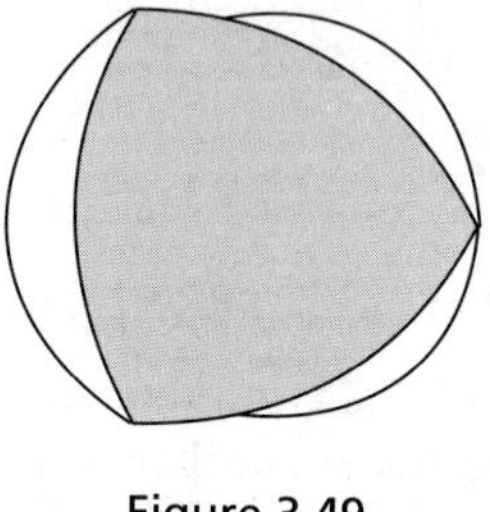

Figure 3.49

A second application of a Reuleaux triangle is in the gear for driving a movie film. The action desired is a brief, quick movement, followed by a momentary stoppage. See Figure 3.50a. A third application is the construction of a drill that makes a square hole. This is possible because a Reuleaux triangle can be moved around (but not rotated) inside a square with sides equal to the width of the Reuleaux triangle; hence, the shape of the drill bit is a Reuleaux triangle, and the bit moves in an eccentric path (Figure 3.50b). Another application is the shape of the top of a "jimmy-proof" fire hydrant that cannot be opened with an ordinary wrench.

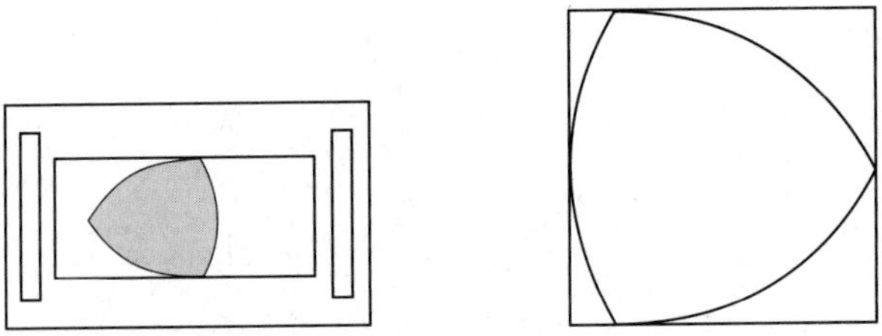

(a) (b)

Figure 3.50

An interesting possible or theoretical application is this: sets of constant width make safe manhole covers, since they can't slip down through their holes as could covers which are not of constant width, such as squares.

Some noncircular sets of constant width do not have corner points. Let ABC be an equilateral triangle with edge d, and extend each side a distance h past the vertices, as in Figure 3.51. With each vertex as center and with radius h, construct the arcs $A'A''$, $B'B''$, and $C'C''$. Then, with each vertex as center and with radius $d+h$, construct the arcs $A''B'$, $B''C'$, and $C''A'$. The six arcs are the boundary of a set of constant width $d+2h$.

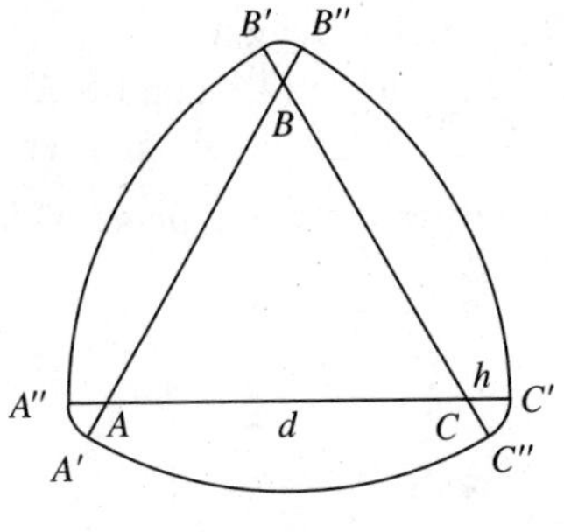

Figure 3.51

Sets of constant width of still another type may be constructed in a manner similar to that used for a Reuleaux triangle. Sets of this type use regular polygons with an odd number of sides and are sometimes called *Reuleaux polygons*. A Reuleaux polygon with five sides is shown in Figure 3.52. Any set of constant width may be considered as the orthogonal trajectory of sets of intersecting lines. The concept of *orthogonal trajectory* is illustrated in part for a Reuleaux triangle in Figure 3.53. The orthogonal trajectory of a set of lines may be thought of as a curve intersecting each line of the set at right angles. Each ray with endpoint A that intersects $\widehat{BC}$ does so at right angles. A similar statement may be made for $\widehat{AC}$ using rays from B, and $\widehat{AB}$ using rays from C.

Note that these sets of rays are the tangent cone for the triangular region ABC, with A, B, C as endpoints.

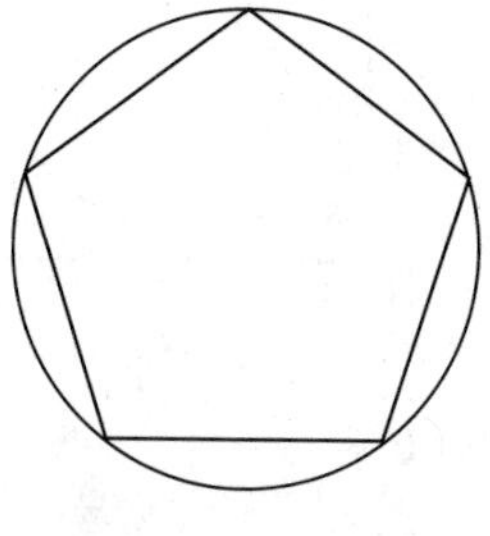

Figure 3.52

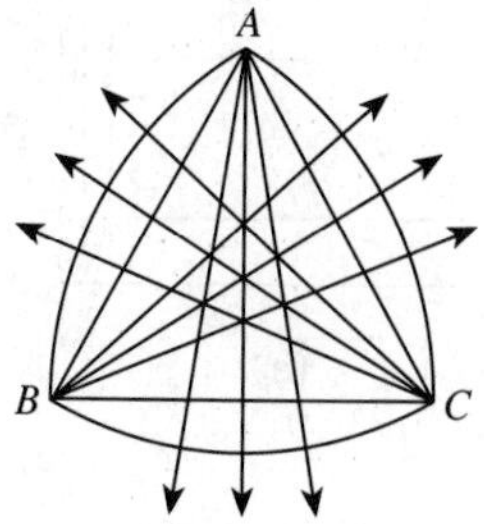

Figure 3.53

Sets of constant width are still being studied intensively by leading research mathematicians. Several of the simpler properties are given here, but the interested reader may pursue the topic in recent professional journals.

THEOREM 3.23 ➤ If K is a convex body of constant width and if $A \notin K$, then the diameter of $K \cup \{A\}$ is greater than the diameter of K.

The intuitive meaning of this theorem is that other outside points cannot be joined to a convex body of constant width without increasing the width. The theorem probably seems self-evident, at least for a circular region.

The proof is outlined. In Figure 3.54, suppose K has constant width w and $A \notin K$. Let B be a point of K closest to A, and let π be the plane through B perpendicular to $\overleftrightarrow{AB}$. Then π is a supporting plane of K, and $BB' = w$. But then $AB + BB' > BB' = w$.

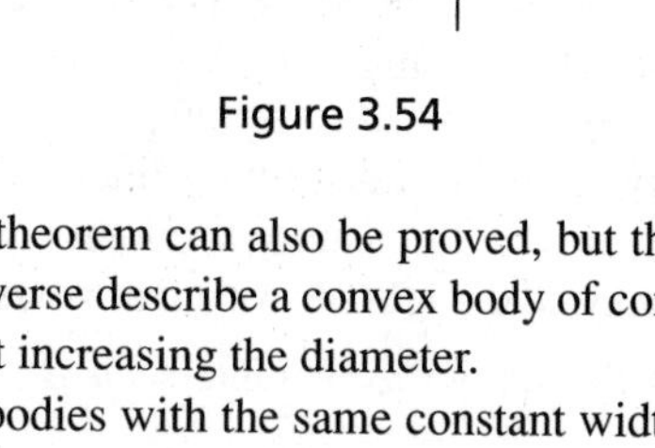

Figure 3.54

The converse of this theorem can also be proved, but the proof is not given here. Theorem 3.23 and its converse describe a convex body of constant width as containing all points possible without increasing the diameter.

Consider all convex bodies with the same constant width w. It is easy to find the perimeter (the length of the boundary) of several of these. Thus, for a circular region, the perimeter is πw. Interestingly enough, this is also the perimeter of the Reuleaux triangle with width w. The following theorem, not proved in this text, describes the general situation.

THEOREM 3.24 ➤ *Barbier's theorem.* The perimeter of a plane convex body of constant width w is πw.

Certainly, all convex sets with the same constant width do not have the same measure of area. The set with maximum area is the circular region. The set with minimum area is named in the following theorem.

THEOREM 3.25 ➤ *Blaschke–Lebesque theorem.* The set of constant width, with a given width specified, with the least area is the Reuleaux triangle.

The Reuleaux triangle has some properties of a circular region, but it does not have a center (a point equidistant from each point on the boundary). In fact, the only convex bodies of constant width having a center are circular regions in the plane and spherical regions in three dimensions.

One of the current areas of investigation in convexity is the concept of universal cover. A *universal cover* is defined as a plane region that can be used to cover every set whose diameter is 1. Any set with a diameter of 1 can be located in the plane in such a way that it is a subset of a universal cover.

A set that will cover every convex set of diameter 1 will cover every set of diameter 1. The smallest square universal cover is a unit square. The smallest equilateral triangle that is a universal cover has an incircle of diameter 1. The general problem of the smallest universal cover of any given shape has not been completely solved.

▲ EXERCISES 3.6 ▲

Find the minimum width and the diameter of each set in Exercises 1–4.

1. Square region with edge 1 cm

2. Rectangular region 2 cm by 3 cm

3. Reuleaux triangle constructed on equilateral triangle with edge 1 cm

4. Isosceles triangle with edges 7 cm, 7 cm, and 3 cm.

Exercises 5–14 are designed to further develop the concept of a Reuleaux triangle. For each statement about a Reuleaux triangle, answer true or false.

5. The boundary is a simple closed curve.

6. It has three corner points.

7. It is a universal cover.

8. The area is equal to the area of a circular region with the same width.

9. The perimeter is equal to the circumference of a circle with the same diameter.
10. The intersection of two Reuleaux triangles is a Reuleaux triangle.
11. It is a triangle.
12. It is a polygon.
13. The boundary is the union of three circular arcs.
14. It has a center point equidistant from each point on the boundary.
15. Picture nonconvex sets of constant width, other than those shown in the text.
16. Draw a Reuleaux polygon of seven sides.
17. Construct a set of constant width as in Figure 3.51, but begin with a pentagon instead of a triangle.
18. Find the perimeter of a Reuleaux triangle in terms of the constant width, and prove that your answer is correct.
19. Find the maximum and minimum areas of plane sets of constant width 4.
20. Prove that the smallest square universal cover is a unit square.
21. Find the length of the edge of the smallest equilateral triangle that is a universal cover.
22. Prove that the smallest equilateral triangle that is a universal cover has an incircle of diameter 1.
23. Prove that if P and Q are any two points of a plane closed set K of constant width w, and if $PQ = w$, then the lines at P and Q perpendicular to $\overline{PQ}$ are both supporting lines of K.
24. Prove Theorem 3.22.
25. Give an example to show that Theorem 3.23 is not necessarily true for a nonconvex body of constant width.
26. Give an example to show that Theorem 3.23 is not necessarily true for a convex body not of constant width.
27. Find a reference and study the proofs of Theorems 3.24 and 3.25.
28. Read and discuss portions of the original text by Franz Reuleaux [*The Kinematics of Machinery*, translated and edited by Alex.B.W. Kennedy London: Macmillan and Company,1876] in which he introduces the figure that now bears his name.
29. Study the article by J. Chris Fisher, "Curves of Constant Width from a Linear Viewpoint," listed in the bibliography, which shows how calculus and linear algebra can be used to prove theorems about curves of constant width.

▲ 3.7 Helly's Theorem and Applications

The central theorem of this section is named for the Austrian mathematician Eduard Helly (1884–1943). Helly studied at the University of Vienna and at Gottingen. He discovered his theorem in 1913 and published it in 1923. Interestingly enough, Helly, like Poncelet, who discovered projective geometry, spent several years as a prisoner of the Russians. In 1938, he and his family moved to the United States. His theorem and the wealth of material related to it make a substantial contribution to the recent discoveries in the geometry of convexity.

THEOREM 3.26 ➤ *Helly's theorem.* Let K_1, K_2, ..., K_N be N convex sets of points, lying in n-space, with $N \geq n+1, n = 1, 2$, or 3, and such that any collection of $n+1$ of these sets has a nonempty intersection. Then the intersection of all the sets is not empty.

Figure 3.55 is an illustration of Helly's theorem for four convex sets in a plane. You may verify that each set of three of the four convex sets has a nonempty intersection, and also that the intersection of all four sets, indicated by the darkest shading, is not empty.

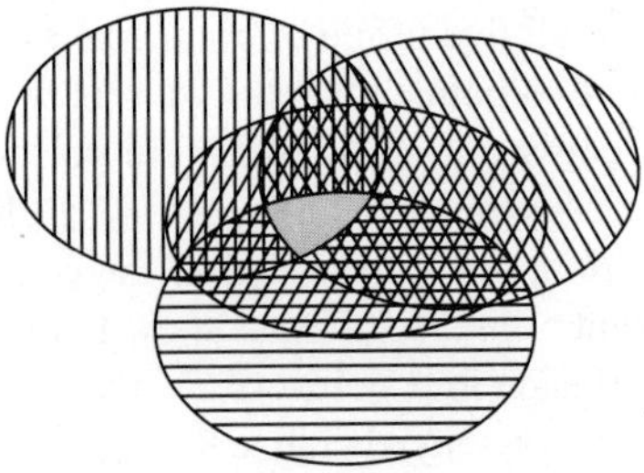

Figure 3.55

The theorem depends on a lemma called Radon's theorem.

THEOREM 3.27 ➤ *Radon's Theorem.* For $P = \{P_1, \ldots, P_N\}$ a finite set of points in n-space ($n = 1, 2, 3$), if $N \geq n+2$, then P can be partitioned into two disjoint sets whose convex hulls intersect.

A proof of this lemma is given in Benson, *Euclidean Geometry and Convexity*.

Example ➤ Consider the set of 6 points in a plane shown in Figure 3.56a. One way this set can be partitioned so the convex hulls intersect is 1, 2, 5 and 3, 4, 6. The convex hulls are intersecting triangular regions, as shown in Figure 3.56b.

The outline of the proof of Helly's Theorem for $n = 3$ follows.

1. Helly's theorem is obvious for $N = 4$.

2. For $N = 5$, construct all possible sets of points $P = \{P_1, P_2, P_3, P_4, P_5\}$, such that P_i is an element of the intersection of K_j for all values of j from 1 through 5, with the restriction $j \neq i$. For example, P_2 is an element of the intersection of K_1, K_3, K_4, K_5.

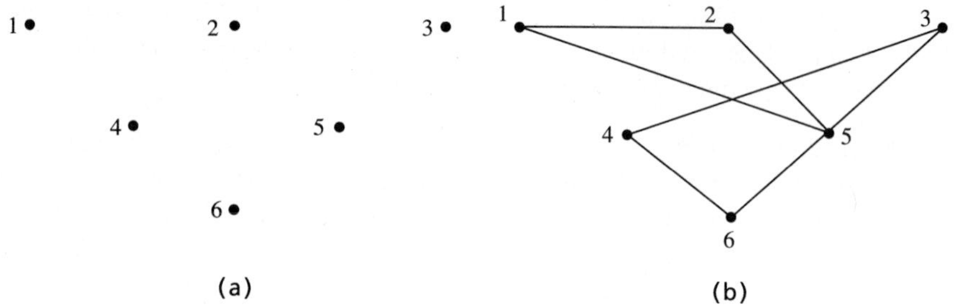

Figure 3.56

From Radon's theorem, each set P can be partitioned into two sets S_1 and S_2 whose convex hulls intersect. Now it needs to be shown that any point in the intersection of these convex hulls belongs to all five of the original convex sets. P can be partitioned into two sets with three and two points, or into two sets with four and one points. P was constructed so that the convex cover of every four points lies entirely in one of the sets of K. For example, the convex cover of P_1, P_2, P_3, P_4 lies in K_5.

For the three–two partition, you may visualize the convex hulls as a triangle and segment with a common point of intersection, as shown in Figure 3.57a. This common point is in the convex hull of each four points of P, and, therefore, lies in every set of K. For the four-one partition, you may visualize the convex hulls as a tetrahedron with the fifth point inside, as shown in Figure 3.57b. Point P_5 is in the convex cover of every set of four points of P.

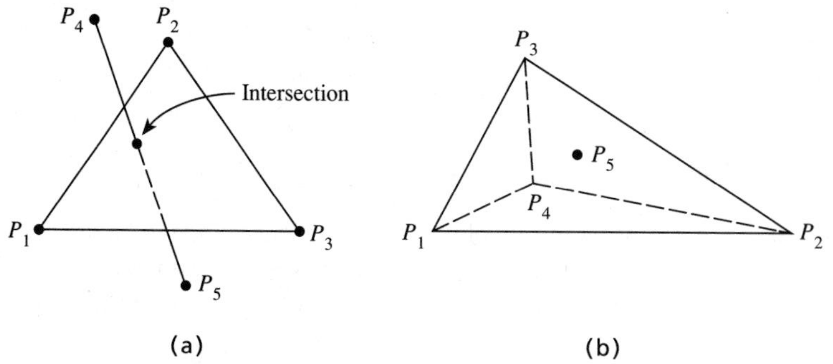

Figure 3.57

3. The proof is completed by mathematical induction using N, the number of convex sets. Assume that the theorem holds for m convex sets. Then we must show that it holds for $m+1$ sets. Let $K_1, \ldots, K_m, K_{m+1}$ be $m+1$ convex sets, such that each four have a nonempty intersection. Now create a set K_I, that is the intersection of K_m and K_{m+1}. We know that this set and any other three sets of K have a common intersection. This means that $K_1, K_2, \ldots, K_{m-1}$, K_I, are m convex sets, each four of which have a common point. According to our assumption, then, all of these sets have a nonempty intersection. This common intersection belongs to all the sets $K_1, \ldots, K_m, K_{m+1}$.

Many applications of Helly's theorem depend on a more general form stated here without proof.

THEOREM 3.28 ➤ For any collection of convex bodies in n-space, if every $n+1$ of them have a common point, then all the convex bodies have a common point.

One of the interesting applications of Helly's theorem shows the existence of points that behave somewhat like the center of symmetry of a set of points even when the set of points is not symmetric.

THEOREM 3.29 ➤ Let $S = \{P_1, \ldots, P_n\}$ be any finite set of points in space. Then there is a point A such that every closed half-space formed by a plane through A contains at least $n/4$ points of S.

As an illustration, suppose S consists of six points in space. The theorem says that there is a point A such that every closed half-space formed by a plane through A contains at least $6/4$ points of S. This means each half-plane must actually contain at least two points of S, since it contains more than one. It is also important to point out that A does not have to be a member of S. The theorem does not explain how the point A is actually found.

Proof: Since S is a finite set of points, there exists a spherical region B containing all the points P_i. Consider the set of all closed half-spaces containing more than $3n/4$ points of S. Let H_1, H_2, H_3, H_4 be any four of these closed half-spaces. In a course in set theory, it can be proved that

$$(H_1 \cap H_2 \cap H_3 \cap H_4)' = H_1' \cup H_2' \cup H_3' \cup H_4'$$

where H_i' is the complement of H_i. Since each H_i' contains fewer than $n/4$ points of S, $H_1' \cup H_2' \cup H_3' \cup H_4'$ does not contain all n of the points of S. Then $H_1 \cap H_2 \cap H_3 \cap H_4$ must contain at least one point of S.

Since every four of the closed half-spaces have a common point of S, Helly's theorem (in the form of Theorem 3.28) can be applied to the intersections of these half-spaces with B to conclude there is a point A common to all such half-spaces.

This point A is the desired point in the theorem. If it were not, then there would be a plane π through A that is the boundary of a closed half-space containing less than $n/4$ points of S. The opposite open half-space H will contain more than

$3n/4$ points of S. Let π' be the plane parallel to π passing through the points of $S \cap H$ closest to π. The closed half-space H' with boundary π' and lying in H contains more than $3n/4$ points of S, but H' does not contain A. This is a contradiction, since A was assumed to be in the intersection of all half-spaces containing more than $3n/4$ of the points.

The applications of Helly's theorem are surprisingly numerous; only a few are listed here. A theorem similar to Theorem 3.29, but involving volume, is the following, stated without proof.

THEOREM 3.30 ➤ Let S be a bounded set of points in space having volume V. Then there is a point A such that every closed half-space formed by a plane through A intersects S in a set with volume at least $V/4$.

Helly's theorem is rather closely related to other geometric concepts, including the ideas of universal cover and width of a set, as is evident by the final four applications. Drawing a figure to illustrate the last three will help you better understand these applications.

THEOREM 3.31 ➤ If each three of n points in a plane can be enclosed in a circle of radius 1, then all n points can be enclosed in such a circle.

PROOF: The proof consists of showing that a point (the center of the desired circle) exists that is not more than 1 unit from each of the points. A typical figure for three points is shown in Figure 3.58.

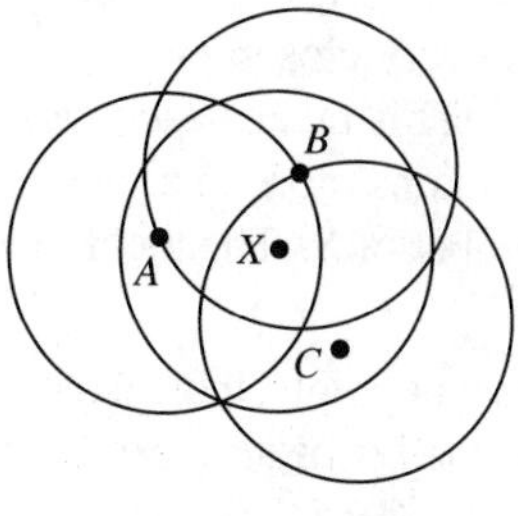

Figure 3.58

Let a circle with center X be a unit circle enclosing three of the n points, A, B, C. The three circles with centers A, B, C all contain X. By Helly's theorem, since any three of the unit circles with the given points as center have a nonempty intersection, all of the circles have a nonempty intersection. A point in this common intersection will serve as the center of a unit circle enclosing all the points.

THEOREM 3.32 ➤ *Jung's theorem.* For n points of the plane, such that each two are less than 1 unit apart, all can be enclosed in a circle of radius $\sqrt{3}/3$.

THEOREM 3.33 ➤ *Blaschke's theorem.* Every bounded convex figure of constant width 1 contains a circle of radius $1/3$.

THEOREM 3.34 ➤ Given n parallel line segments in the same plane, if there exists a line that intersects each three of them, then there is a line intersecting all the line segments.

The geometry of convexity is a rapidly expanding area of modern geometry. Many of the familiar figures of Euclidean geometry have been shown to possess new properties when they are considered as convex sets. This chapter is deliberately open-ended in the hope that readers will seek out recent research articles on convexity, many of which are fairly elementary.

▲ EXERCISES 3.7 ▲

1. Draw a picture with six convex sets in a plane to illustrate Helly's theorem.
2. Give an example in one dimension to show that Helly's theorem does not hold for nonconvex sets.
3. Carefully state Helly's theorem for two dimensions. Give specific numbers where appropriate.
4. Give an example in two dimensions to show that Helly's theorem does not hold for nonconvex sets.
5. Draw a model for Helly's theorem for $n = 1$ and $N = 4$.
6. Outline the proof of Helly's theorem for one dimension.
7. Outline the proof of Helly's theorem for two dimensions.
8. For the eight vertices of a cube, does Theorem 3.29 apply for A the center of the cube?
9. Suppose that S is a set of exactly 24 distinct points in space. Then there is a point such that every closed half-space formed by a plane through this point contains at least how many points of S?
10. State and prove a theorem for the plane that is analogous to Theorem 3.29.
11. Verify Theorem 3.33 for a Reuleaux triangle of constant width 1.
12. Draw a picture to verify Theorem 3.34.
13. Prove Theorem 3.30.
14. ▷ Take various examples, starting with the vertices of a unit triangle, to develop Theorem 3.32 intuitively. Then study a formal proof.
15. ▶▷ Make your own conjecture of possible applications, then consult a reference text on convexity and list additional applications of Helly's theorem besides those mentioned in this section.

▲ CHAPTER 3 REVIEW EXERCISES ▲

For Exercises 1–4, tell which statements are true for all convex sets consisting of more than one point.

1. The intersection of any two of these sets is a convex set.

2. Each point of the set has a supporting line containing it.

3. Removing exactly one boundary point would result in a convex set.

4. Every point belonging to the set is an interior point.

For Exercises 5–10, tell whether each set is always convex.

5. The union of two angles and their interiors

6. The interior of a triangular region

7. A simple closed curve and its interior

8. A set that is closed, bounded, and nonempty

9. $\{(x, y) : y > |x|\}$

10. $\{(x, y) : x^2 + y^2 \geq 1\}$

11. Describe the boundary of this set: $\{(x, y) : x$ and y are rational numbers$\}$.

For each set in Exercises 12 and 13, describe the regular points and corner points, and tell whether or not the line is a supporting line for the set.

12. Set: Hexagonal region; Line: Line intersecting set at one vertex only

13. Set: 50° angle and its interior; Line: One of the rays forming the angle

For Exercises 14–20, indicate which sets are always convex bodies.

14. Reuleaux triangle

15. Intersection of two convex bodies

16. Convex hull of an ellipse

17. Simple closed curve and its interior

18. Circular region with one boundary point missing

19. Rectangular right prism and interior

20. Interior of ellipsoid

For Exercises 21–23, describe the convex hull of each set.

21. Three distinct points, not collinear

22. An acute angle

23. Triangular right prism

24. Find the maximum width of a rectangular region that is 2 cm by 4 cm.

For Exercises 25–28, which statements are always true for the boundary of a Reuleaux triangle?

25. It is a simple closed curve.

26. It has three corner points.

27. It is a universal cover.

28. The perimeter is greater than the circumference of a circle with the same diameter.

29. Suppose that K is a convex body of constant width and suppose that A is a point not in K. What can you conclude about the diameter of the union of K and A?

30. Is it possible for a two-dimensional convex set to have no extreme points?

31. In order to make two kinds of items, P and Q, two machines, I and II, are required. To make item P, machine I runs for 1 hour and machine II runs for $\frac{8}{5}$ hours. To make item Q, machine I runs 1 hour and machine II runs $\frac{4}{5}$ hour. Each machine can run 16 hours or less per day. On item P, the company realizes a profit of \$30 and on item Q, a profit of \$20. How many of each item should be produced per day for a maximum profit?

Additional chapter review exercises may be selected from any exercises not already assigned from previous exercise sets in this chapter.

4 Euclidean Geometry of the Polygon and Circle

4.1 Fundamental Concepts and Theorems

Chapter 2 emphasized Euclidean geometry as a study of the invariant properties of sets of points under Euclidean or similarity transformations. Chapter 3 concentrated on convexity, one of the properties preserved by similarity transformations (as well as by some more general transformations). This chapter begins with some fundamental concepts and theorems concerning polygons (particularly triangles) and circles—concepts and theorems that have long been a part of Euclidean geometry. Then it progresses rapidly to material that has been discovered since 1800. The last two sections include many applications of Euclidean geometry: the golden ratio, tessellations, packing problems, Pick's theorem, extremum problems, geometric probability and fuzzy sets. Much of the material of the first part of this chapter, and part of the next, was once a major portion of the course called "college geometry." If necessary, you should quickly review the beginning theorems in Appendix 5 before continuing.

Among the key concepts and theorems from elementary geometry are the concurrence theorems that identify four significant points connected with a triangle.

THEOREM 4.1 ➤ The perpendicular bisectors of the sides of a triangle are concurrent at a point called the *circumcenter*.

Theorem 4.1 is illustrated in Figure 4.1. It is significant when any *three* lines have a common point in a geometry, because any three lines generally intersect by pairs to determine three distinct points, rather than a single one.

Proof: The proof of Theorem 4.1 depends on showing that the point of intersection of two of the perpendicular bisectors, say $\overleftrightarrow{B'O}$ and $\overleftrightarrow{C'O}$ in Figure 4.1, also lies on the perpendicular bisector of the third side. Point O is equidistant from A and C because it is on $\overleftrightarrow{B'O}$; it is also equidistant from A and B because it is on $\overleftrightarrow{C'O}$. Therefore, O is equidistant from B and C on the perpendicular bisector of $\overline{BC}$.

Since O is the point in a triangle equidistant from the three vertices, it is called the circumcenter to show that it is the center of the circumscribed circle (also called the circumcircle), the unique circle containing all three vertices of a triangle.

In Figure 4.2, triangle $A'B'C'$ is formed by joining the midpoints of the sides of the original triangle. Since the segments joining the midpoints of two sides of a triangle are parallel to the third side, the perpendicular bisectors of the sides of triangle ABC are also perpendicular to the sides of triangle $A'B'C'$. This means that $\overleftrightarrow{A'O}$ is perpendicular to $\overleftrightarrow{B'C'}$ and so on; thus, the perpendicular bisectors of the sides of triangle ABC are the altitudes of triangle $A'B'C'$. Reversing this statement leads to the claim that the altitudes of a triangle are the perpendicular bisectors of the sides of another triangle and, hence, are concurrent (by Theorem 4.1).

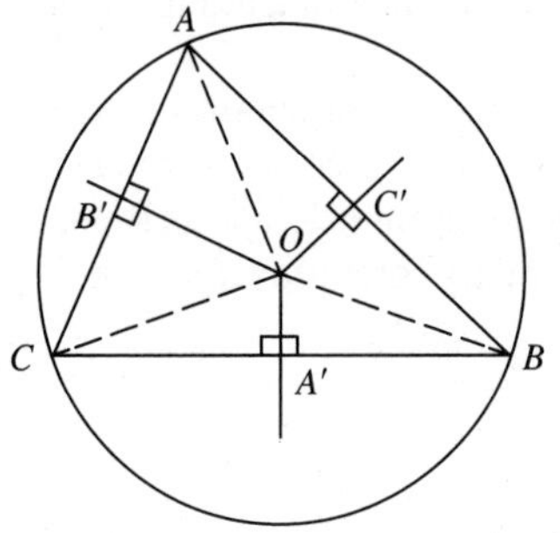

Figure 4.1

THEOREM 4.2 ➤ The altitudes of a triangle are concurrent at a point called the *orthocenter*.

Figure 4.3 shows triangle ABC and its orthocenter H. When they are distinct, the four points A, B, C, and H constitute an *orthocentric set* of four points, so named because each of the four points is the orthocenter of the triangle formed by the other three. For example, B is the orthocenter of the triangle with vertices A, C, and H.

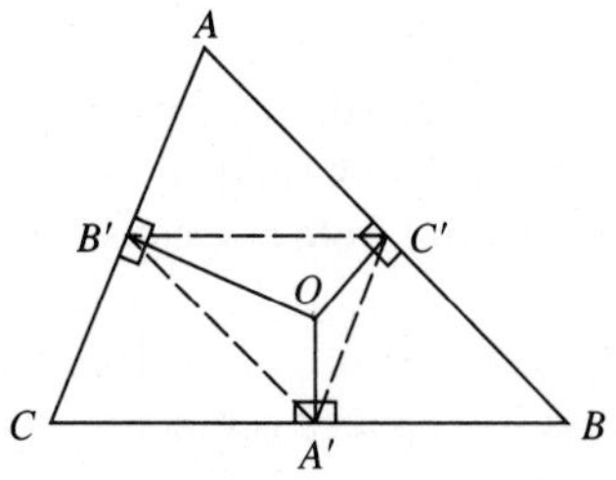

Figure 4.2

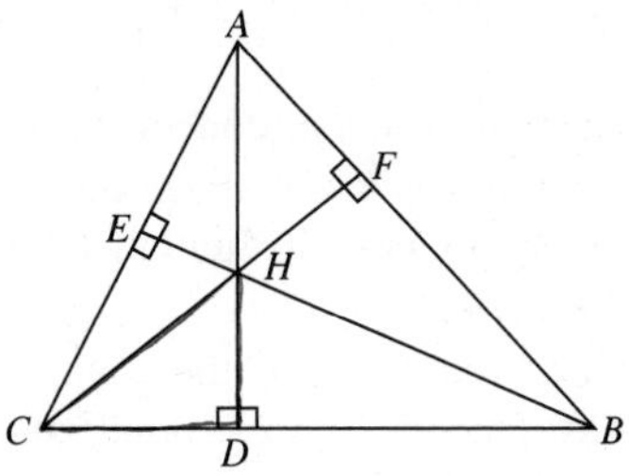

Figure 4.3

Two additional theorems about concurrency involve the internal angle bisectors and the medians of a triangle.

THEOREM 4.3 ➤ The internal bisectors of the angles of a triangle meet at a point called the *incenter*.

The proof of Theorem 4.3 (the details are left as Exercise 15, Exercises 4.1) depends on the idea that every point on the internal bisector of an angle is equidistant from the adjacent sides of the angle. For example, in Figure 4.4a, if I is on the angle bisector of angle B, then $IY = IX$. Since I is equidistant from all three sides of the triangle, it is the center of the *incircle*, a circle inscribed in the triangle. This means that the three sides of the triangle are tangent to the incircle.

THEOREM 4.4 ➤ The medians of a triangle meet at a point called the *centroid*.

Medians join the vertex and the midpoint of the opposite side of a triangle. In Figure 4.4b, triangles CBG and $GC'B'$ are similar, with a ratio of similarity of 2. Completion of the details of the proof is left as an exercise. The centroid is a point that is also called the center of gravity for a triangle.

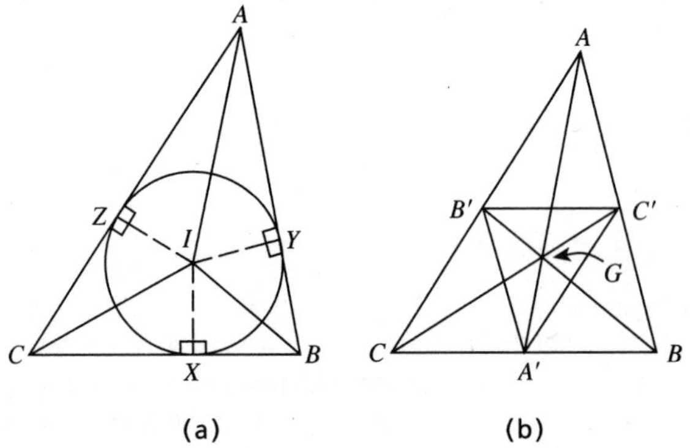

Figure 4.4

So far, four points of concurrency have been introduced. It is natural to wonder whether the four new points constitute a significant set of points in their own right. This matter will be discussed in Section 4.3, but you may profit from conjecturing about the location of these points now.

Much of the study of the triangle in Euclidean geometry involves work with proportions. One important proportion used as a tool in college geometry is connected with internal angle bisectors of the angles of a triangle, as stated in Theorem 4.5.

THEOREM 4.5 ➤ The internal bisector of an angle of a triangle divides the opposite side into two segments proportional to the sides of the triangle adjacent to the angle.

Proof: In Figure 4.5, assume that $\overline{AD}$ is the internal angle bisector of angle A of triangle ABC. Draw $\overline{CE}$ parallel to $\overline{AD}$, where E is on $\overleftrightarrow{AB}$. Because of the parallelism, $\angle ECA \cong \angle CAD \cong \angle CEA$. This means that triangle ECA is isosceles and that $\overline{EA} \cong \overline{AC}$. Now think of the figure as formed by two transversals from B intersecting a pair of parallel lines, $\overleftrightarrow{EC}$ and $\overleftrightarrow{AD}$, so that the following holds:

$$\frac{CD}{EA} = \frac{DB}{AB} \quad \text{or} \quad \frac{CD}{CA} = \frac{DB}{AB}$$

so that

$$\frac{CD}{DB} = \frac{CA}{AB}$$

as was to be established.

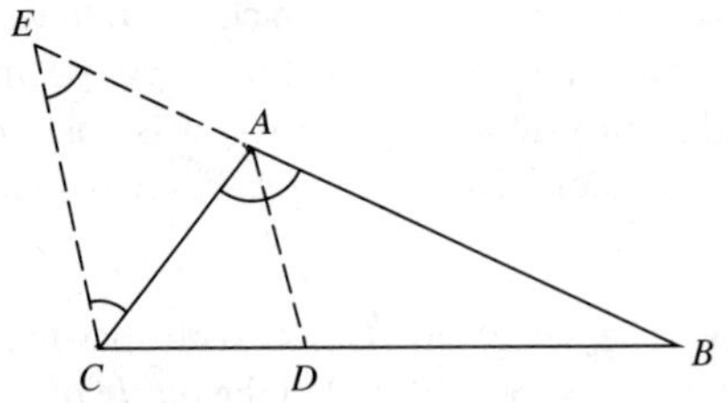

Figure 4.5

Since Theorem 4.5 is used frequently in connection with directed segments in Section 4.2 and its exercises, we are assuming in the preceding proof that the sides of the triangle must be placed so that the ratio is positive since the internal bisector divides the opposite side internally, making that ratio positive.

The proof of the analogous theorem for *external bisectors* is left as Exercise 20, Exercises 4.1. The external bisectors of angles of a triangle have properties corresponding to those of the internal bisectors. Corresponding to the concept of incenter, for example, is the concept of excenter given in the statement of Theorem 4.6.

THEOREM 4.6 ➤ The external bisectors of two angles of a triangle meet the internal bisector of the third angle at a point called an *excenter*.

Of course, when talking about external bisectors of a triangle, one must think of the sides as "extended" or consider the sides as lines rather than segments. In Figure 4.6, assume that E is an excenter. Since E is equidistant from all three sides of the triangle, it is the center of a circle externally tangent to all three sides of the triangle. This circle is one of the *excircles* of the given triangle.

One other property of internal and external angle bisectors—the internal and external bisectors of an angle of a triangle are perpendicular (Exercise 22)—is needed to help prove Theorem 4.7.

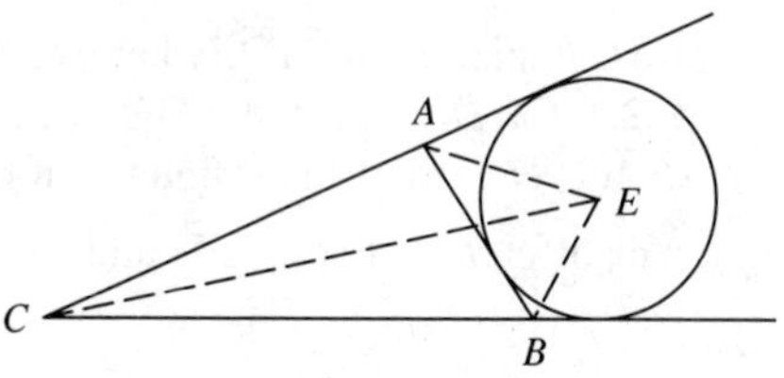

Figure 4.6

In geometry, a set of points can sometimes be described as satisfying some particular condition. For example, the set of points in a plane the same distance from two fixed points is the perpendicular bisector of the segment having the two fixed points as endpoints. The set of all points in the plane the same distance from the sides of an angle is the bisector of the angle. The set of points that subtend a right angle with a given segment is a circle with the segment as diameter.

Originally, the concept of *locus* was used in connection with finding points having a certain property. The word *locus* implied that the curve desired was traced by a path of a moving point. Today, in synthetic geometry, it is customary to use the concept of *set of points* rather than locus. The following theorem states another property that determines a set of points.

THEOREM 4.7 ➤ The set of all points, such that the distances to two fixed points have a given ratio, is a circle, called the *circle of Apollonius*.

The circle of Apollonius is named after the Greek mathematician Apollonius, who wrote a comprehensive treatment on conic sections prior to 200 B.C. Apollonius (262–190 B.C.) is often called "The Great Geometer." His work, *Conic Sections*, consisted of 8 books and 400 propositions. He is given credit for naming the conic sections using *ellipse*, *parabola*, and *hyperbola*. In the proof of Theorem 4.7, a set of points is shown to be a circle because the points are vertices at the right angle of a right triangle whose hypotenuse is the diameter. The proof depends on the fact that a triangle inscribed in a semicircle is a right triangle.

PROOF: Assume, using the notation of Figure 4.7a, that $PA/PB = c$, a given constant greater than one. It must be shown that the set of all locations for point P is a circle. There are two points on $\overleftrightarrow{AB}$ whose distances from A and B have the desired ratio. These are indicated by points C and D in the figure. Then, in triangle APB, $\overline{PC}$ and $\overline{PD}$ are internal and external angle bisectors of the angle at P. This can be shown for point C, for example, assuming the validity of the converse of Theorem 4.5, since

$$\frac{AC}{CB} = \frac{PA}{PB}$$

But the internal and external angle bisectors at a vertex of a triangle are perpendicular, so $\angle CPD$ is a right angle. Triangle CPD is inscribed in a semicircle, so P lies on a circle with $\overline{CD}$ as diameter. See Exercises 20 and 22. Conversely, it can be shown that any point P on the circle with $\overline{CD}$ as diameter has the same ratio of distances from A and B. In the special case where $c = 1$, the set of points is a line. To preserve generality, we can consider this line as a degenerate circle.

Figure 4.7b shows a specific example of the circle of Apollonius, for which $P'A'/P'B' = 5/3$. Each point on the circle is 5/3 as far from A' as it is from B', and the points on the circle are the only points in the plane with this property.

In the proof of Theorem 4.7 and elsewhere, even though parallelism (perpendicularity) is a relationship usually defined for lines, use of the concept is extended to segments and rays without difficulty. For example, although it is technically imprecise to say that $\overline{AB}$ is parallel (perpendicular) to $\overline{CD}$, that expression is understood to mean that $\overleftrightarrow{AB}$ of which $\overline{AB}$ is a subset, is parallel (perpendicular) to $\overleftrightarrow{CD}$, of which $\overline{CD}$ is a subset.

Although directed segments were avoided in the proof of Theorem 4.7, they will be needed in the next section and elsewhere. For points on a line, one direction may be indicated as positive and the opposite direction as negative. When discussing a directed segment, one must distinguish between the first and second endpoint. The length of a particular segment is positive if one must go in a positive direction from the first to the second endpoint and negative if one goes in a negative direction from the first to the second endpoint. In Figure 4.7a, if right is designated as positive, then AB and BD are examples of positive lengths, while BA and DB are examples of negative lengths.

Ratios of distances also may be indicated as positive or negative. If a third point divides the segment joining two other points, the ratio of division is considered positive if the segment is divided internally and negative if it is divided externally. In Figure 4.7a, AC/CB is positive, while AD/DB is negative.

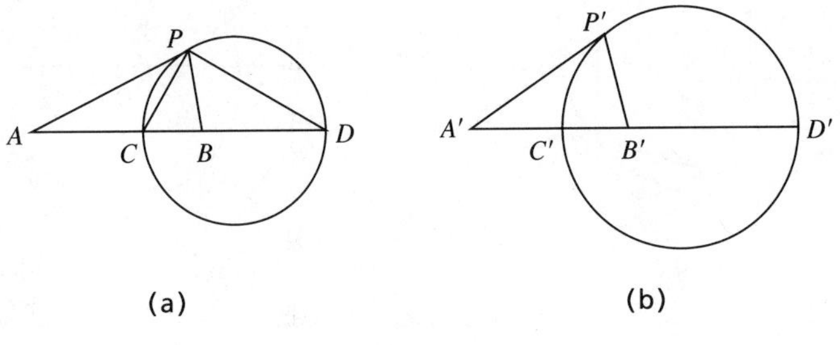

Figure 4.7

Two final theorems about the segments related to a circle are connected to the previous theorem and also are useful in proving more advanced theorems.

THEOREM 4.8 ➤ A quadrilateral is inscribed in a circle if and only if the opposite angles are supplementary.

The terms *cyclic quadrilateral* and *concyclic* are also used for a quadrilateral inscribed in a circle. Study Figure 4.8. You should recall that the measure of an inscribed angle in a circle is half that of its intercepted arc. For example, the measure of angle B is half that of arc ADC. The ancient Babylonians first divided the circumference of the circle into 360 parts, now called degrees. Since angles B and D together intercept arcs with a total measurement of 360°, the angles are supplementary since the sum of their measures is 180° (π radians).

Opposite sides of an inscribed quadrilateral are sometimes called *antiparallel* with respect to the remaining pair of sides. The prefix "anti" suggests across from or opposite. In Figure 4.8, if angles D and A were supplementary, then $\overline{DC}$ and $\overline{AB}$ would be parallel. But instead, it is the angle opposite or across from A that is supplementary to D; hence, the segments are antiparallel instead of parallel.

THEOREM 4.9 ➤ The product of the lengths of the segments from an exterior point to the points of intersection of a secant with a circle is equal to the square of the length of the tangent from the point to the circle.

Recall that a secant is a line intersecting a circle in two points.

PROOF: Using the notation of Figure 4.9, the theorem states that $AC \cdot AD = (AB)^2$. Recall that an inscribed angle and the angle between a tangent and a chord are measured by half the intercepted arc. Triangles ABC and ADB are similar because corresponding angles are congruent; therefore, $AC/AB = AB/AD$, from which the theorem follows.

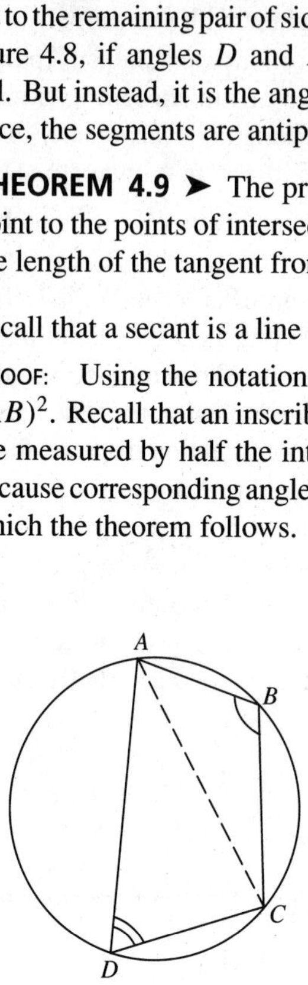

Figure 4.8

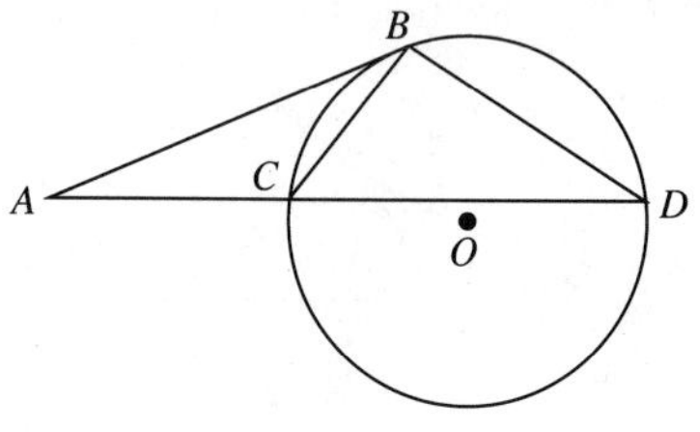

Figure 4.9

Cyclic quadrilaterals often play an important role in geometric proofs. The proof of Theorem 4.15 in the next section is an excellent example.

A final practical consideration involving triangles and cyclic quadrilaterals is a historical note explaining how to find the areas, given only the lengths of the sides.

Heron of Alexandria, in his geometrical work called *Metrica*, used the formula $K = \sqrt{s(s-a)(s-b)(s-c)}$ (now called Heron's formula) for the area of a triangular region. In the formula, s is the semiperimeter and a, b, c are the lengths of the sides.

Example ➤ Find the area of the triangular region if the lengths of the sides are 6, 8, 9 inches.

Let $a = 6, b = 8, c = 9$. Then $s = \frac{1}{2}(6 + 8 + 9) = 11.5$ and the area is $K = \sqrt{11.5\,(11.5 - 6)(11.5 - 8)(11.5 - 9)} \approx 24$ sq. inches.

The Hindu mathematician Brahmagupta (A.D. 628) gave the correct formula $K = \sqrt{(s - a)(s - b)(s - c)(s - d)}$ for the area of a cyclic quadrilateral, for s the semiperimeter and a, b, c, d the lengths of the sides.

▲ EXERCISES 4.1 ▲

1. Use the notation of Figure 4.3 to name all of the triangles whose vertices are three of the given points and whose orthocenter is the fourth distinct point.
2. Could the centroid be outside the triangle?
3. Could the incenter be outside the triangle?
4. Could the circumcenter be outside the triangle?
5. Could the incircle and circumcircle have points in common?
6. Where is the orthocenter of a right triangle?
7. Under what conditions would the orthocenter of a triangle lie outside the triangle?
8. How many excenters does a triangle have?
9. Could antiparallel segments also be parallel?

For Exercises 10 and 11, find the area of the triangular regions if the sides have the given lengths.

10. $a = 7, \quad b = 12, \quad c = 15$
11. $a = 4, \quad b = 8, \quad c = 5$

For Exercises 12 and 13, find the area of the cyclic quadrilateral if the sides have the given lengths.

12. $a = 16, \quad b = 15, \quad c = 23, \quad d = 29$
13. $a = 40, \quad b = 28, \quad c = 15, \quad d = 13$
14. Prove that the segment joining the midpoints of two sides of a triangle is parallel to the third side.
15. Complete the proof of Theorem 4.3.
16. Prove that the angle between the segments from the incenter to two vertices of a triangle has a radian measure equal to $\pi/2$ plus one-half the measure of the angle of the triangle at the third vertex.
17. Complete the proof of Theorem 4.4.

For Exercises 18 and 19, paraphrase the proof of Theorem 4.5 by using the angle indicated, rather than angle A.

18. Angle B, Figure 4.5

19. Angle C, Figure 4.5

20. Prove that the external bisector of an angle of a triangle (not isosceles) divides the opposite side (externally) into two segments proportional to the sides of the triangle adjacent to the angle.

21. Prove Theorem 4.6.

22. Prove that the internal and external angle bisectors at a vertex of a triangle are perpendicular.

23. Prove the converse of Theorem 4.8.

24. Prove that a quadrilateral is concyclic if and only if one side subtends equal angles at the other two vertices.

25. Prove that two vertices of a triangle and the feet of the altitudes to the sides adjacent to the third vertex can be inscribed in a circle. (The feet are the points of intersection of the altitudes with the opposite sides of the triangle.)

26I. Use a paper model of a triangular region and demonstrate, by paper folding, an intuitive proof of (a) Theorem 4.1, (b) Theorem 4.2, (c) Theorem 4.3, (d) Theorem 4.4.

▷ **27.** Try to discover a proof of Heron's formula for a triangular area; then consult a reference text, if necessary.

▷ **28.** Investigate how some of the theorems in this section might be proved using motions, by beginning with a set of postulates found in a secondary school text on transformational geometry.

▷ **29.** A paradox that results from an incorrect assumption in Figure 4.10 is the proof that any triangle is isosceles. Let ABC be any triangle, with D the intersection of the angle bisector at B and the perpendicular bisector of $\overline{AC}$. $\overline{DE}$ and $\overline{DF}$ are perpendicular to the sides.

a. Use congruent triangles to prove $\triangle ABC$ is isosceles, with $AB = BC$.

b. Find the false assumption on which the proof is based.

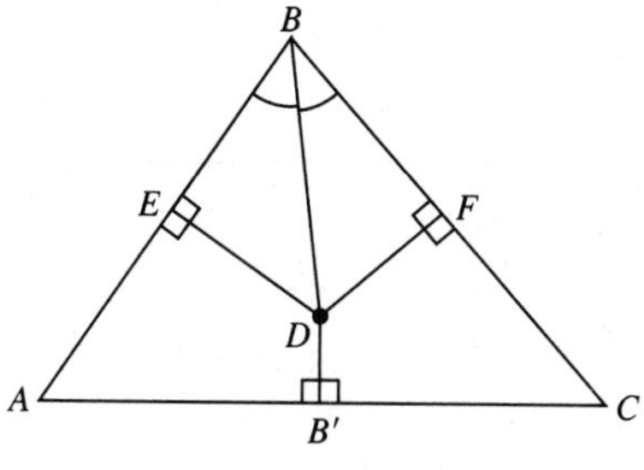

Figure 4.10

▲ 4.2 Some Theorems Leading to Modern Synthetic Geometry

Two theorems usually studied together, though proved in different ages, have great significance because they concern only collinearity and concurrence and hence are of value in the study of projective geometry, where distance is no longer an invariant. The first of these is credited to the Greek mathematician Menelaus, who lived in Alexandria about 100 A.D.

THEOREM 4.10 ➤ *Menelaus' theorem.* If three points, one on each side of a triangle (extended if necessary) are collinear, then the product of the ratios of division of the sides by the points is negative one (-1), if internal ratios are considered positive and external ratios are considered negative.

For directed segments, a positive direction along a line is established arbitrarily. In Figure 4.11, if the direction from A to C is positive, then AC indicates a segment with a positive length and $AC = -CA$. Thus, AG and GC are both positive, and the internal ratio AG/GC is positive. If C to A had been considered positive, then AG and GC would both be negative, but the ratio would still be positive. Using this convention of directed segments, the two segments for an external ratio would be opposite in sign, and the ratio would be negative.

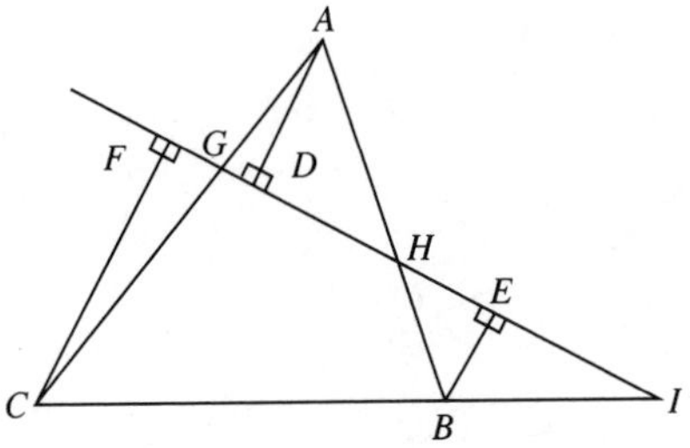

Figure 4.11

PROOF: Suppose, using the notation of Figure 4.11, that points G, H, I are the three given collinear points, one on each side of the triangle ABC. We need to show

$$\frac{AG}{GC} \cdot \frac{CI}{IB} \cdot \frac{BH}{HA} = -1$$

The following pairs of similar triangles can be observed:

$$\triangle ADG \sim \triangle CFG$$
$$\triangle BEI \sim \triangle CFI$$
$$\triangle BEH \sim \triangle ADH$$

From the three pairs of similar triangles are derived the three proportions, since pairs of corresponding sides of similar triangles have the same ratio. Also recall that the notation AG indicates the measure of $\overline{AG}$, which is a number.

$$\frac{AG}{GC} = \frac{AD}{CF} \qquad \frac{CI}{IB} = -\frac{CF}{BE}$$

(Considered negative because I divides CB externally)

$$\frac{BH}{HA} = \frac{BE}{AD}$$

Then

$$\frac{AG}{GC} \cdot \frac{CI}{IB} \cdot \frac{BH}{HA} = -\frac{AD}{CF} \cdot \frac{CF}{BE} \cdot \frac{BE}{AD} = -1$$

The three ratios may be named by starting at a vertex (say A) and proceeding around the triangle along directed segments (say in a counterclockwise direction), going from a vertex to a point of division, then to the next vertex, then to the next point of division, and so on. This same pattern may be used when all three points of division are external, as in Figure 4.12:

$$\frac{AE}{EC} \cdot \frac{CF}{FB} \cdot \frac{BD}{DA} = -1$$

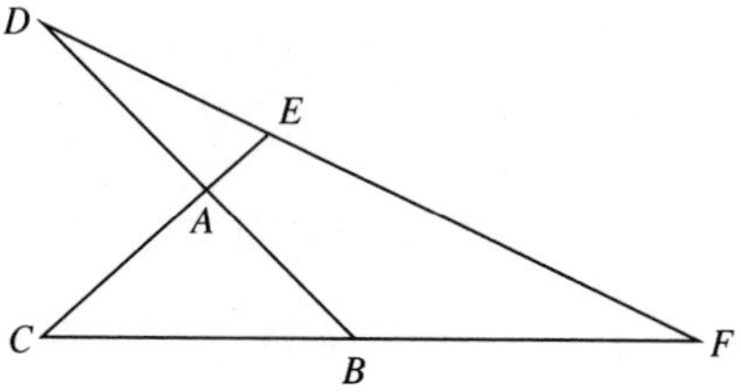

Figure 4.12

The line cutting the sides of the triangle (line GHI in Figure 4.11 and DEF in Figure 4.12) is called a transversal.

Example ➤ Use Figure 4.12, but consider triangle ADE the basic triangle with CBF the transversal. Write the product of the three ratios equal to negative one by the theorem of Menelaus.

If we think of beginning at vertex A and proceeding in a counterclockwise direction around the triangle, the product is $AC/CE \cdot EF/FD \cdot DB/BA = -1$.

In case the transversal goes through a vertex, the statement of the theorem can be modified to avoid a zero in the denominator (see Exercise 6, Exercises 4.2).

The converse of Menelaus' theorem provides an important tool for proving that three points are collinear.

THEOREM 4.11 ➤ *Converse of Menelaus' theorem.* If the product of the ratios of division of the sides by three points, one on each side of a triangle, extended if necessary, is negative one, then the three points are collinear.

PROOF: Suppose, using the notation of Figure 4.12, that D, E, and F are three points on each side of the triangle and that the product of the ratios of division is negative one. Suppose, furthermore, that $\overleftrightarrow{EF}$ meets $\overleftrightarrow{AB}$ at a distinct point D'. It must be shown that $D' = D$. Using Menelaus' theorem for E, F, and D',

$$\frac{AE}{EC} \cdot \frac{CF}{FB} \cdot \frac{BD'}{D'A} = -1$$

But this means

$$\frac{AE}{EC} \cdot \frac{CF}{FB} \cdot \frac{BD'}{D'A} = \frac{AE}{EC} \cdot \frac{CF}{FB} \cdot \frac{BD}{DA}$$

or

$$\frac{BD'}{D'A} = \frac{BD}{DA}$$

$$\frac{BD' + D'A}{D'A} = \frac{BD + DA}{DA}$$

and since $BA = BD' + D'A$ or $BD + DA$,

$$D'A = DA$$

Therefore, D and D' are the same point, which means that D, E, and F are collinear.

For the proof of Theorem 4.11, it has been assumed that line DEF is not parallel to any side of triangle ABC, so that D, E, and F are ordinary points. The theorem could be extended to take care of the special case in which D, E, or F is an ideal point (see Chapter 7). For example, if D is ideal, then you could assume that $BD/DA = -1$ in the modified proof.

Theorem 4.11 can be used to prove several other theorems, one of which is stated below but proved as an exercise. For Theorem 4.12, assume that the lines intersect in ordinary points.

THEOREM 4.12 ➤ The internal bisectors of two angles of a triangle and the external bisector of the third angle intersect the opposite sides of the triangle in three collinear points.

The theorem often paired with the theorem of Menelaus, Ceva's theorem, was discovered about 1678 by the Italian mathematician Ceva. Menelaus' theorem deals with three collinear points, whereas the theorem of Ceva concerns three concurrent lines.

THEOREM 4.13 ➤ *Ceva's theorem and its converse.* Three lines that join three points, one on each side of a triangle, to the opposite vertices are concurrent if and only if the product of the ratios of division of the sides is 1.

PROOF: Using the notation of Figure 4.13, with $\triangle ABC$ as the basic triangle, Menelaus' theorem applied to triangle ABD and $\overleftrightarrow{CF}$ shows that

$$\frac{AG}{GD} \cdot \frac{DC}{CB} \cdot \frac{BF}{FA} = -1$$

For triangle ACD and $\overleftrightarrow{BE}$,

$$\frac{AE}{EC} \cdot \frac{CB}{BD} \cdot \frac{DG}{BA} = -1$$

Multiplying left and right members of these equations results in

$$\frac{AG}{GD} \cdot \frac{DC}{CB} \cdot \frac{BF}{FA} \cdot \frac{AE}{EC} \cdot \frac{CB}{BD} \cdot \frac{DG}{GA} = 1$$

or

$$\frac{AE}{EC} \cdot \frac{CD}{DB} \cdot \frac{BF}{FA} = 1$$

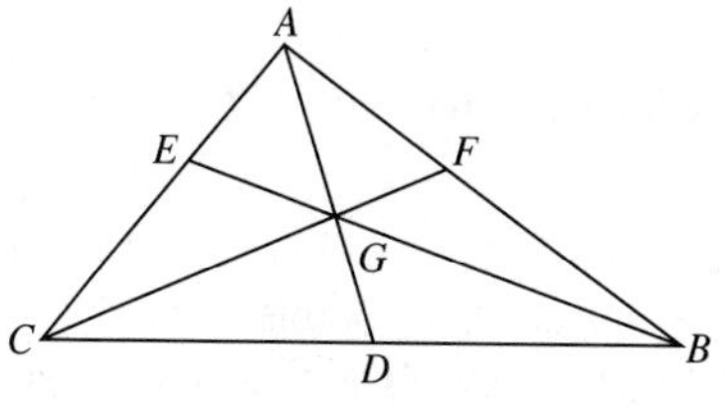

Figure 4.13

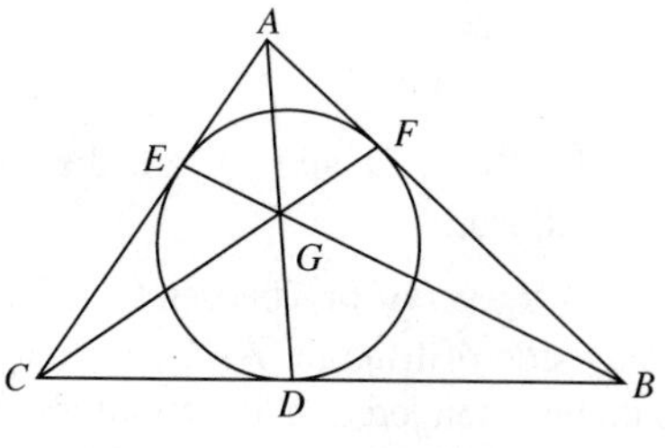

Figure 4.14

The three ratios can be written quickly by following the same pattern described for the theorem of Menelaus: Start at a vertex, go to the point of division, then the next vertex, and so on. Similar triangles may be used to prove Ceva's theorem directly, without using the theorem of Menelaus (Exercise 11, Exercises 4.2).

The converse of Ceva's theorem, proved indirectly in a way similar to Theorem 4.11, is a useful tool for proving three lines concurrent. For example, it can now be used to give very simple proofs that the medians and the internal bisectors of the angles of a triangle are concurrent (see Exercises 12–14, Exercises 4.2).

A more modern application of the converse of Ceva's theorem in ordinary Euclidean geometry is in the proof of the following theorem:

THEOREM 4.14 ➤ The segments from the vertices of a triangle to the points of tangency of the incircle with the opposite sides of the triangle are concurrent.

This theorem was proved in the early nineteenth century by J. D. Gergonne, a French mathematician, and the point of concurrency is called the *Gergonne point.*

PROOF: Using the notation of Figure 4.14, since the two tangents from a point to a circle are congruent, $AE = AF, CE = CD, BD = BF$, and

$$\frac{AE}{EC} \cdot \frac{CD}{DB} \cdot \frac{BF}{FA} = 1$$

Thus, the three segments are concurrent by the converse of Ceva's theorem.

Many other theorems about concurrency and collinearity have interested geometers throughout the history of geometry. One additional example, credited to the English mathematician Robert Simson (1687–1768), refers to a triangle and a point on its circumcircle.

THEOREM 4.15 ➤ The three perpendiculars from a point on the circumcircle to the sides of a given triangle intersect the sides in three collinear points. The line on which the three points lie is called the *Simson line.*

PROOF: Use the notation of Figure 4.15, with P the point on the circumcircle and D, E, F the feet of the perpendiculars. By the converse of Theorem 4.8, points P, D, A, and E are cyclic (lie on a circle). Similarly, P, A, C, and B are on another circle, and P, D, B, and F are on yet a third circle.

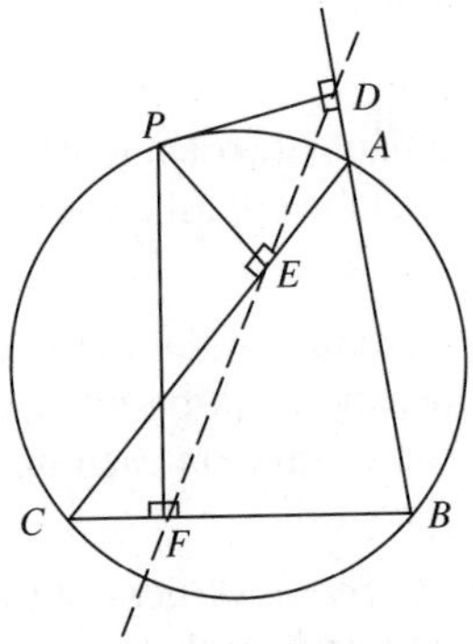

Figure 4.15

In circle P, D, A, E	$\angle PDE \cong \angle PAE$
	$\angle PAE \cong \angle PAC$
In circle P, A, C, B	$\angle PAC \cong \angle PBC$
	$\angle PBC \cong \angle PBF$
In circle P, D, B, F	$\angle PBF \cong \angle PDF$

Since this establishes the fact that $\angle PDE \cong \angle PDF$, points D, E, F are collinear.

For a given point P on the circumcircle, one Simson line is determined. The special case when P coincides with a vertex is considered in the exercises.

▲ EXERCISES 4.2 ▲

1. Can a transversal intersect one side of a triangle internally and the other two sides externally?

For Exercises 2–5, use Figure 4.16. For the triangle and the transversal named in the exercise, write the product of the three ratios equal to negative one by the theorem of Menelaus.

2. Triangle ABF and transversal E, D, C
3. Triangle BCD and transversal A, E, F
4. Triangle EFD and transversal A, B, C
5. Triangle AEC and transversal F, D, B
6. Write an alternative form of the theorem of Menelaus, showing the product of three segments equal to the negative of the product of three other segments. What is the value of the product if one of the points of intersection coincides with a vertex of the triangle?
7. Prove Theorem 4.12.
8. Use the converse of Menelaus' theorem to prove that the external bisectors of the angles of a triangle meet the opposite sides of the triangle in three collinear points.
9. Altogether there are six bisectors of the angles of a triangle, three internal and three external. Prove that these six bisectors meet the opposite sides in six points that are on four lines, three on each line. (Compare this with a finite geometry studied in Chapter 1.)
10. In Figure 4.17, let A, B, C be the vertices of the triangle, with P the point of concurrency of three segments from the vertices. Use Ceva's theorem to write three ratios whose product is equal to 1.

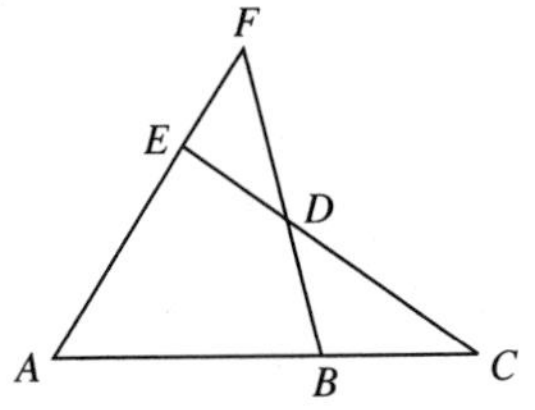

Figure 4.16

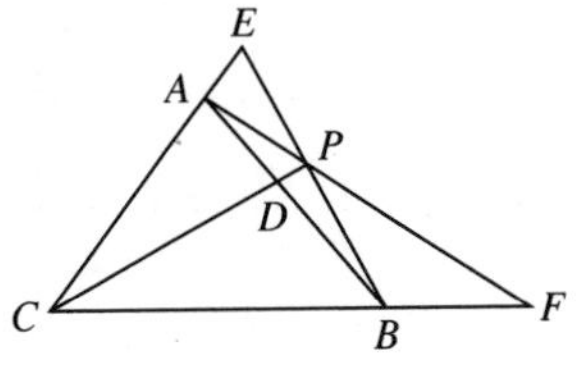

Figure 4.17

11. Prove Ceva's theorem directly, without using the theorem of Menelaus.

12. Use Theorem 4.13 to prove that the medians of a triangle are concurrent.

13. Use Theorem 4.13 to prove that the internal bisectors of the angles of a triangle are concurrent.

14. Use Theorem 4.13 to prove that the external bisectors of two angles of a triangle and the internal bisectors of the third angle are concurrent.

15. How many Simson lines are associated with one triangle?

16. Describe the location of the Simson line if the point on the circumcircle is a vertex of the triangle.

▷ **17.** Construct a paper model and use paper folding to demonstrate the intuitive proof of (a) Theorem 4.14 and (b) Theorem 4.15.

▲ 4.3 The Nine-Point Circle and Early Nineteenth-Century Synthetic Geometry

Probably the most significant of the discoveries leading to renewed interest in the classical geometry of the triangle and the circle early in the nineteenth century was the nine-point circle, sometimes credited to the German mathematician Feuerbach in 1822.

THEOREM 4.16 ➤ The midpoints of the sides of a triangle, the points of intersection of the altitudes and the sides, and the midpoints of the segments joining the orthocenter and the vertices all lie on a circle called the *nine-point circle*.

A nine-point circle is shown in Figure 4.18. Except for special cases, the circle intersects each side of the triangle in two distinct points.

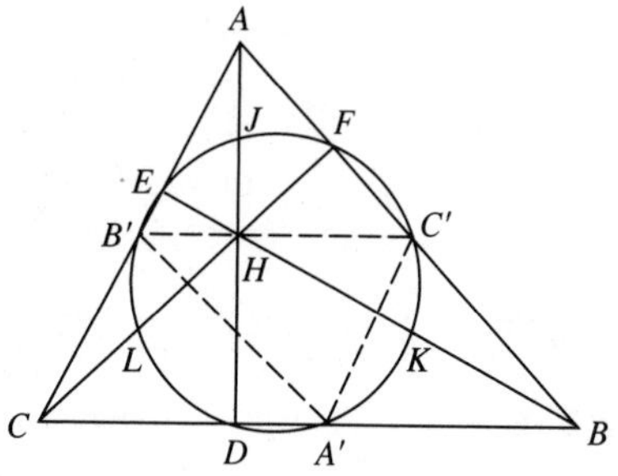

Figure 4.18

Proof: The proof of the theorem follows this pattern:

a. There is a circle through A', B', C', the midpoints of the sides, because any three noncollinear points are on a circle.

b. Show that points D, E, F are on this same circle.

c. Show that points J, K, L are on this same circle.

To prove that D is on the same circle as A', B', C', it is possible to show that $DB'C'A'$ is an isosceles trapezoid (see Exercise 5) and, hence, inscribed in the circle. $\overline{A'C'} \cong \overline{DB'}$ because $\overline{A'C'}$ connects the midpoints of the sides of triangle ABC and, therefore, has a measure equal to half the measure of the base; $\overline{DB}$ connects the vertex and the midpoint of the hypotenuse of right triangle ADC, which implies that DB' is half the hypotenuse (see Exercise 4).

For part c of the proof, we can begin by considering the circle with $\overline{JA'}$ as diameter. $\angle JB'A'$ is a right angle (see Exercise 6); hence, B' is on this circle because $\triangle JB'A'$ is a right triangle inscribed in a semicircle with $\overline{JA'}$ as diameter. Similarly, $\angle JC'A'$ is a right angle and C' is on the same circle. Therefore, J lies on the same circle as $A'B'C'$. Repeating this entire process, by considering the circle with KB' as diameter and then the circle with $\overline{LC'}$ as diameter, we can show that points K and L are also on the same circle as $A'B'C'$.

There is a wealth of additional information about the nine-point circle and its relationships with other sets of points in Euclidean geometry. One of the important ideas is the location of the center of the nine-point circle in reference to several other points previously mentioned. This information depends on a theorem established by the German mathematician Euler in 1765.

THEOREM 4.17 ➤ The centroid of a triangle trisects the segment joining the circumcenter and the orthocenter.

The line containing the three points is called the *Euler line*. The word "trisect" as used here means that the distance along the line from the circumcenter to the centroid is one-third of the distance along the same line from the circumcenter to the orthocenter.

PROOF: In Figure 4.19, let O be the circumcenter, G the centroid, and H the orthocenter. The measure of $\overline{AH}$ is twice that of $\overline{OA'}$. This is true because triangles CBI and COA' are similar with a ratio of similarity of 2 to 1, and $IB = AH$ because $AHBI$ is a parallelogram. See Exercise 7, Exercises 4.3. Now triangles GOA' and GHA are similar with a ratio of 1 to 2; therefore, $OG - 1/2GH$, and G trisects $\overline{OH}$. Points O, G, H are collinear since $\angle AGH \cong \angle OGA'$.

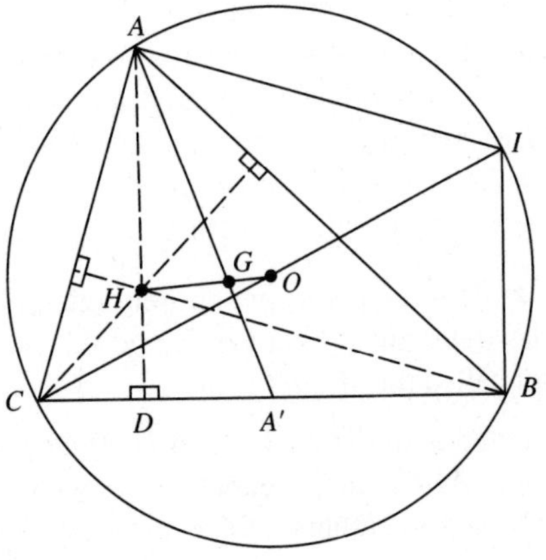

Figure 4.19

It may be surprising to find that the Euler line also contains the center of the nine-point circle, as indicated in the following theorem.

THEOREM 4.18 ➤ The center of the nine-point circle bisects the segment joining the orthocenter and the circumcenter of a triangle (see Figure 4.20).

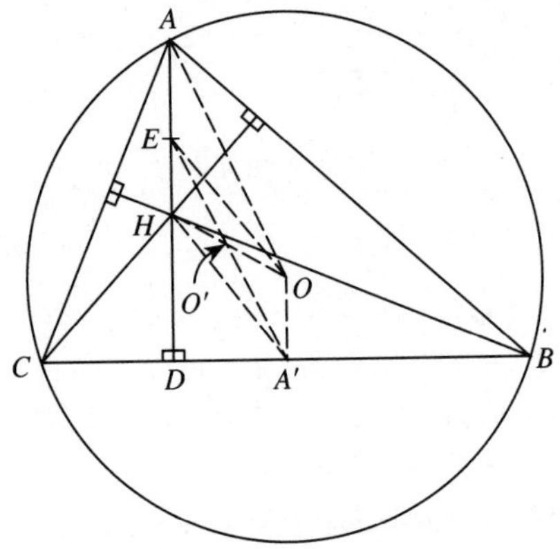

Figure 4.20

Proof: From Theorem 4.17, $\overline{EH} \cong \overline{OA'}$, and the two segments are also parallel. If O' is the center of the nine-point circle, it is the midpoint of the diameter $\overline{A'E}$. But $OA'HE$ is a parallelogram, and the diagonals of a parallelogram bisect each other. This means that O' is the midpoint of $\overline{OH}$.

The nine-point circle is the circumcircle of $\triangle A'B'C'$. If a nine-point circle were then to be constructed for $\triangle A'B'C'$, N would be its circumcenter. If this process were continued, the centroid would remain the same for all triangles.

All the theorems about concurrence considered so far relate to the concurrence of three lines. *Miquel's theorem* is significant in part because it considers the concurrence of sets of three circles associated with any triangle. The theorem is named for Auguste Miquel, who included it in "Théorèms de géométrie," published in the *Journal de mathématiques pure et appliquées*, Vol. iii in 1838.

THEOREM 4.19 ➤ If three points are chosen, one on each side of a triangle, then the three circles determined by a vertex and the two points on the adjacent sides meet at a point called the *Miquel point*.

Proof: Using the notation of Figure 4.21, let D, E, and F be the arbitrary points on the sides of triangle ABC. Suppose that circles with centers I and J intersect at point G. In circle J, $\angle EGD$ and $\angle ECD$ are supplementary, and in circle I, $\angle FGD$ and $\angle DBA$ are supplementary. In the rest of the proof, the notation m means the measure of the angle. Since

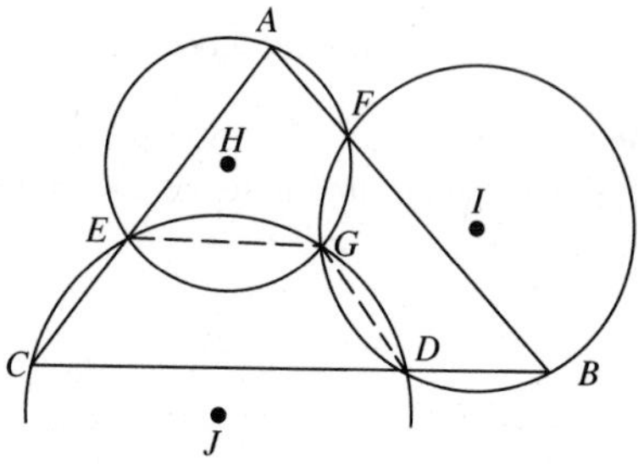

Figure 4.21

$$m\angle EGD + m\angle DGF + m\angle EGF = 2\pi$$
$$\pi - m\angle C + \pi - m\angle B + m\angle EGF = 2\pi$$
$$m\angle EGF = m\angle C + m\angle B = \pi - m\angle A$$

This means that $\angle A$ and $\angle EGF$ are supplementary, so A, F, G, E are on a circle and all three of the circles are concurrent. It is possible that the Miquel point could be outside the triangle, in which case the proof must be modified slightly.

▲ EXERCISES 4.3 ▲

1. A triangle has how many:
   a. Nine-point circles? **b.** Euler lines? **c.** Miquel points?
2. In general, a triangle and its nine-point circle have how many points in common?
3. Give the maximum and minimum number of points a triangle and its nine-point circle might have in common.

In the proof of Theorem 4.16 (for Exercises 4–6),

4. Prove that the segment connecting the vertex of the right angle of a right triangle with the midpoint of the hypotenuse has a measure half that of the hypotenuse.
5. Prove that $DB'C'A'$ is an isosceles trapezoid.
6. Prove that $JB'A'$ is a right angle.
7. Complete the details of the proof of Theorem 4.17 not included in the text.
8. Prove that the radius of the nine-point circle is half that of the circumcircle.
9. Prove that the nine-point circle bisects every segment connecting the orthocenter to a point on the circumcircle.
10. Prove that the four triangles formed by the points of an orthocentric set of points have the same nine-point circle.
11. **a.** Draw a figure showing an example in which the Miquel point is outside the triangle.
    b. Modify the proof as necessary so that the three circles are concurrent.
12. Draw a figure showing an example in which the Miquel point is on a side of the triangle and two of the circles are tangent. Modify the proof as necessary.
13. Show that, on the Euler line, the centroid and the orthocenter divide internally and externally in the same ratio the segment whose endpoints are the circumcenter and the center of the nine-point circle.
14. Prove that the Miquel point is a point on the circumcircle if the three points on the sides of the triangle are collinear.
15. Prove that if the Miquel point is on the circumcircle, then the three points on the sides of the triangle are collinear.
16. ▷ Investigate additional applications of the theory involving the Miquel point in Davis, *College Geometry*.

▲ 4.4 Isogonal Conjugates

Three major contributions to the modern synthetic geometry of the triangle are discussed in this section and in the next. The first of these, credited to Lemoine (about 1873), is the concept of symmedians.

Consider Figure 4.22. If $\overline{AE}$ is the bisector of angle A and if $\angle DAE \cong \angle FAE$, then $\overline{AD}$ and $\overline{AF}$ are called *isogonal lines* and one is called the *isogonal conjugate* of the other. In other words, $\overline{AD}$ is the isogonal conjugate of a segment $\overline{AF}$ if $\angle DAC \cong \angle FAB$. The bisector of the angle is the bisector of the angle between two isogonal conjugates. According to this definition, the adjacent sides of the triangle are themselves isogonal conjugates.

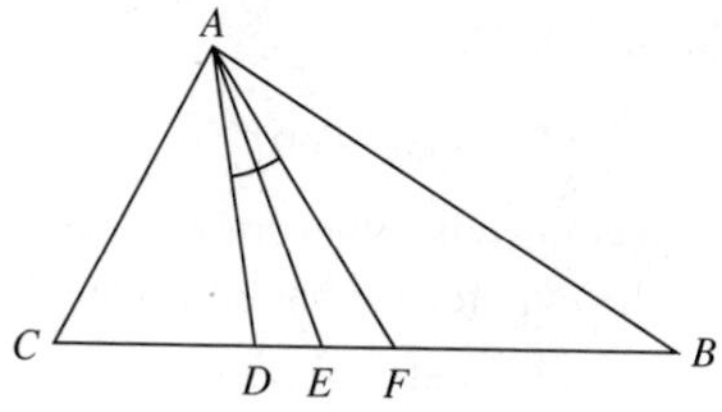

Figure 4.22

DEFINITION ➤ A *symmedian* is the isogonal conjugate of a median.

Suppose, in Figure 4.22, that $\overline{AF}$ is a median. Then $\overline{AD}$ is a *symmedian*. Before reading ahead, see what you can discover for yourself about the set of symmedians. The three symmedians of a triangle have been investigated in great detail by Lemoine and other, more recent geometers. They have properties analogous in many ways to the medians themselves. Only a few examples are included in this section. The next theorem includes symmedians as a special case.

THEOREM 4.20 ➤ The isogonal conjugates of a set of concurrent segments from the vertices to the opposite sides of a triangle are also concurrent.

PROOF: Ceva's theorem should suggest itself as a tool to use in proving Theorem 4.20, but the form of the theorem needs to be changed so that ratios of angles are involved.

In Figure 4.23, assume first that G is the point of concurrency of three segments and that $\overline{AD}$, $\overline{BE}$, and $\overline{CF}$ are the isogonal conjugates of these segments. In this proof, it is not assumed that A', B', C' are midpoints.

Consider triangles $AA'C$ and $AA'B$. By the law of sines,

$$\frac{CA'}{\sin CAA'} = \frac{CA}{\sin CA'A}$$

so that

$$CA' = \frac{CA \sin CAA'}{\sin CA'A}$$

Similarly,

$$A'B = \frac{AB \sin BAA'}{\sin BA'A}$$

Since the sines of the supplementary angles at A' are equal,

$$\frac{CA'}{A'B} = \frac{CA \sin CAA'}{AB \sin BAA'}$$

Note that Theorem 4.5 is a special case of this equality for $\overline{AA'}$ an angle bisector.

Expressing each ratio of division of the sides of triangle ABC in this form leads to a trigonometric form for the theorem of Ceva and its converse:

$$\frac{\sin CAA'}{\sin BAA'} \cdot \frac{\sin BCC'}{\sin ACC'} \cdot \frac{\sin ABB'}{\sin CBB'} = 1$$

Now, in Figure 4.23, because $\overline{AD}, \overline{BE}$, and $\overline{CF}$ are isogonal conjugates of the three original segments, various angles can be substituted. For example, $\angle BAA' \cong \angle CAD$ and so on. When all of the appropriate substitutions are made,

$$\frac{\sin BAD}{\sin CAD} \cdot \frac{\sin ACF}{\sin BCF} \cdot \frac{\sin CBE}{\sin ABE} = 1$$

and $\overline{AD}, \overline{BE}$, and $\overline{CF}$ are concurrent by the converse of Ceva's theorem.

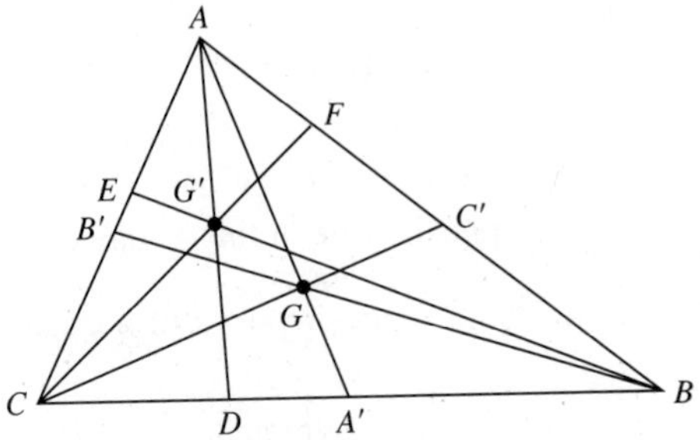

Figure 4.23

If $\overline{AA'}, \overline{BB'}$ and $\overline{CC'}$ are medians, then their isogonal conjugates are symmedians; thus a corollary of Theorem 4.20 is:

THEOREM 4.21 ➤ The symmedians of a triangle are concurrent.

DEFINITION ➤ The point of concurrency of the symmedians of a triangle is the *symmedian point* or the *Lemoine point*.

In Figure 4.24, $\overline{AA'}$, $\overline{BB'}$, $\overline{CC'}$ are medians and $\overline{AD}$, $\overline{BE}$, $\overline{CF}$ are symmedians. Point S is the symmedian point.

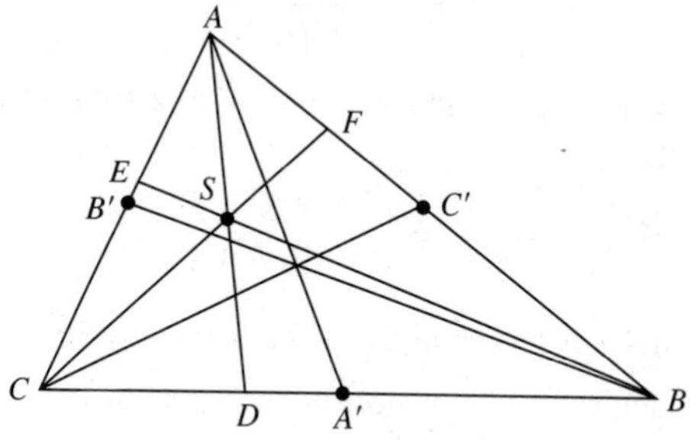

Figure 4.24

The median from a vertex is the bisector of a segment with endpoints on sides adjacent to the vertex and parallel to the side of the triangle opposite the vertex. For example, in Figure 4.25, if segment $\overline{DE} \parallel \overline{CB}$ and if $\overline{AA'}$ is a median, then $DF = FE$. An analogous property for the symmedian is given in the next theorem.

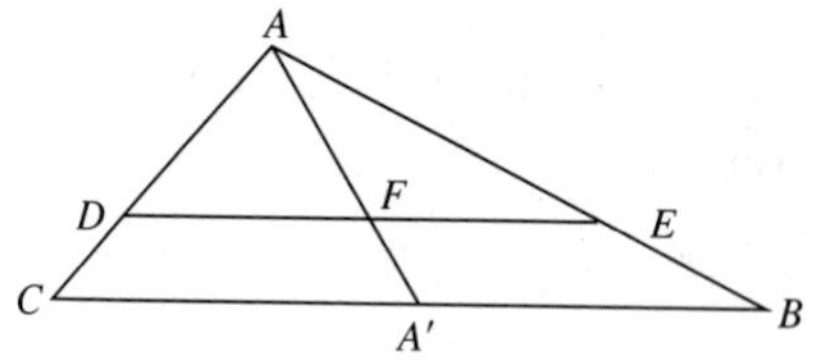

Figure 4.25

THEOREM 4.22 ➤ The symmedian from one vertex of a triangle bisects any segment antiparallel to the opposite side of the triangle.

Recall that antiparallel segments are opposite sides of a quadrilateral inscribed in a circle.

PROOF: In Figure 4.26, let $\overline{AA''}$ be a symmedian and $\overline{DE}$ a segment antiparallel to $\overline{BC}$ so that quadrilateral $DEBC$ can be inscribed in a circle.

$$AD \cdot AC = AE \cdot AB,$$

by Theorem 4.9. Also, in the proof of Theorem 4.20, it was established that

$$\frac{CA'}{A'B} = \frac{CA \sin \angle CAA'}{AB \sin \angle BAA'}$$

If A' is the midpoint of $\overline{BC}$, then

$$\frac{\sin \angle CAA'}{\sin \angle A'AB} = \frac{BA}{CA}$$

so that

$$\begin{aligned}\frac{DF}{FE} &= \frac{AD \sin DAF}{EA \sin EAF} \\ &= \frac{AB}{AC} \cdot \frac{CA}{BA} \\ &= 1\end{aligned}$$

and F is the midpoint of $\overline{DE}$.

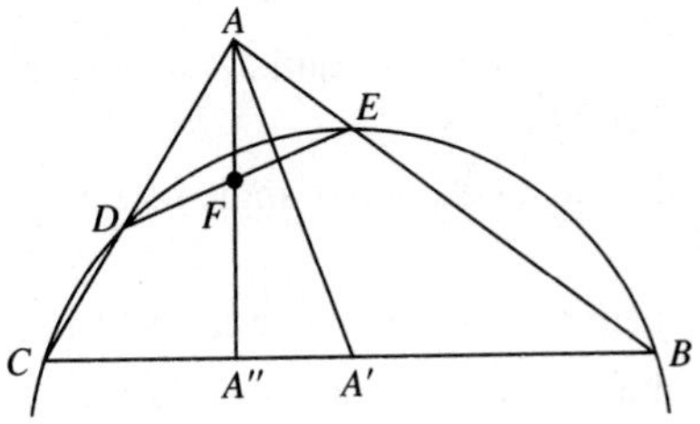

Figure 4.26

A somewhat different sort of property, one that could be used to describe a symmedian as a set of points satisfying a certain condition, is considered next. The set of points equidistant from the two sides of a triangle lies on a bisector of the angle. There is also a constant ratio of distances from points on a symmedian to the sides.

THEOREM 4.23 ➤ The ratio of the distances from a point on the symmedian to the adjacent sides of the triangle equals the ratio of the lengths of those sides.

PROOF: Use the notation of Figure 4.27, with $\overline{AD}$ the symmedian and $\overline{AA'}$ the median.

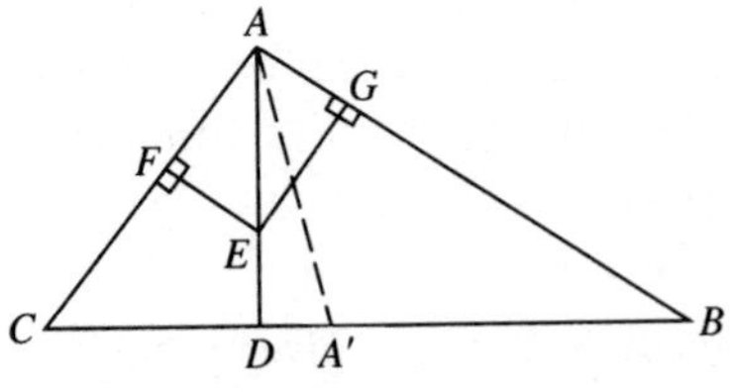

Figure 4.27

From the law of sines,

$$\frac{EF}{EG} = \frac{\sin \angle CAD}{\sin \angle BAD}$$

$$= \frac{\sin \angle BAA'}{\sin \angle CAA'} = \frac{AC}{AB} \qquad \text{since } A' \text{ is the midpoint of } \overline{CB}$$

▲ EXERCISES 4.4 ▲

1. What is the isogonal conjugate of a symmedian?
2. Name the point of concurrency for the isogonal conjugates of the symmedians of a triangle.
3. For an isosceles triangle that is not equilateral, how many symmedians coincide with their isogonal conjugates?
4. Prove that the median to a side of a triangle bisects any segment in the triangle parallel to that side.
5. Where is the symmedian point of an equilateral triangle?
6. Could the symmedian point be outside the triangle?
7. Use Theorem 4.23 to answer this question. Assume that the adjacent sides of a triangle are not congruent. Is a point on the symmedian nearer the longer or shorter of the adjacent sides of the angle?
8. Apply Theorem 4.23 to the special case of a right triangle with legs 3 and 4 units in length. For each symmedian, state the ratio of the distances from a point on it to the adjacent sides of the triangle.
9. Prove that the ratio of the segments into which a symmedian divides the opposite side of a triangle equals the ratio of the squares of the measures of the adjacent sides.
10. Prove that the altitudes of a triangle are the isogonal conjugates of the circumradii to the vertices of the triangle.
11. Use Theorem 4.20 and Exercise 10 to give an alternative proof that the altitudes of a triangle are concurrent.
12. Prove that a segment joining the feet of the perpendiculars from a point on a symmedian to the adjacent sides of the triangle is perpendicular to the corresponding median.
13. Prove the converse of Theorem 4.22.
14. Let S be the symmedian point of a given triangle, and D, E, F be the feet of the perpendicular segments from S to the sides of the triangle. Prove that S is the centroid of $\triangle DEF$.

▲ 4.5 Recent Synthetic Geometry of the Triangle

A second major contribution to the recent geometry of the triangle has to do with the *Brocard points* and the *Brocard circle*, named after Henri Brocard (1845–1922), who proved some theorems about these topics about 1875. Like the Miquel point, the Brocard points are defined as special points of intersection of three circles.

THEOREM 4.24 ➤ The three circles, each with one side of a triangle as a chord and tangent to an adjacent side, taken in order around the figure, meet in a point called a *Brocard point*.

In Figure 4.28, circle 1 is tangent to $\overline{AC}$, circle 2 is tangent to $\overline{BC}$, and circle 3 is tangent to $\overline{AB}$.

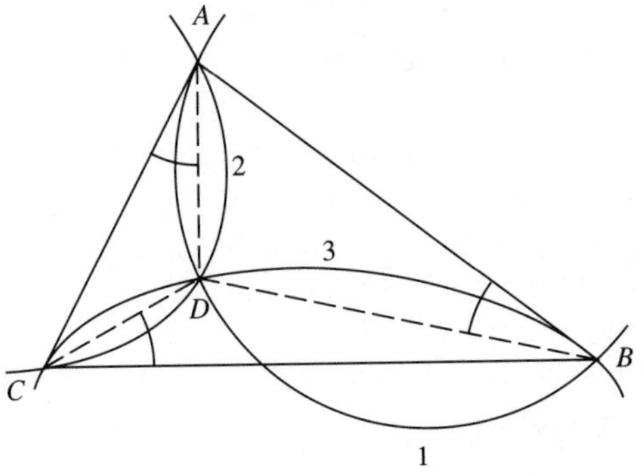

Figure 4.28

Assuming that circles 1 and 2 meet at D, $\angle ABD \cong \angle CAD \cong \angle BCD$, since the angle between a tangent and a chord is, like an inscribed angle, measured by half the intercepted arc. Circle 3 is tangent to $\overline{AB}$ at B, and $\overline{BD}$ subtends $\angle DCB$ equal to $\angle ABD$, so this circle also passes through D. The order chosen was counterclockwise, and point D is called the *first* or *positive Brocard point*.

DEFINITION ➤ Points are isogonal conjugates if one is the intersection of lines that are isogonal conjugates of lines intersecting at the other point.

In general, there are two distinct Brocard points, since the order around the triangle can be either clockwise or counterclockwise. Furthermore, it can be proved that the two Brocard points are isogonal conjugates of each other.

Yet another theorem shows a connection between previous theorems about symmedians and Theorem 4.24 about the Brocard points. The circle whose diameter has as endpoints the circumcenter and the symmedian point of a triangle is called the Brocard circle.

THEOREM 4.25 ➤ The Brocard points are on the Brocard circle.

PROOF: In Figure 4.29, let O and S be the circumcenter and the symmedian point, and let the perpendicular bisectors of the sides of the triangle meet the Brocard circle again at D', E', and F'. It can be shown that the lines AF', BE', and CD' are concurrent at a Brocard point.

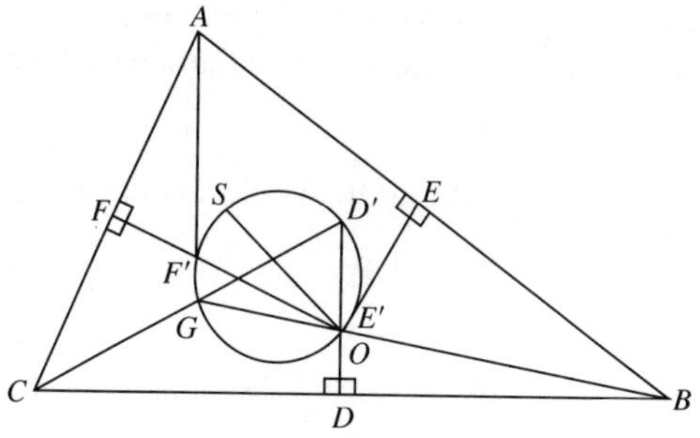

Figure 4.29

Because right angles are inscribed in a semicircle, $\overline{SD'}$ is parallel to $\overline{BC}$, $\overline{SF'}$ is parallel to $\overline{AC}$, and $\overline{SE'}$ is parallel to $\overline{AB}$. This means that the distances from S to the sides of the original triangle are equal to the distances along the perpendicular bisectors from the points D', E', F' to points D, E, F.

Because of Theorem 4.23,

$$\frac{DD'}{CD} = \frac{EE'}{EB} = \frac{FF'}{FA}$$

so that $\triangle CDD' \sim \triangle BEE' \sim \triangle AFF'$. As a result,

$$\angle BCD' \cong \angle CAF' \cong \angle ABE'$$

A comparison with Figure 4.28 shows that the three sides of the angles meet at a Brocard point G. Also, since $\angle AF'F \cong \angle GF'O \cong \angle GD'O$, points F', D', O, G all lie on a circle, the Brocard circle, as was to be proved. For an equilateral triangle, the circumcenter and symmedian point coincide so that the Brocard circle reduces to a single point.

The second major topic in this section is a theorem discovered about 1899 by Frank Morley, father of the author Christopher Morley. Its significance is that it concerns trisectors of angles rather than bisectors. Angle bisectors of a triangle meet at a point, but it seems that mathematicians have only recently considered what happens to sets of three adjacent trisectors.

THEOREM 4.26 ➤ *Morley's theorem.* The adjacent trisectors of the angles of a triangle are concurrent by pairs at the vertices of an equilateral triangle.

Before reading further, you will find it profitable to use a protractor and carefully draw several triangles of various shapes to verify the reasonableness of Morley's theorem.

Since 1900, many different proofs have been given for this theorem, but the one here is an indirect approach, starting with an equilateral triangle and ending with the original given triangle.

Proof: Figure 4.30a shows what is meant by adjacent trisectors. For example, $\overline{CD}$ and $\overline{BD}$ are a pair of adjacent trisectors. In Figure 4.30b, assume that $\triangle DEF$ is an equilateral triangle. Isosceles triangles DGE, DHF, and FIE can be determined such that the measures X, Y, Z of the angles are each less than 60 and $X + Y + Z = 120$. Points A, B, and C are determined by extending the sides of these isosceles triangles.

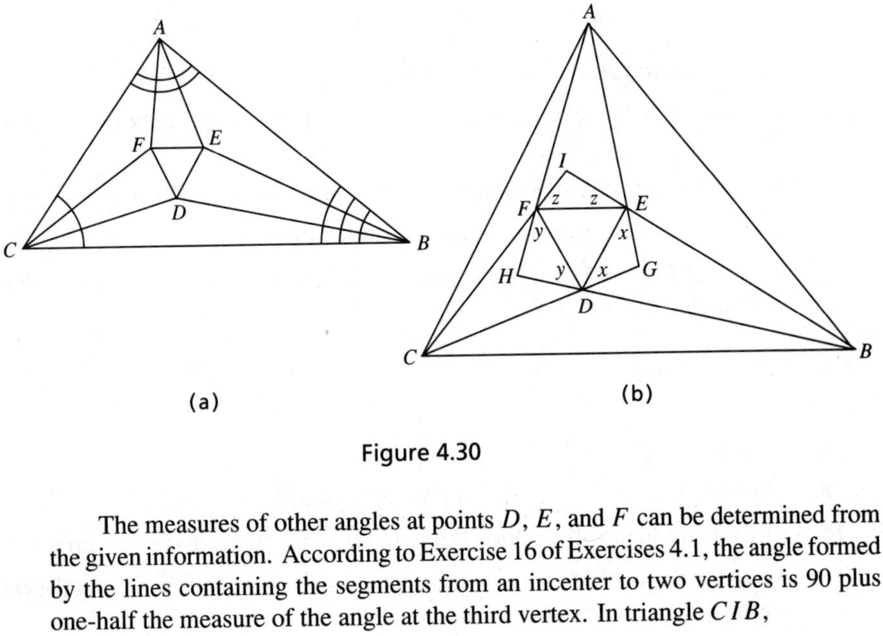

Figure 4.30

The measures of other angles at points D, E, and F can be determined from the given information. According to Exercise 16 of Exercises 4.1, the angle formed by the lines containing the segments from an incenter to two vertices is 90 plus one-half the measure of the angle at the third vertex. In triangle CIB,

$$m\angle CIB = 180 - 2Z$$

whereas $m\angle CDB = 360 - (X + Y + 2Z + 60)$.

$$\begin{aligned} 360 - (X + Y + 2Z + 60) &= 360 - (X + Y + Z) - Z - 60 \\ &= 360 - 120 - Z - 60 \\ &= 180 - Z \\ &= 90 + \frac{1}{2}(180 - 2Z) \end{aligned}$$

From this discussion, along with Exercise 16 of Exercises 4.1, and the fact that D is on the angle bisector of $\angle CIB$, point D is the incenter of $\triangle CIB$. Similarly, F is the incenter of $\triangle AGC$ and E is the incenter of $\triangle AHB$. Thus, the three angles at C are congruent as are the three at A and at B. These angles have measures

$$60 - Y, \qquad 60 - Z, \qquad 60 - X$$

The base angles of the isosceles triangle can be found in terms of the angles of the original triangle; for example, $X = 60 - \frac{1}{3}\angle B$ so that a triangle ABC can always be found similar to any given triangle.

▲ EXERCISES 4.5 ▲

1. Does a Brocard point always lie within a triangle?
2. Draw a figure locating the second Brocard point for the triangle shown in Figure 4.28.
3. Prove that the two Brocard points are isogonal conjugates of each other.
4. Name four significant points on the Brocard circle.
5. Triangle $D'E'F'$ in Figure 4.29 is called the first Brocard triangle. Prove that this triangle is similar to the original triangle.

In Exercises 6–9, draw a figure showing the equilateral triangle of Morley's theorem if the measurements of two of the angles of the original triangle are:

6. 90° and 40°
7. 60° and 20°
8. 30° and 80°
9. 25° and 75°
10. Prove that, in Figure 4.30b, the segments $\overline{DI}$, $\overline{FG}$, and $\overline{EH}$ are concurrent.
11. Investigate the figure formed by the intersections of the four pairs of adjacent trisectors of the angles of a square.
12. Investigate the figure formed by the intersections of the four pairs of adjacent four-sectors (or quadrisectors) of the angles of a square, if four-sectors are defined to be the rays partitioning the angle and its interior into four congruent angles and their interiors.
13. ▷ Investigate the figure formed by the six pairs of adjacent angle trisectors of a regular hexagon.
14. ▷▷ Make various large-scale drawings of triangles, measure carefully, then make a conjecture about whether or not a theorem analogous to Morley's theorem holds for the four-sectors of the angles of a triangle.

▷ **15.** For an answer to the truth or falsity of the conjecture in Exercise 14IG, and for a general theorem, consult D. J. Kleven, "Morley's Theorem and a Converse," *The American Mathematical Monthly*, February 1978.

▷ **16.** For a more general form of Morley's theorem, study Cletus O. Oakley and Justine C. Baker, "The Morley Trisection Theorem," *The American Mathematical Monthly*, January 1979.

▲ 4.6 Golden Ratio, Tessellations, Packing Problems and Pick's Theorem

Applications of synthetic Euclidean geometry to be discussed briefly in this section include the golden ratio, tessellations, packing problems, and Pick's theorem. The interested reader may wish to extend the study of various topics introduced.

With their sense of beauty and proportion, the Greeks came to regard certain shapes as more pleasing than others and to build many of their buildings in those shapes. The most famous example of this is what is called the *golden ratio*, the ratio of the lengths of the sides of a rectangle with the most pleasing proportion.

In Figure 4.31, suppose that $AB/BC = AC/AB$. In this case, point B is said to divide AC in *extreme and mean ratio*, and, if the length of $\overline{BC}$ is 1, the length of $\overline{AB}$ is the *golden ratio*.

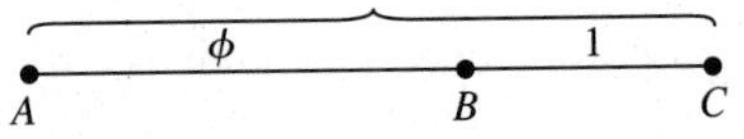

Figure 4.31

Let φ represent the numerical value of the golden ratio. From the previous proportion,

$$\frac{\varphi}{1} = \frac{\varphi + 1}{\varphi}$$

$$\varphi^2 = \varphi + 1$$

$$\varphi^2 - \varphi - 1 = 0$$

$$\varphi = \frac{1 \pm \sqrt{5}}{2}$$

The positive value of φ,

$$\frac{1 + \sqrt{5}}{2}$$

is the numerical value of the golden ratio. This, of course, is an irrational number. An approximate decimal value is 1.62. The set of Fibonacci numbers, $1, 1, 2, 3, 5, 8, 13, 21, \ldots$ can be extended by using the rule that the next number is the sum of the last two previous numbers. Mathematicians have also shown that the ratios of the $(n + 1)$ to the nth term in the set of Fibonacci numbers $1, 1, 2, 3, 5, 8, 13, 21, \ldots$

approach the golden ratio as a limit. The reader should express 3/2, 5/3, 8/5, and so on, as decimals to see that this statement is reasonable.

In Figure 4.32, $ABCD$ is a golden rectangle, with $AB = \varphi$ the golden ratio and $\overline{BC}$ the unit segment. If square $AEFD$ is removed, the remaining rectangle $EBCF$ is also a golden rectangle.

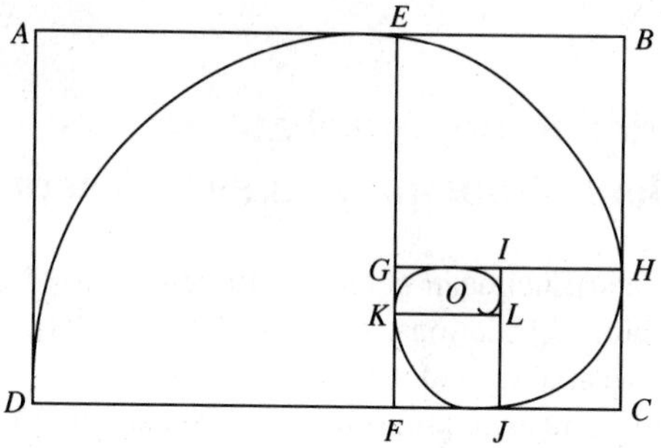

Figure 4.32

It is necessary to show that $CB/EB = \varphi$. Since $CB = 1$ and $BH = \varphi - 1$,

$$\frac{1}{\varphi - 1} = \frac{CB}{BH}$$

Or, since $BH = EB$, the proportion may be written as

$$\begin{aligned} \frac{CB}{EB} &= \frac{1}{\varphi - 1} \\ &= \frac{1}{\frac{1+\sqrt{5}}{2} - 1} \\ &= \frac{1+\sqrt{5}}{2} \\ &= \varphi \end{aligned}$$

This process of removing squares may be continued, and each time the result is a smaller golden rectangle. As shown in Figure 4.32, the pattern suggests a spiral, called the *golden spiral*, whose equation in polar coordinates is $r = \varphi^{2\theta/\pi}$, where φ is the golden ratio.

One additional application of the golden ratio is its connection to a regular pentagon.

THEOREM 4.27 ➤ The diagonals of a regular pentagon divide each other in the golden ratio.

PROOF: The notation of Figure 4.33 is used to illustrate Theorem 4.27 by showing AF/FD is the golden ratio. In isosceles trapezoid $ACDE$, $\triangle ACF \sim \triangle DEF$, so $AF/FD = AC/ED$. But $ED = BC$ and $BC = AF$, since $ABCF$ is a rhombus (see Exercise 4). Thus,

$$\frac{AF}{FD} = \frac{AC}{AF} \quad \text{or} \quad \frac{AF}{FD} = \frac{AD}{AF}$$

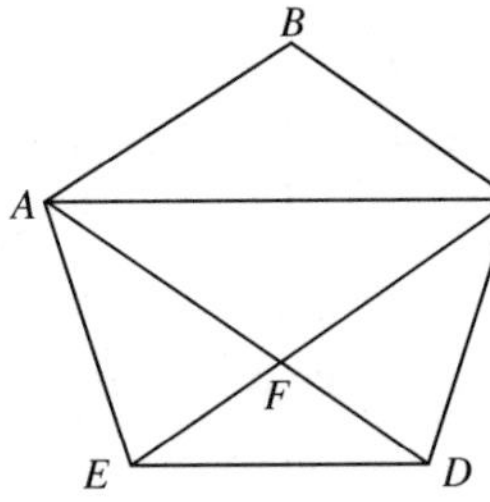

Figure 4.33

so AF/FD is the golden ratio. If the sides of a regular pentagon are 1 unit, then the length of the diagonals is the numerical value of the golden ratio.

For additional information about the golden ratio, see Garth E. Runion, *The Golden Section and Related Curiosa*, published by Scott, Foresman & Company, Glenview, Illinois, 1972.

The second application of synthetic Euclidean geometry to be discussed is tessellations (sometimes called tilings).

DEFINITION ➤ A *tessellation* is a pattern of polygons fitted together to cover an entire plane without overlapping. *Regular tessellations* are those in which all of the polygons are congruent and regular, with common vertices.

Figure 4.34 shows regular tessellations with squares, equilateral triangles, and regular hexagons. That these are the only three regular tessellations is established in the proof of the following theorem.

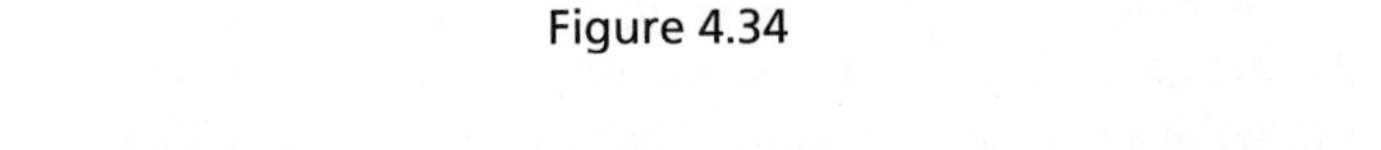

(a) (b) (c)

Figure 4.34

THEOREM 4.28 ➤ The only three regular tessellations of the plane are those with the square, the equilateral triangle, and the regular hexagon.

PROOF: Let x be the number of sides for each polygon and y the number of polygons coming together at each vertex. Then the two expressions for the measure of each angle at a vertex can be set equal.

$$\frac{\pi(x-2)}{x} = \frac{2\pi}{y}$$
$$\pi xy - 2\pi y = 2\pi x$$
$$xy - 2x - 2y = 0$$
$$xy - 2x - 2y + 4 = 4 \qquad \text{add 4 to both members}$$
$$(x-2)(y-2) = 4$$

There are only three ways to factor 4 (counting the change in order of factors), and these three ways lead to the three tessellations previously named. See Table 4.1.

Table 4.1

$x-2$	$y-2$	x	y	**Tessellation**
4	1	6	3	Figure 4.34c
2	2	4	4	Figure 4.34a
1	4	3	6	Figure 4.34b

If we consider nonregular polygons, then we can show that any triangle or any quadrilateral will tessellate the plane.

Example ➤ Figures 4.35a and 4.35b show one way this can be done.

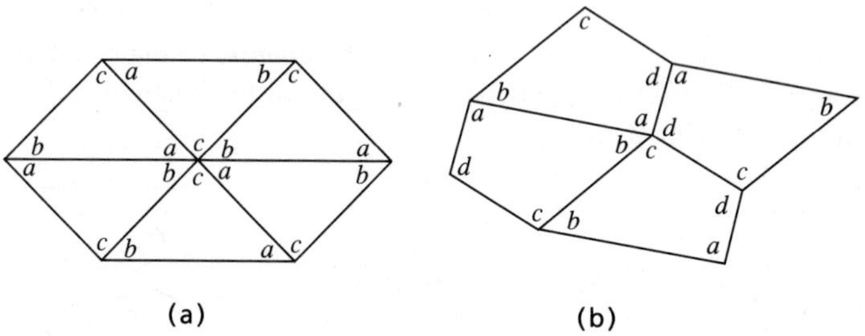

Figure 4.35

For the triangle, 6 angles (2 of each) will fit around each vertex and for the quadrilateral, the four angles will fit around each vertex. One way to explain the figures is to think of them as a series of half turns about the midpoint of each side of the triangle or quadrilateral.

Two other ways to use a triangle to tessellate a plane are to use only translations or to use a combination of translations and reflections. In the first case, shown in Figure 4.36a, triangle ABC is translated to the right so that segment AB is moved to segment BB'. Triangle ABC is also translated in the direction $\overrightarrow{AC}$ so that segment AC is moved to CC'. A repetition of this pattern of translating in the direction of two sides of the triangle accomplishes the tessellation. The third method to use a triangle to tessellate a plane is shown in Figure 4.36b. The triangle is reflected about a side to form a quadrilateral, then the quadrilateral is translated in the directions along the two diagonals.

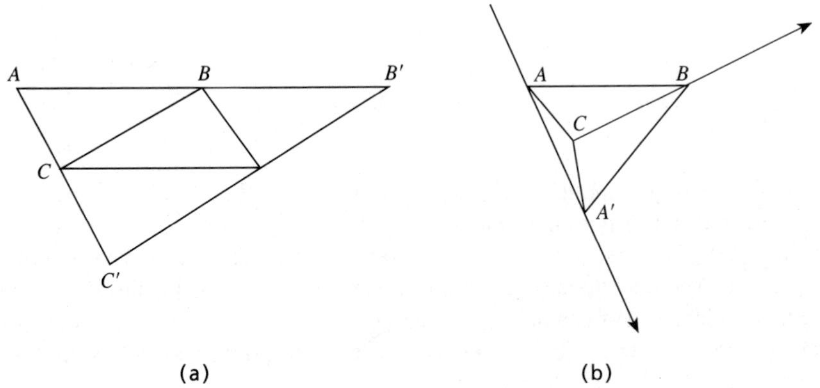

Figure 4.36

Additional explorations with tessellations are suggested in the exercises.

Other tessellations are also a fruitful area to examine. What other types of tessellations can you discover?

Sometimes, the definition of tessellation is broadened to include figures other than polygons. The art of M.C. Escher, mentioned in Chapter 2, provides examples of varied plane-filling drawings.

Another application closely related to regular tessellations is that of packing of congruent circles in a plane. The problem is to pack as many congruent circles as possible by choosing the maximum density. The density of the packing is defined to be the ratio of the area of the circular region to the area of the polygon in which it is inscribed. Since the polygons must cover the entire plane without overlapping, the only possibilities for regular packing are to use the three regular tessellations, as shown in Figure 4.37.

By using formulas for the area of a circular region and the regular polygons, one can show that the density is greatest for the packing in Figure 4.37c, and that the numerical value is greater than 0.9. Actually, the packing with the hexagon is still the densest possible even when the restriction of using regular polygons is removed. The analogous problem in three dimensions, the packing of spheres, is much more complex, and is a subject of continued research.

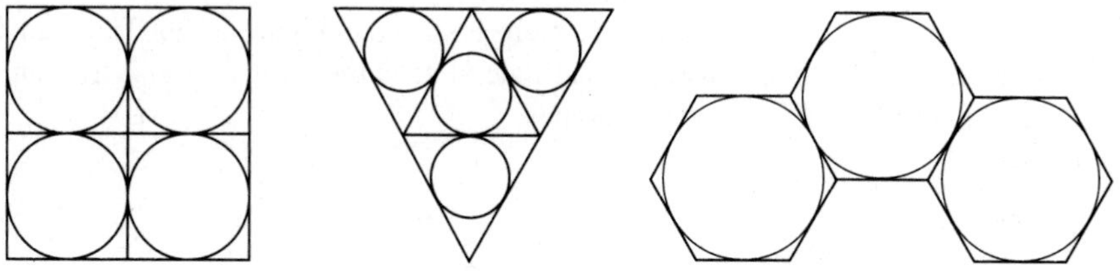

(a) (b) (c)

Figure 4.37

A numerical type application related to polygons is a theorem for finding the area of a polygon if the vertices have integral coordinates. The array of dots on the plane in Figure 4.38a is a special case of a set of points called a lattice, and each point is a lattice point. You may think of these points as the points in the Euclidean plane with integer coordinates. For example, (3, 7) and (6, 1) are two examples of lattice points in this special case, while (3.2, 7.5) is not. The lattice points shown are in rows and columns, theoretically infinite in extent.

THEOREM 4.29 ➤ *Pick's theorem* (1899). If a simple polygon has vertices that are lattice points, then the formula for the area is $\frac{1}{2}p + q - 1$, where p is the number of lattice points on the boundary and q is the number of lattice points inside the polygon. A simple polygon has one interior only, with no sides crossing each other.

A reference containing the proof is provided in the exercises. In Figure 4.38b, the formula provides areas of $\frac{1}{2}(20) + 6 - 1 = 15$ and $\frac{1}{2}(16) + 13 - 1 = 20$ for the two polygons shown.

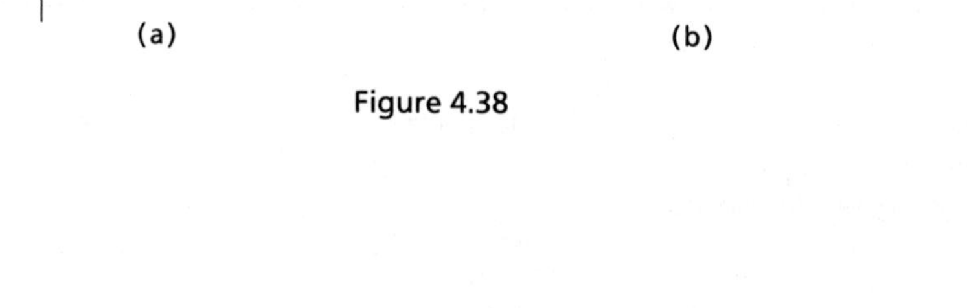

(a) (b)

Figure 4.38

▲ EXERCISES 4.6 ▲

1. Find the approximate value of the golden ratio correct to four decimal places.

2. Use Fibonacci numbers to give a series of approximations to the golden ratio until one is reached that is correct to three places past the decimal point.

3. In Figure 4.32, if $BC = 1$, find the measure of segment $\overline{IL}$.

4. In Figure 4.33, prove that $ABCF$ is a rhombus.

5. Show by a drawing and a numerical explanation why regular octagons cannot be used for a regular tessellation.

Use the triangle in Figure 4.39a to tessellate the plane:

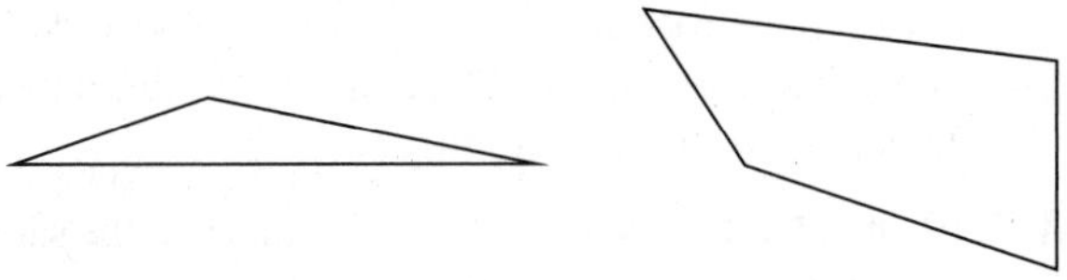

(a) (b)

Figure 4.39

6. Using half-turns about the midpoints of the sides.

7. Using only translations.

8. Using translations and reflections.

Use the quadrilateral in Figure 4.39b to tessellate the plane:

9. Using half-turns about the midpoints of the sides.

10. Using translations along the diagonals.

For Exercises 11–14, use Pick's theorem to find the area of each polygon shown in Figure 4.40.

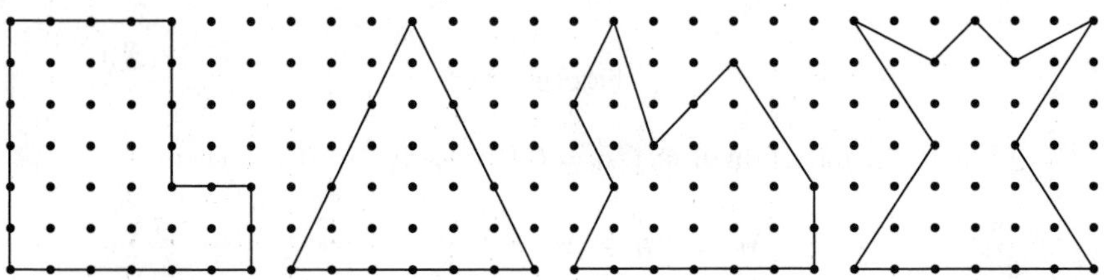

11. **12.** **13.** **14.**

Figure 4.40

▷ **15.** Investigate the concept and properties of semiregular tessellations.

▷ **16.** Develop a general formula for finding the density of the three regular packings of circles in a plane and calculate the three numerical ratios.

▷ **17.** Investigate the current status of research on the sphere packing problem.

▷ **18.** Outline the proof of Pick's theorem found in Coxeter, *Introduction to Geometry* or another reference.

▲ 4.7 Extremum Problems, Geometric Probability, and Fuzzy Sets

One of the ancient yet very modern applications of the Euclidean geometry of similar triangles is today called *caroms*. A carom is an impact followed by a rebound. Historically, applications of this sort are based on an extremum property (involving maxima or minima) stated as Heron's theorem. Heron of Alexandria was the author of *Catoptrica*, which included the elementary properties of mirrors.

THEOREM 4.30 ➤ *Heron's theorem.* For two points on the same side of a line, the shortest path from the first point to the line and then to the second point is by way of the point of intersection of the line and the segment from the first point to the reflection of the second point.

Proof: In Figure 4.41, we need to prove that $AC + CB$ is a minimum. This means that $AC + CB < AC' + C'B$ for any other point C' on the given line.

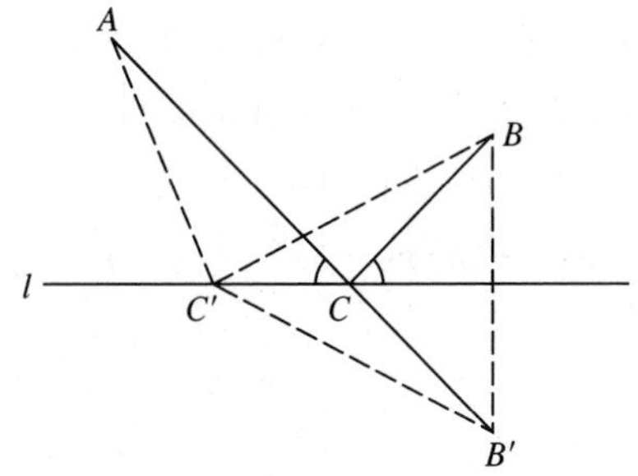

Figure 4.41

For B' the reflection of B, $CB = CB'$ and $BC' = B'C'$. Then

$$AC + CB = AC + CB' = AB'$$

and

$$AC' + C'B = AC' + B'C'$$

But $AB' < AC' + B'C'$, since $AC'B'$ is a triangle, so $AC + CB < AC' + C'B$, as was to be established.

Physical applications of Heron's theorem depend on the assertion that the angles $\overline{AC}$ and $\overline{CB}$ make with line l are congruent. For example, the shortest path from A to B by way of a point on line l is the path of a light ray reflected in a mirror at line l. A corner reflector, or angle mirror, shown in Figure 4.42, uses the physical principle of reflection to allow a person to "see around a corner."

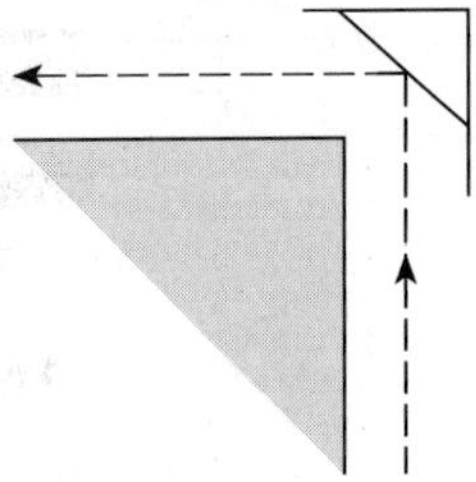

Figure 4.42

The problem of finding the shortest path can be generalized to more than one line, thus increasing the usefulness of the theory. For example, Figure 4.43 shows the shortest path from A to B, touching l_1 and then l_2. B' is the reflection of B in l_2 and B'' is the reflection of B' in l_1. The path leads from A toward B'', then from C toward B'. A simple periscope makes use of this theory.

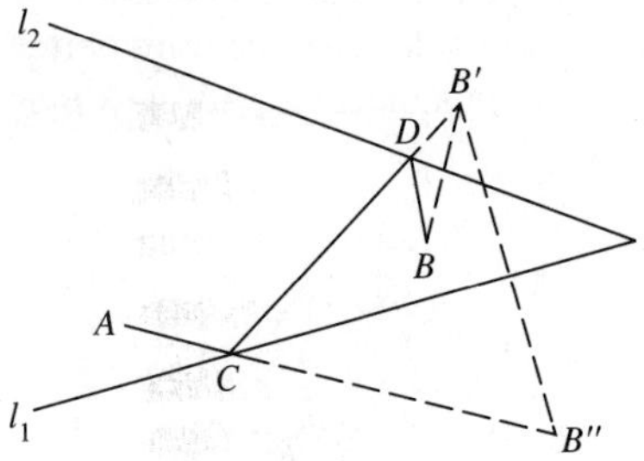

Figure 4.43

An application of the theory of caroms that is of current interest in mathematics as well as in everyday life is in connection with a pool table. Figure 4.44a shows how the ball at A can be aimed to hit ball B by caroming off one, two, or three sides of the table.

In Figure 4.44a, B' is the image of B with respect to the side containing point C. In Figure 4.44b, B' is the image of B with respect to the side containing point D, and B'' is the image of B' with respect to the side containing point C. In Figure 4.44c, B' is the image of B with respect to the side containing point E, B'' is the image of B' with respect to the side containing point D, and B''' is the image of B'' with respect to the side containing point C.

In Figure 4.44b, $\overline{AC}$ and $\overline{BD}$ are parallel and in Figure 4.44c, $\overline{AC}$ and $\overline{ED}$ are parallel and $\overline{CD}$ and $\overline{BE}$ are parallel.

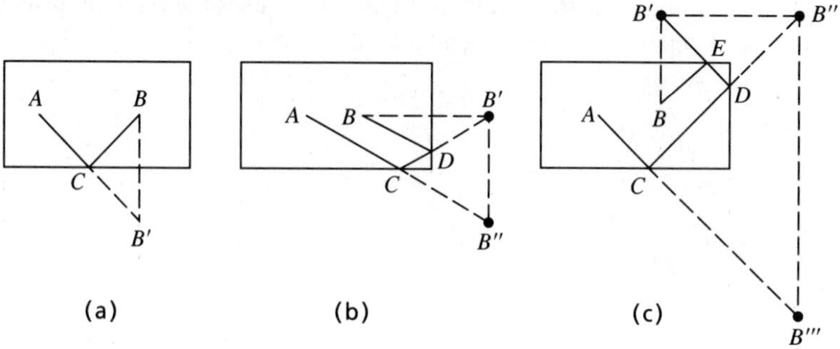

Figure 4.44

Heron's theorem and the transformation of reflection also provide the basis for other applications of geometry to *extremum problems*. Two examples, solved without the use of calculus, are included.

THEOREM 4.31 ➤ For a triangle with a given area and side, the sum of the measures of the other two sides is a minimum if and only if the triangle is isosceles.

The proof is outlined, but the details are left as Exercises 9 and 10, Exercises 4.7. In Figure 4.45, if $\overline{AB}$ is the fixed side, then the third vertex X must be on a line $\overleftrightarrow{CD}$ parallel to $\overline{AB}$. Why? By Heron's theorem, $AX + XB$ is a minimum if $\triangle AXB$ is isosceles. Why?

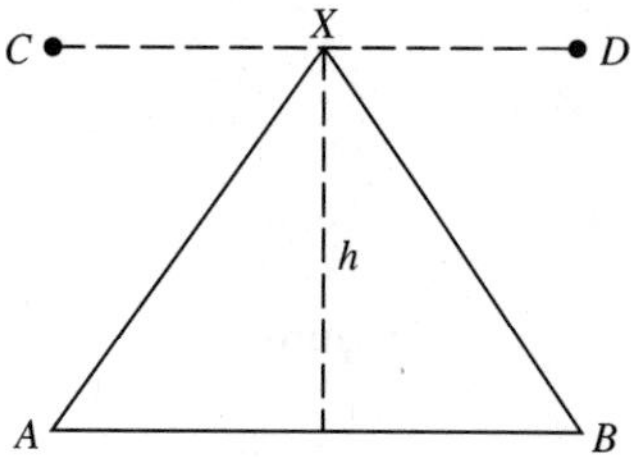

Figure 4.45

A second example, called *Fagnano's problem* after J. F. Fagnano, who proposed it in 1775, is stated as the last theorem of the chapter.

THEOREM 4.32 ➤ In a given acute-angled triangle, the inscribed triangle with minimum perimeter is the orthic triangle, whose vertices are the feet of the altitudes.

Details of the proof are again left as an exercise (Exercise 11, Exercises 4.7). In Figure 4.46, if $\triangle DEF$ is an arbitrary inscribed triangle and if G and H are the images of D under reflection with respect to $\overleftrightarrow{AC}$ and $\overleftrightarrow{AB}$, then the distance $GE + EF + FH$ is the perimeter of the inscribed triangle. Why? This distance will be a minimum if G, E, F, H are collinear. The measure of $\angle GAH = 2m\angle CAB$. Why? $\triangle GAH$ is isosceles. Why? Because the measure of $\angle GAH$ is independent of the choice of point D, the base $\overline{GH}$ will have a minimum measure when the congruent sides are a minimum. But this condition is met when $\overline{AD}$ has a minimum length, and this occurs when $\overline{AD}$ is an altitude. Why?

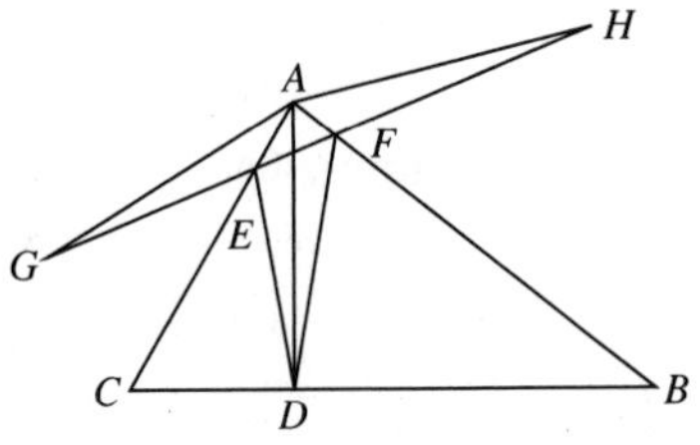

Figure 4.46

Geometric probability is the name given to an application in which geometry is used to calculate probabilities. For a typical problem in geometric probability, outcomes are represented by points in a plane region. The sample space is represented by a plane region, and the event is represented by a subregion.

Example ➤ What is the probability that a point chosen at random in the larger rectangular region in Figure 4.47 will be in the shaded region?

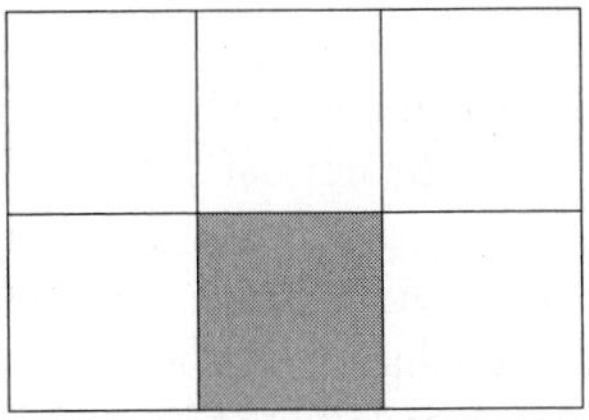

Figure 4.47

The larger region is partitioned into 6 congruent regions, so the area of each is $1/6$ of the total area. Thus the probability is $1/6$.

Example ➤ What is the probability that a point chosen at random in at least one of the circular regions in Figure 4.48a will be in the shaded region? (Assume the radius is 1.) The triangles CAE and DAE are 30, 60, 90 triangles. $\angle CAD = 120°$. The shaded area is $\frac{2}{3}\pi$, minus the area of the two small circular segments shown cross hatched. See Figure 4.48b for more detail.

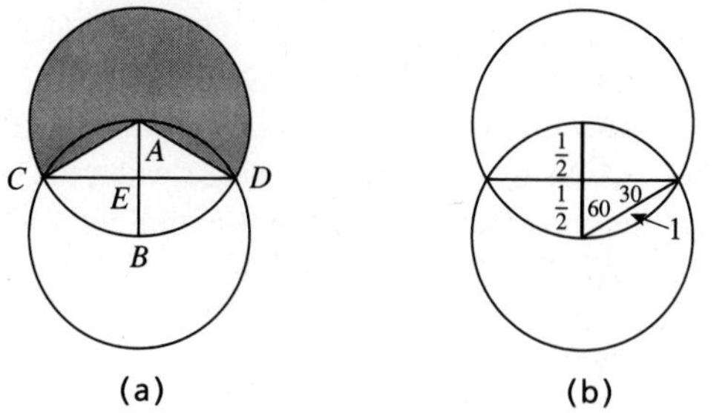

Figure 4.48

The area of each of these is $\frac{1}{2}\left(\frac{\pi}{3} - \sin\frac{\pi}{3}\right) = \frac{1}{2}\left(\frac{\pi}{3} - \frac{\sqrt{3}}{2}\right)$, so that the total shaded area is $\frac{2}{3}\pi - \left(\frac{\pi}{3} - \frac{\sqrt{3}}{2}\right) = \frac{\pi}{3} + \frac{\sqrt{3}}{2}$. The area for the entire figure is $\pi + \frac{\pi}{3} + \frac{\sqrt{3}}{2} = \frac{4}{3}\pi + \frac{\sqrt{3}}{2}$, so that the probability is

$$\frac{\dfrac{\pi}{3} + \dfrac{\sqrt{3}}{2}}{\dfrac{4\pi}{3} + \dfrac{\sqrt{3}}{2}} = \frac{2\pi + 3\sqrt{3}}{8\pi + 3\sqrt{3}}$$

From one point of view, geometric probability is an application of the mensuration formulas of geometry. Practical applications of geometric probability include search patterns in rescue operations and military applications involving bombing of targets whose exact positions in a geometric region are not known.

Two additional geometric applications, not usually considered a part of geometric probability, use the concept of comparison of areas to derive a probability. The first is statistical theory having to do with the normal curve. If the area between the normal curve and the x axis is one, then the area under the curve between two vertical lines represents the probability that an individual score would be in that region. For example, the probability is approximately .68 that an individual score would be between -1 and $+1$ standard deviations from the mean (See Figure 4.49).

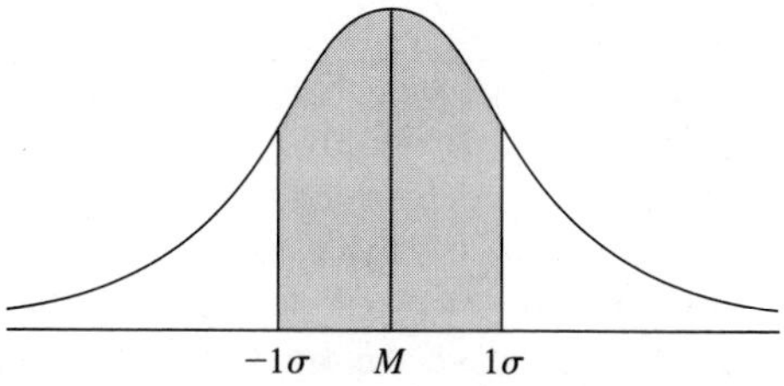

Figure 4.49

The second application, a common method of finding area in computer science, is called the Monte Carlo method. Suppose the problem is to find the shaded area under the curve in Figure 4.50, assuming the area of the entire rectangle is known. Large numbers of random coordinates are generated to find points in the rectangle, and the percent of those points lying in the shaded region gives an approximation of the area. The average of 3 samples of 25,000 points each, for example, can easily be provided by a computer, and gives results that are precise enough in many cases.

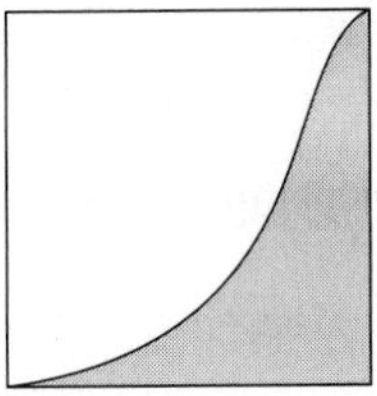

Figure 4.50

An application of geometry that is related to probability, but which does not depend on randomness, is the use of *fuzzy sets* described geometrically. Intuitively, a fuzzy set is a set that has the possibility of partial membership. For example, for each point of an ordinary set of points, the value 0 could be used to mean the point is not in the set, and the value 1 to mean the point is in the set. For a fuzzy set, the grade of membership can be any real number in the range from 0 to 1.

Example ➤ Suppose a fuzzy set over the real numbers greater than 1 consists of the numbers much greater than one. The formula for grade of membership might be defined by the formula $(x - 1)/x$ for x greater than one. Figure 4.51 shows the grade of membership for various points on a number line. The grade is not random, but increases to the right.

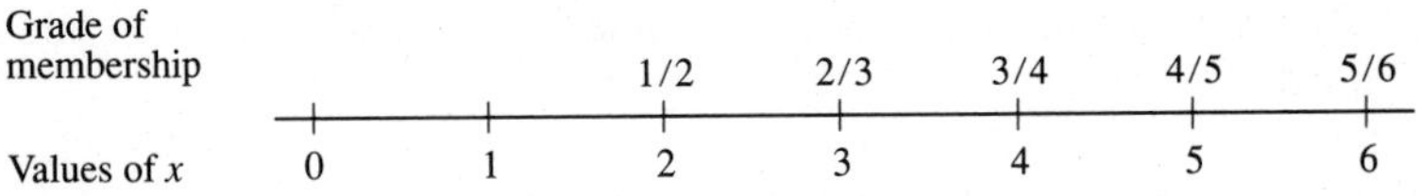

Figure 4.51

The theory of fuzzy sets was introduced in 1965 by L. A. Zadeh. The idea of fuzziness comes from the name of sets without exactly defined boundaries. Other examples of fuzzy sets that might be represented geometrically include the set of real numbers close to 10, the set of small real numbers, and the set of integers much smaller than 100. Today there are an enormous number of reference articles on fuzzy sets, and they are also applied in many branches of mathematics besides geometry.

New branches of science continue to suggest new applications of polygons. A subject of current research is motion planning in the engineering application called robotics. One of the stated problems is to plan a continuous path for a rod moving in two dimensions with polygonal obstacles (Figure 4.52 shows a simplified drawing of the idea) and the three-dimensional analogue involving polyhedral bodies.

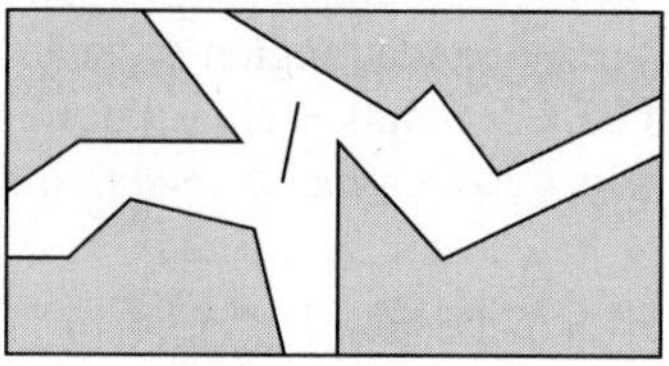

Figure 4.52

Section 8.5 discusses polyhedra and Euler's formula relating the number of vertices, faces, and edges.

▲ EXERCISES 4.7 ▲

For Exercises 1–6, draw figures similar to Figure 4.53 for the following circumstances.

1. Where should ball A hit side $\overline{CF}$ in Figure 4.53a in order to carom and hit ball B?
2. Where should ball A hit side $\overline{DE}$ in Figure 4.53a in order to carom and hit ball B?
3. Where should ball A hit side $\overline{CD}$ in Figure 4.53a in order to carom and hit ball B?
4. Ball A in Figure 4.53b is to carom off $\overline{DE}$, then $\overline{EF}$, and then hit ball B.
5. Ball A in Figure 4.53b is to carom off $\overline{EF}$, then $\overline{CF}$, and then hit ball B.
6. Ball A in Figure 4.53c is to carom off side $\overline{CD}$, then $\overline{DE}$, then $\overline{EF}$, and finally hit ball B.

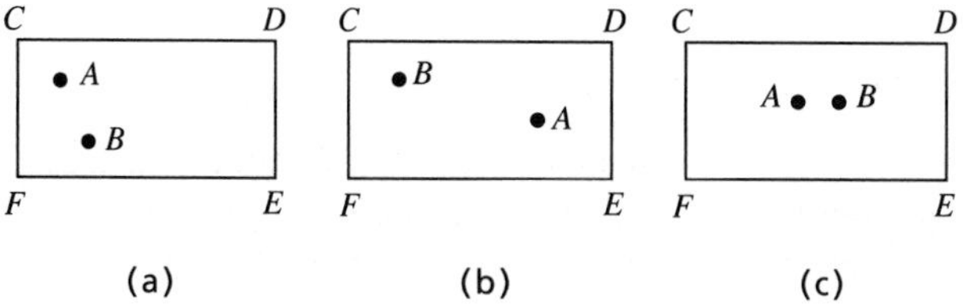

Figure 4.53

7. Cutting a diamond for maximum brilliance or sparkle is an application related to caroms. In Figure 4.54 it is desirable to have the rays parallel so that light will

be reflected back in the same direction. What angle θ will make this reflection possible?

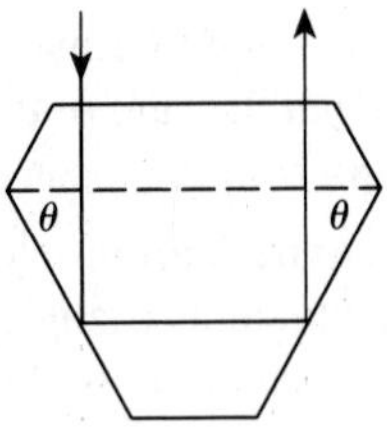

Figure 4.54

8. Using the notation of Figure 4.43, prove that the shortest path from A to B, by way of l_1 and then l_2, is the path shown.

9. Complete the "if" part of the proof of Theorem 4.31.

10. Complete the "only if" part of the proof of Theorem 4.31.

11. Complete the proof of Theorem 4.32.

12. What is the probability that a point chosen at random in the larger rectangular region in Figure 4.55a will be in the shaded region?

13. Figure 4.55b shows an equilateral triangle and its medians. What is the probability a point chosen at random in the triangle will be in the shaded region?

14. In Figure 4.55c, what is the probability a point chosen at random in the region will be in the semicircle?

15. Figure 4.55d shows an equilateral triangle inscribed in a circle. What is the probability a point chosen at random in the circle will be in the triangle?

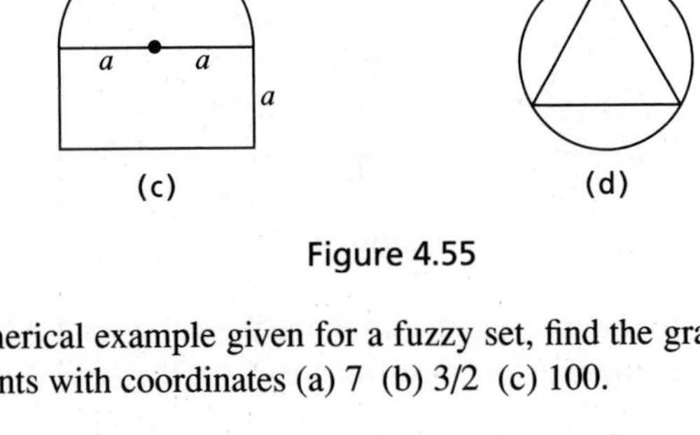

Figure 4.55

16. In the numerical example given for a fuzzy set, find the grade of membership for the points with coordinates (a) 7 (b) 3/2 (c) 100.

17. Suppose the grade of membership in the set or points with coordinates much greater than 9 is defined by the formula $(x-9)/x$. Find the grade of membership for points with coordinates (a) 10 (b) 18 (c) 36.

▷ 18. Another extremum problem, called Fermat's problem, is to find a point in an acute-angled triangle such that the sum of the distances from the three vertices is a minimum. How can this point be located?

▷ 19. A further extension of the extremum problem of Fermat is sometimes called Napoleon's theorem. If equilateral triangles are constructed outward on the sides of a triangle, then the circumcircles of these three triangles are concurrent at Fermat's point. Investigate this property and a property of the three centers of the equilateral triangles.

▷ 20. Make additional sketches of figures and use geometric probability to find the percent of total area shaded.

▷ 21. Study or review the connection between geometric probability and the area under a normal curve.

▷ 22. Find or write an example of a computer program for using the Monte Carlo method of finding area under a curve.

▲ CHAPTER 4 REVIEW EXERCISES ▲

1. What is the name of the point that is the same distance from all three sides of a triangle?
2. What is the name of the point that is the same distance from all three vertices of a triangle?
3. An excircle is tangent to how many sides of a triangle?
4. In general, the nine-point circle of a triangle has how many points in common with the triangle?
5. In Figure 4.56, find the length of segment CE if segments AD, BE, and CF are concurrent, using the given measures.

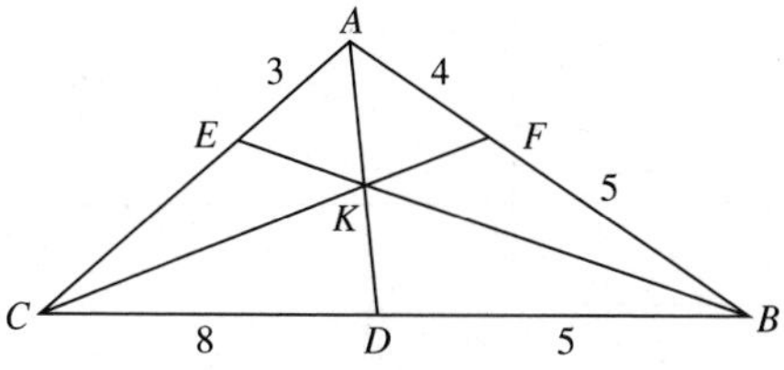

Figure 4.56

6. In Figure 4.56, for the triangle CFB and the transversal AD, write the product of the three ratios equal to negative one by the theorem of Menelaus.

7. Each symmedian of a triangle has what minimum number of points in common with the Brocard circle?

8. Make a sketch for Figure 4.57 showing how ball A can be made to hit ball B by first caroming off side $\overline{EF}$, then off side $\overline{FC}$.

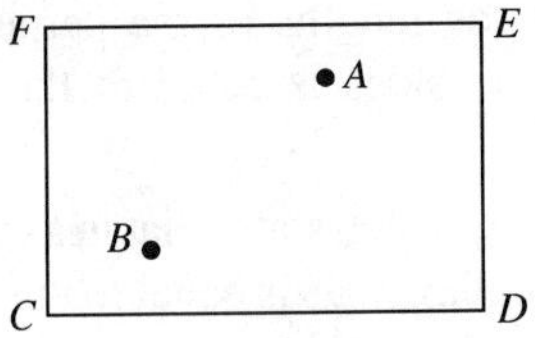

Figure 4.57

9. For each given point on the circumcircle of a given triangle, how many Simson lines are determined?

10. Name a line segment that is its own isogonal conjugate.

11. Which two of these points do not always lie on the Euler line: incenter, circumcenter, centroid, symmedian point?

12. Which of these points always lie on the Euler line: incenter, centroid, Brocard point, Lemoine point, orthocenter, excenter, Gergonne point?

13. Use Pick's theorem to find the area of the polygon in Figure 4.58.

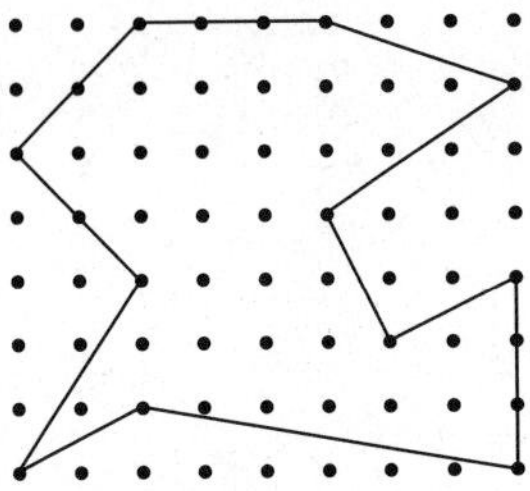

Figure 4.58

14. Sketch a quadrilateral with consecutive angles 60°, 70°, 80°, and use it to tessellate the plane in two different ways.

15. Suppose a square has a circle circumscribed around it and a second circle inscribed in it. Find the probability that a point selected at random in the larger circular region is also in the smaller circular region.

16. Prove that if an altitude of a triangle is extended to meet the circumcircle, the side it intersects bisects the segment between the orthocenter and the circumcircle.

17. Prove that the distance from a vertex of a triangle to the orthocenter is twice the distance from the circumcenter to the opposite side.

18. Prove that the three triangles formed by joining the centroid of a triangle to the three vertices have equal areas.

19. Prove that if an equilateral triangle is constructed on each side of a given triangle and the third vertex of each of these triangles is joined to the opposite vertex of the original triangle, the three segments determined are congruent.

20. Prove that the sum of the lengths of the perpendiculars from a point in an equilateral triangle to its sides is equal to the length of the altitude of the triangle.

21. Prove Ptolemy's theorem, which states that in a cyclic quadrilateral, the product of the lengths of the two diagonals is equal to the sum of the products of lengths of the opposite sides.

22. Prove that the 6 feet of the perpendiculars from the two Brocard points to the sides of a triangle are concyclic.

Additional chapter review exercises may be selected from any exercises not already assigned from previous exercise sets in this chapter.

5 Constructions

▲ 5.1 The Philosophy of Constructions

The concept of a construction was basic to the axiomatic system of Euclid, as can be seen by reviewing his axioms and postulates, which are stated in Chapter 1. The concept of a construction remains significant in geometric thought. Constructions are studied in various modern geometries not only because they are interesting in their own right but also because they provide applications of other geometric concepts.

Plato, the great Greek philosopher who lived from 427 to 347 BC, about a century before Euclid, is given credit for specifying the use of the *straightedge* and the *compass* as the only two permissible instruments for performing constructions. Technically, Plato specified the use of *dividers*, or a *collapsing compass*, not the modern compass we use today. When the instrument had been used to construct one circle, the measurement was lost because the instrument could not be moved to construct another circle with the same radius. It will be shown, however, that the divider (collapsing compass) is equivalent to the modern compass, so the straightedge and the compass are still the only two instruments allowed in the classical mathematical constructions of Euclidean geometry (and even the straightedge can be discarded).

The word *construction* is used in at least three ways:

1. To describe the geometric problem to be solved
2. To describe the process of solving the problem
3. To describe the completed drawing that results from solving the problem

The result of a construction is a drawing that shows certain relationships among lines and circles. Philosophically, constructions may be explained as methods for solving certain geometrical problems according to a fixed set of rules.

The problem in a construction is not simply that of drawing a figure to satisfy certain conditions but whether, by using a compass and straightedge only, a theoretically exact solution can be obtained. The first three postulates of Euclid provide the axiomatic basis for the constructions of the Greeks, who even used constructions to solve algebraic problems such as the solution of a quadratic equation. The first two of these postulates make it possible to construct any portion of a straight line through two points, whereas the third makes it possible to construct a circle if the center and the measure of the radius are given.

An example of a simple construction problem solved with the modern compass and then with the collapsing compass will illustrate the difference in the methods that must be used. Figure 5.1a shows the familiar method of finding the midpoint of a segment by construction. $\overline{AC}$ and $\overline{BC}$ must be congruent, $\overline{AD}$ and $\overline{BD}$ must also be congruent, but $\overline{AC}$ and $\overline{AD}$ are not necessarily congruent. Figure 5.1b shows the same problem solved with a collapsing compass. In this case, the measure of $\overline{A'B'}$ is used for the radius since its length can be determined using either endpoint as the center of the arc, whereas the arbitrary radius $\overline{AC}$ in Figure 5.1a cannot be reproduced again with the second endpoint B as the center of the circle.

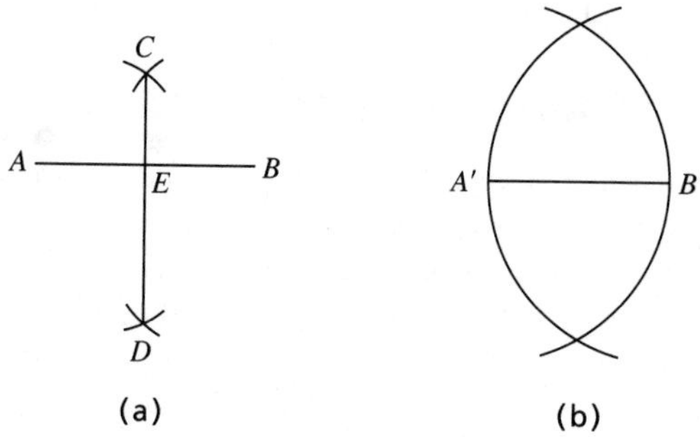

Figure 5.1

THEOREM 5.1 ➤ The compass and the collapsing compass (dividers) are mathematically equivalent.

The proof of this theorem consists of showing that a circle can be constructed with the collapsing compass, given the center and two other points that determine the length of the radius. The steps in performing this construction are stated, using the notation of Figure 5.2. The problem is to construct a circle with center A and with radius BC.

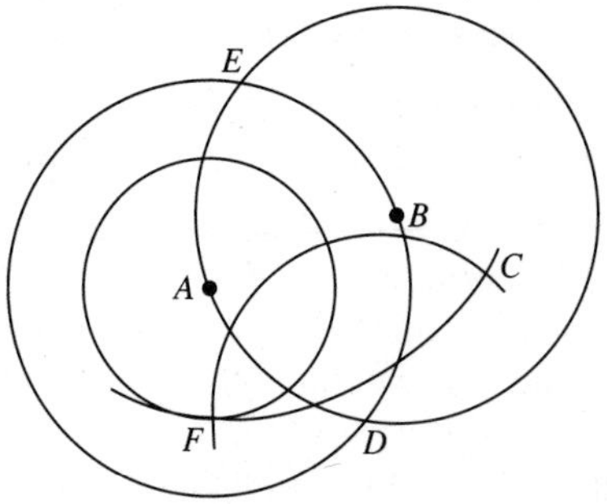

Figure 5.2

1. Construct the circle with center A and passing through B.
2. Construct the circle with center B and passing through A. These two circles meet at D and E.
3. Construct the circle with center E and passing through C.
4. Construct the circle with center D and passing through C.
5. The circles in steps 3 and 4 intersect again in a point F, and the circle with center A and radius AF is the required circle.

The proof, using congruent triangles, that $\overline{AF}$ is congruent to $\overline{BC}$ is left as an exercise.

The basic constructions of Euclidean geometry include transferring a segment, bisecting a segment, constructing a perpendicular to a line at a certain point, constructing a perpendicular from a point to a line, constructing an angle bisector, copying an angle, constructing a triangle given an angle and the two adjacent sides, constructing a triangle given the three sides, constructing a tangent from a point to a circle, and constructing a line through a point parallel to a given line. Each of these constructions is illustrated in Appendix 3 for students unfamiliar with them, or who need to review them before going ahead with this chapter.

Several additional basic constructions are introduced or reviewed here so that they may be used easily.

Partition of a segment into n congruent segments, for n a positive integer greater than 1, is the first of these constructions and is illustrated in Figure 5.3, for $n = 5$. The problem is to partition $\overline{AB}$ into five congruent segments. $\overline{AC}$ is constructed through A at any convenient angle. Then a unit AD is chosen to determine five congruent segments along $\overleftrightarrow{AC}$. Next $\overline{HB}$ is constructed, and the four segments $\overline{DI}$, $\overline{EJ}$, $\overline{FK}$, and $\overline{GL}$ are constructed parallel to $\overline{HB}$, through the established points on $\overleftrightarrow{AC}$ determining the four required points I, J, K, and L.

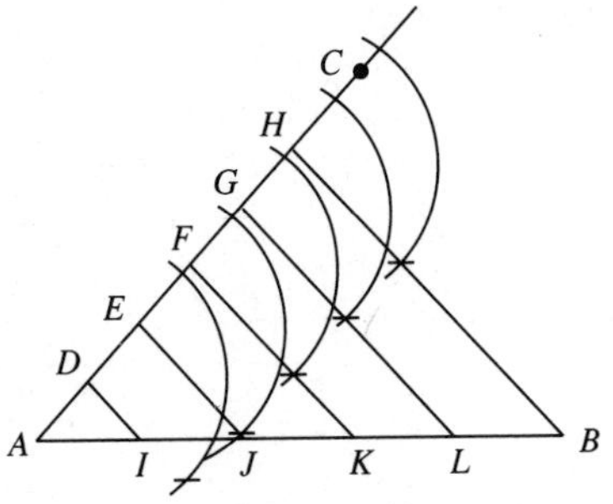

Figure 5.3

A second problem related to the first is to *partition a segment into a given ratio*. The problem is to find a point B on $\overleftrightarrow{AC}$ such that AB/BC has a given constructible numerical value. One use of this construction is in the theory of harmonic sets of points in projective geometry. Figure 5.4a is an *analysis figure*, showing the completed solution. It is assumed that B and G divide $\overline{AC}$ internally and externally in the desired ratio. If $\overline{AE}$ and $\overline{DF}$ are parallel, then $\triangle AEB \sim \triangle CDB$ and $\triangle AEG \sim \triangle CFG$, and the ratios of similarity for the pairs of similar triangles are the given ratio of division. Then the given ratio is also the ratio of AE to CD and of AE to CF; this makes possible the following construction, illustrated in Figure 5.4b for a ratio of 5/2. Through A' and C', construct parallel lines. Using an arbitrary unit, find E' so that $A'E' = 5$ and find D' and F' so that $C'D' = C'F' = 2$. Connect E', D' and E', F' to locate the required points B' and G'.

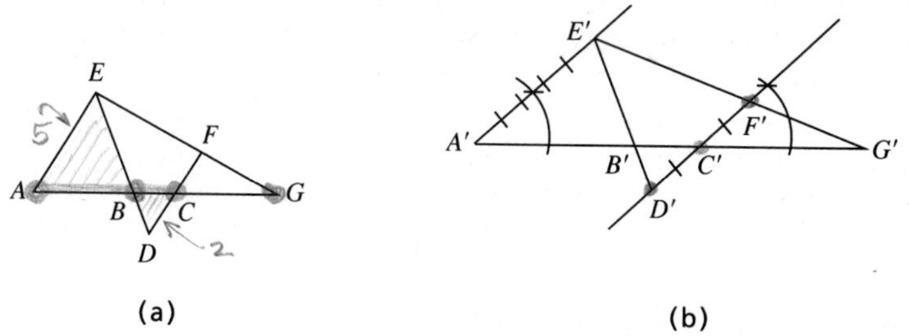

Figure 5.4

In the problem of partitioning a segment into a given ratio, the known information might be given entirely in the form of segments. For example, $\overline{AC}$ could be given along with a unit segment and a third segment, whose length, in terms of the unit, represents the given ratio. In this case, the unit segment could be used for $C'D'$, whereas the third segment would determine the length $A'E'$. A specific example of this type of construction is given in Figure 5.5. The problem is to partition $\overline{AC}$ internally and externally in the given ratio.

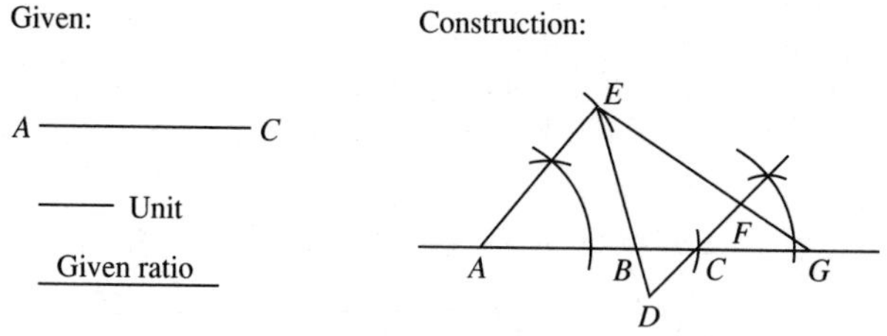

Figure 5.5

The use of the basic constructions of this section leads to the solution of more complicated construction problems. Constructions involving the partitioning of segments, and other constructions of figures from Chapter 4, are given in the exercises at the end of this section. More difficult constructions are found in Section 5.3.

In a formal study of constructions, four distinct steps are required in the solution of any construction problem.

1. *Analysis.* In this step, the solver assumes that the construction has been performed, then analyzes the completed picture of the solution to find the needed connections between the unknown elements in the figure and the given facts in the original problem.
2. *Construction.* The result of this step is the drawing itself, made with straightedge and compass and showing the construction marks.
3. *Proof.* It is necessary to prove that the figure constructed is actually the required figure.
4. *Discussion.* The number of possible solutions and the conditions for any possible solution are explained in this step.

In this text, it will not be necessary to carry through all four steps in complete detail, although each will be illustrated.

▲ EXERCISES 5.1 ▲

For Exercises 1 and 2, explain how a collapsing compass can be used to:

1. Bisect an angle
2. Transfer a segment
3. Complete the proof of Theorem 5.1.

For Exercises 4–8, carry out the construction indicated.

4. Partition a given segment into seven congruent segments.
5. Partition a given segment internally and externally in the ratio of 3 to 2, given a unit segment.
6. Carry out the same construction as in Exercise 5, but use a ratio of 3 to 4.
7. Carry out the same construction as in Exercise 5, but use a ratio of 5 to 7.
8. Partition a given segment internally and externally in the ratio of length of two given arbitrary segments, neither of which is the unit segment.

For Exercises 9–18, begin with a triangle and locate the required point or set of points by construction, using the basic Euclidean constructions as needed.

9. The circumcenter
10. The orthocenter
11. The incenter
12. The centroid

13. An excenter

14. The Gergonne point

15. The nine-point circle

16. The Euler line

17. The symmedian point

18. The two Brocard points

19. Investigate how the Greeks solved quadratic equations by means of geometric constructions.

20. Show how additional constructions can be made with a collapsing compass.

5.2 Constructible Numbers

A unit segment represents the number one. What other numbers can also be represented by segments, beginning with this unit segment and using only the straightedge and compass to construct other segments? The answer to this question defines what is known as the set of *constructible numbers*. Figure 5.6 shows a geometric interpretation of the four rational operations on whole numbers, as well as the construction of some irrational numbers by the process of extracting the square root of a positive rational number. The given information consists of the three segments in Figure 5.6a.

The diagrams in Figure 5.6b and c for addition and subtraction should be self-explanatory. For multiplication, the proof of the construction in Figure 5.6d depends on the proportion

$$\frac{1}{b} = \frac{a}{AE}$$

so that $AE = ab$.

For division, the proof of the construction in Figure 5.6e also depends on a proportion,

$$\frac{1}{AD} = \frac{b}{a} \quad \text{or} \quad AD = \frac{a}{b}$$

The proof that BD is equal to $\sqrt{a}$ in Figure 5.6f likewise depends on a proportion that, in turn, is derived from similar right triangles in the figure.

$$\triangle ADB \sim \triangle DCB$$

so that

$$\frac{AB}{BD} = \frac{BD}{BC}$$
$$AB = (BD)^2$$
$$\sqrt{AB} = BD$$

The first four constructions of addition, subtraction, multiplication, and division make it possible to construct a segment representing any number in the field of rational numbers, given the unit segment.

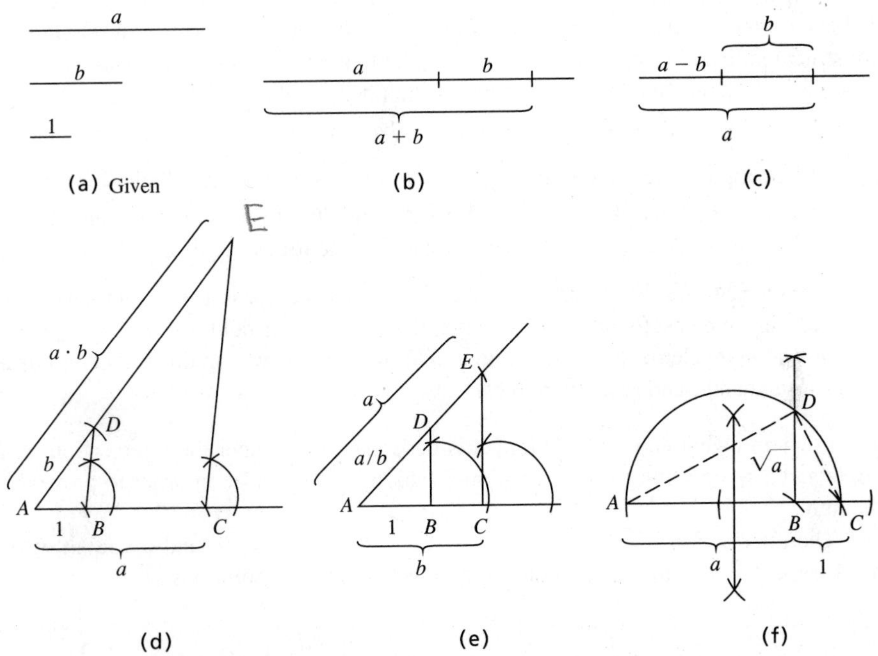

Figure 5.6

Since lengths are considered positive, directed segments with a negative sign in front of the length may be used to represent negative numbers.

Suppose that all the numbers in some ordered number field F can be constructed. The next two theorems show how the straightedge and compass are involved in determining which members are constructible.

THEOREM 5.2 ➤ The use of a straightedge alone can never yield segments for numbers outside the original number field.

Proof: The equations for any two pairs of lines through distinct pairs of points with coordinates (a, b), (c, d), (e, f), (g, h) in a field are

$$y - b = \frac{d-b}{c-a}(x-a) \qquad \text{and} \qquad y - f = \frac{h-f}{g-e}(x-e)$$

The point of intersection of these two lines has coordinates obtained by rational operations on elements of the field F; hence, the use of the straightedge alone did not result in a number outside the original field. It is naturally assumed that division by zero is avoided.

Now select an element a of F such that $\sqrt{a}$ is not an element of F. All numbers of the type $b+c\sqrt{a}$ (b and c also in F) can be constructed with the use of the straightedge and compass. Numbers of the form $b+c\sqrt{a}$ themselves constitute a field, and this is called an extension field of F. The construction of the square root makes it possible to construct numbers in *extension fields* having the field of rational numbers as a subset. Examples of extension fields of constructible numbers are the sets of numbers $a+b\sqrt{2}$, $a+b\sqrt{3}$, $\sqrt{a+b\sqrt{2}}$, for a and b rational numbers and the entire radicand positive in the third example. The number $5+\sqrt{3+\sqrt{7}+\sqrt{2}}$ is a constructible number, while $\sqrt[3]{7}$ is not. (See Exercise 18, Exercises 5.4.) A single use of the compass cannot lead from F beyond an extension field of F, as stated in the following theorem.

THEOREM 5.3 ➤ A single application of the compass using numbers of a number field results only in elements of the extension field $l+m\sqrt{n}$, where l, m, and n are elements of the original field, with n positive. In this theorem, n is a constant and l and m are variables.

From an algebraic point of view, it is necessary to consider the intersection of a circle and a straight line and then the intersection of two circles in order to prove the theorem.

The intersection of a circle, $x^2+y^2+ax+by+c=0$, and a straight line, $dx+ey+f=0$, with all coefficients in F, is given by the solutions of

$$x^2+\left(\frac{-f-dx}{e}\right)^2+ax+b\left(\frac{-f-dx}{e}\right)+c=0$$

This equation may be written in the form

$$\left(1-\left[\frac{d}{e}\right]^2\right)x^2+\left(\frac{2df-bd}{e}+a\right)x+\left(\frac{f^2-bf}{e}+c\right)=0$$

Each of the coefficients of this equation is an element of F, so they may be indicated as g, h, k, so that $gx^2+hx+k=0$. The quadratic formula yields the solutions

$$x=\frac{-h\pm\sqrt{h^2-4gk}}{2g} \qquad x \text{ has real values for } h^2-4gk\geq 0$$

Both of these solutions are of the form $l+m\sqrt{n}$ for certain numbers in F, so that the use of the compass did not lead outside the extension field. The proof of the second part of the theorem involving the intersection of two circles is left as an exercise. As a result of this analysis, constructible numbers can be characterized as those that can be obtained through a sequence of extension fields of the type discussed. For example, the number $5+\sqrt{3+\sqrt{7+\sqrt{2}}}$ can be constructed because it can be obtained by the sequence of extension fields $7+\sqrt{2}$, $3+\sqrt{7+\sqrt{2}}$, and $5+\sqrt{3+\sqrt{7+\sqrt{2}}}$.

In the next section, construction problems are used as a vehicle to learn more about the use of construction instruments and geometric concepts from Chapter 4. An alternate suggestion for some readers is to study the theory of constructions and impossibility proofs from Section 5.4, then to go back to the applied problems of Section 5.3.

▲ EXERCISES 5.2 ▲

For Exercises 1–8, tell whether the number is constructible or not.

1. $3/4$
2. $\sqrt[8]{3}$
3. $3+4\sqrt{7}$
4. π
5. $\sqrt{7/5}$
6. $\sqrt{2+\sqrt{3}+\sqrt{5}}$
7. 8,759
8. $\sqrt[3]{5+\sqrt[3]{7}}$

Take a given unit segment and two other given segments to perform the constructions in Exercises 9–13.

9. Find a segment whose measure is the sum of the measures of the two given segments.

10. Find a segment whose measure is the difference of the two given segments.

11. Find a segment whose measure is the product of the measures of the two given segments.

12. Find a segment whose measure is the quotient of the measures of the two given segments.

13. Find a segment whose measure is the square root of the measure of the longer segment.

For Exercises 14–18, with a given unit segment, construct segments for each of these numbers in the set of constructible numbers. Use the results of one exercise for the next.

14. 3
15. $\sqrt{3}$
16. $2+\sqrt{3}$
17. $\sqrt{2+\sqrt{3}}$
18. $\sqrt{3+\sqrt{2+\sqrt{3}}}$

19. Construct a segment for $\sqrt{2/3}$, given a unit segment.

20. In Theorem 5.2, find the point of intersection of the two lines.

21. In the proof of Theorem 5.2, show that the set of numbers of the form $b+c\sqrt{a}$ is closed under the four rational operations.

22. Complete the proof of Theorem 5.3 for two intersecting circles.

23. Construct a golden rectangle (see Section 4.6).

▷ 24. Investigate the property of closure under various operations for the set of constructible numbers.

▷ 25. Study portions of Charles R. Hadlock's *Field Theory and Its Classical Problems* (Washington, D.C.: Mathematical Association of America, 1978).

▲ 5.3 Constructions in Advanced Euclidean Geometry

The reader of this section should get some idea of the variety of construction problems and should also appreciate the use of constructions in the application of concepts already studied in previous chapters. Developing skill in using constructions is not the desired outcome. Instead, emphasis is on using the analysis portion of a construction problem as an application of concepts from Chapter 4. Remember that constructions provide an interesting vehicle for extending the understanding of concepts from advanced Euclidean geometry.

Example ➤ Construct a triangle, given the length of one side of the triangle and the lengths of the altitude and the median to that side.

The analysis figure is shown in Figure 5.7a. From the given information, right triangle $AA'D$ can be constructed immediately since two sides are known. Then points B and C can be located on $\overleftrightarrow{DA'}$. Each is a distance half the measure of the given side from the determined midpoint A'. The actual construction is shown in Figure 5.7b. Triangle $AA'D$, and hence the required triangle, can always be constructed as long as the segment for the median is at least as long as the segment for the altitude. There is only one possible solution. Triangle $AA'D$ is an example of an *auxiliary triangle* in a construction. An auxiliary triangle, usually a right triangle, is one that can be constructed immediately from the given information.

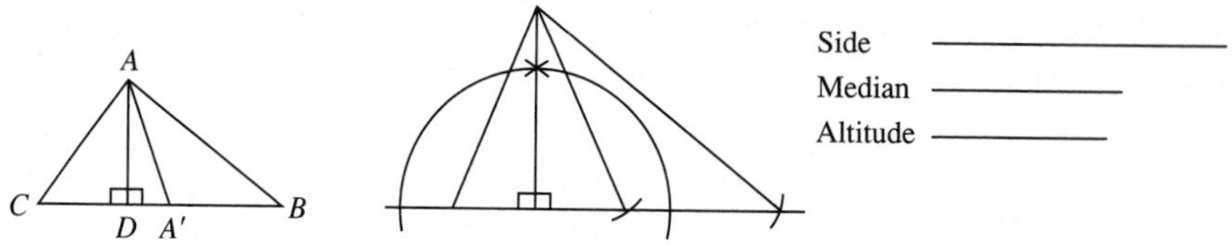

(a) (b)

Figure 5.7

The second example illustrates the important idea of locating one of the vertices of the required triangle as the intersection of two sets of points satisfying given conditions.

Traditionally, this point is described as the intersection of two *loci*, a word meaning "places."

Example ➤ *Construct a triangle, given one angle, the length of the side opposite this angle, and the length of the altitude to that side.* An analysis figure is shown in Figure 5.8a. Assume that $\angle BAC$, AD, and BC are given. Since B and C can be located from the given information, the only remaining problem is to locate point A relative to B and C. The other conditions do not determine A individually, but they can be used together. One condition for the location of A is that it must be on a line parallel to $\overline{BC}$ and at a distance of AD from the opposite side. Vertex A must also lie on the circumcircle, and that can be found from knowing the angle and the opposite side. This *auxiliary construction* is shown in Figure 5.8b. The two conditions determine A, and the triangle can be constructed.

The auxiliary problem in this example is interesting in its own right. In Figure 5.8b, the angle FCE is the given angle, so the center of the circle can be located at the intersection of the perpendicular bisector of $\overline{BC}$ and the perpendicular to $\overline{CE}$ at C.

The discussion of the problem represented by Figure 5.8 consists of determining the number of solutions. There could be two lines parallel to $\overline{BC}$, and each of these might intersect the circle twice. In one case, however, the angle would be the supplement of the given angle, so that case must be discarded. This leaves the possibility for two, one, or no solutions, depending on how many times the line intersects the circumcircle.

The three elements consisting of the measure of one angle of a triangle, the length of the opposite side, and the radius of the circumcircle constitute an example of a *datum*.

DEFINITION ➤ A datum is a set of n elements, any $n - 1$ of which determine the remaining one.

Figure 5.8b shows only one part of the proof that the three elements constitute a datum, since it must be shown that *each two* determine the third.

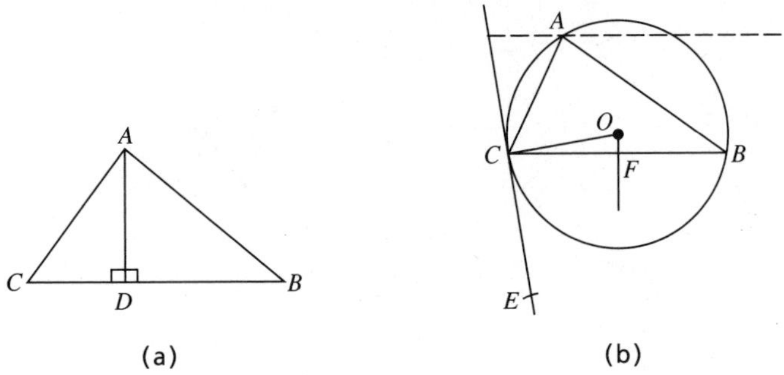

Figure 5.8

Often, the concept of similar triangles can be used in constructions. This is illustrated in the problem of *constructing a triangle, given the measure of two angles and the length of the bisector of the third angle*. Figure 5.9 is the analysis figure. The measures of the three angles of a triangle constitute a datum. From this information, a triangle $AB'C'$, similar to the required triangle, can be constructed. This idea is known as the triangle being determined in *species*. A family of triangles is determined, and the required triangle can be found as one particular member of this family. The required triangle is found by laying off the given length of the bisector along $\overline{AD'}$ to locate point D, and then constructing the parallel to $\overline{C'B'}$ through D. There is always one solution as long as the sum of the measures of the two given angles is less than π.

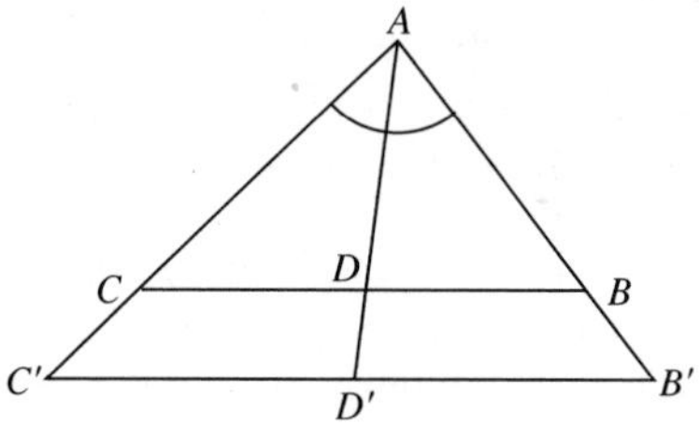

Figure 5.9

Construction of triangles can involve concepts from Euclidean geometry that are more advanced than those used so far in this section. One example of this is the *construction of a triangle, given the circumcenter, the center of the nine-point circle, and the midpoint of one side*.

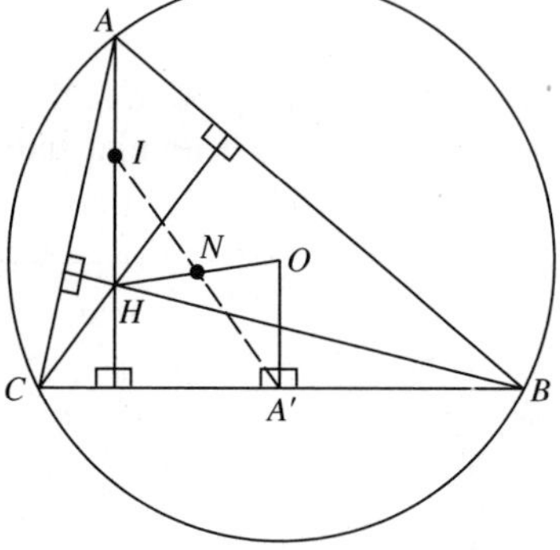

Figure 5.10

The analysis figure is shown in Figure 5.10. The location of H, the orthocenter, is determined from the fact that the nine-point center, N, is the midpoint of the segment joining the orthocenter and the circumcenter O. Since N is also the midpoint of $\overline{A'I}$, where I is the midpoint of $\overline{HA}$ and A' is the midpoint of the side, point I can be determined, then vertex A can be found a known distance on $\overleftrightarrow{HI}$ beyond I. One locus for points B and C is a line through A' perpendicular to $\overline{OA'}$. The second locus is the circumcircle, and this can be determined because the radius OA has been found.

A second example of a construction using a more advanced concept is the *construction of a triangle, given the lengths of the altitude, median, and symmedian from the same vertex.* The analysis figure is shown in Figure 5.11. $\overline{AD}$ is the altitude, $\overline{AE}$ is the symmedian, and $\overline{AA'}$ is the median. The analysis and discussion are left as an exercise.

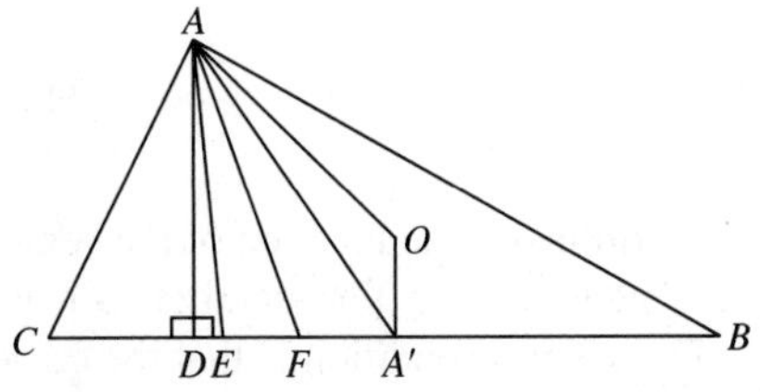

Figure 5.11

The triangle construction exercises at the end of this section apply the ideas given in the examples. The first step is to sketch an analysis figure; the second is to study it and attempt to find a plan for determining the missing data needed to complete the triangle. Some possible questions to ask in reference to the analysis figure are:

1. Is a portion of the figure (a right triangle or other auxiliary triangle) already determined from the given data?
2. Is it possible to determine the third vertex of the required triangle as the intersection of two given loci?
3. Do two of the given elements constitute a datum that gives an additional element?
4. Can the required triangle be constructed in species from the given elements?
5. If the given data involves special points from advanced Euclidean geometry (such as the center of the nine-point circle), what do we know about the location of that point in relation to the vertices of the required triangle or in relation to other given elements?

Construction problems do not always involve constructing a triangle as the completed figure. For example, suppose it is required to *construct a rectangle, given the lengths of one side and one diagonal.* The analysis figure, Figure 5.12, shows that each of the two right triangles BCD and ABD can be constructed from the given information.

A final example is that of *constructing a circle with a known radius that is tangent to a given line and orthogonal to a given circle.* Two curves are *orthogonal* if they meet at right angles. The analysis figure is shown in Figure 5.13. The circle with center O is the required circle, l is the given tangent, and the circle with center O' is the given circle.

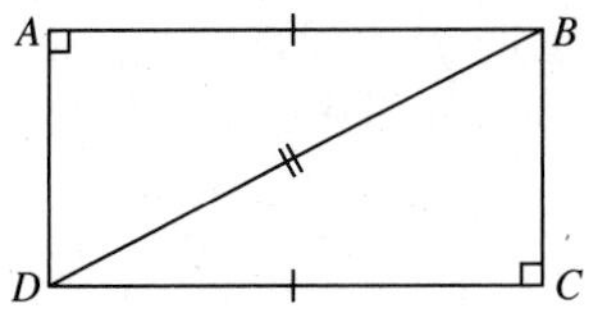

Figure 5.12

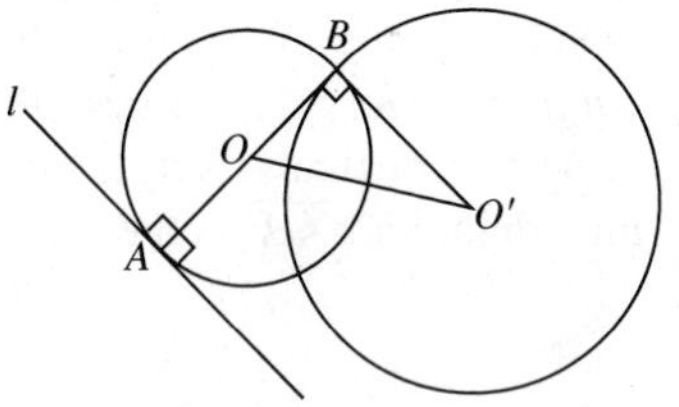

Figure 5.13

The only thing to be determined is the location of the center of the required circle. One condition is that the center lie on a line parallel to l at a given perpendicular distance AO from line l. The second condition is that the center lie a known distance of $OO' = \sqrt{(OB)^2 + (BO')^2}$ from O' so that its position can be fixed. Because the location of O depends on the intersection of a circle with two parallel lines, a complete discussion of the problem would have to consider from four to zero possible solutions.

▲ EXERCISES 5.3 ▲

1. Finish the proof that the measure of one angle of a triangle, the measure of the opposite side, and the radius of the circumcircle constitute a datum.
2. Draw the actual construction figure corresponding to Figure 5.10.
3. Complete the discussion of the number of solutions in Figure 5.13.

For each of the following constructions:

a. Sketch an analysis figure.
b. Give the analysis.
c. Briefly discuss the number of possible solutions.

4. Construct a triangle, given the measure of one angle, the length of an adjacent side, and the length of the altitude to that side.
5. Construct a triangle, given the measure of one angle, the length of that internal angle bisector, and the length of one adjacent side.
6. Construct a triangle, given the length of one side and the lengths of the medians to the other two sides.
7. Construct a triangle, given the length of one side, the length of the altitude to a second side, and the circumradius.
8. Construct a triangle, given the length of one side, the length of the median to that side, and the circumradius.

9. Construct a circle with a given radius and tangent to two given intersecting lines.

10. Construct a circle with a given radius tangent to a given line and tangent to a given circle.

11. Construct a triangle given the length of one side, the length of the median to that side, and the ratio of the lengths of the two remaining sides.

12. Construct a triangle, given the measure of one angle, the measure of the opposite side, and the radius of the incircle.

13. Construct a triangle, given the measure of one angle, the length of the internal bisector of that angle, and the radius of the incircle.

14. Construct a triangle, given the measure of one angle and the length of the altitudes to the two adjacent sides.

15. Construct a triangle, given the length of one side, the length of the median to that side, and the length of one other median.

16. Construct a triangle, given the measure of two angles and the length of the median to the third side.

17. Construct a triangle, given the measure of one angle, the length of the altitude to the opposite side, and the ratio of the two adjacent sides.

18. Construct a triangle, given the orthocenter, the nine-point center, and the foot of one altitude.

19. Construct a triangle, given the lengths of the altitude, median, and symmedian from one vertex.

▷ **20.** Make up and solve three additional triangle construction problems using concepts from Chapter 4.

▷ **21.** Solve additional construction problems found in George Polya, *Mathematical Discovery*, Chapter 1.

▷ **22.** The solution of some construction problems can be facilitated by the use of transformations. Study examples in Eves, *A Survey of Geometry*, then work several problems.

▲ 5.4 Constructions and Impossibility Proofs

Some problems could not be solved by the Greeks using the instruments specified by Plato. Rather than give up, however, they invented new instruments that would make the constructions possible. Not until the nineteenth century, though, did advances in the algebra of the real number system make it possible to identify constructions that were impossible with the straightedge and compass.

In the history of mathematics, three construction problems became so famous that they are called the "Three Famous Greek Problems." These are the problems of *doubling the cube*, of *trisecting the angle*, and of *squaring the circle*. The algebra of the constructible numbers has provided the information necessary to prove that each construction is impossible.

Various legendary stories account for the origin of the problem of doubling the cube. One states that a king wanted to double the size of the cubical tomb of his son. Another tells that the Delians were instructed by their oracle to double the size of the altar erected to Apollo in order to rid the city of a plague. In these problems, it is the volume of the cube, not the edge, that is to be doubled (see Figure 5.14).

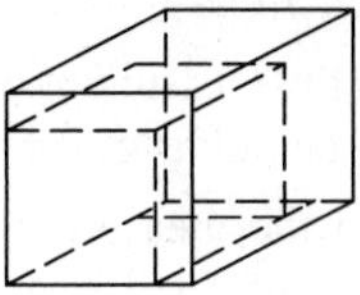

Figure 5.14

If the original edge is 1 unit and the required edge is x units, then the original volume is 1 cubic unit and the volume of the doubled cube is 2 units. This means that x must be a real solution of the equation $x^3 = 2$. A constructible number must be an element of some extension field of the set of rational numbers. The assumption that the solution of $x^3 = 2$ lies in one of these extension fields leads to a contradiction. It can be proved, although the algebraic details are omitted here, that if a number of the form $b + c\sqrt{a}$ is a solution of a cubic equation, then so is $b - c\sqrt{a}$. It has previously been shown that elements of extension fields can always be written in the form $b + c\sqrt{a}$. But the assumption of two real roots for $x^3 = 2$ contradicts the known fact that two of the cube roots of 2 are nonreal. Thus, x is not a constructible number.

This completes the proof of the impossibility of solving the first of the Greek problems.

THEOREM 5.4 ➤ The construction of doubling the volume of a cube cannot be performed by straightedge and compass alone.

The second problem, that of trisecting a general angle, can be disposed of in a way similar to the first. The problem is illustrated in Figure 5.15. If angle AOB has a measurement of c degrees, then the problem is to construct angle AOC with a measurement of $c/3$ degrees.

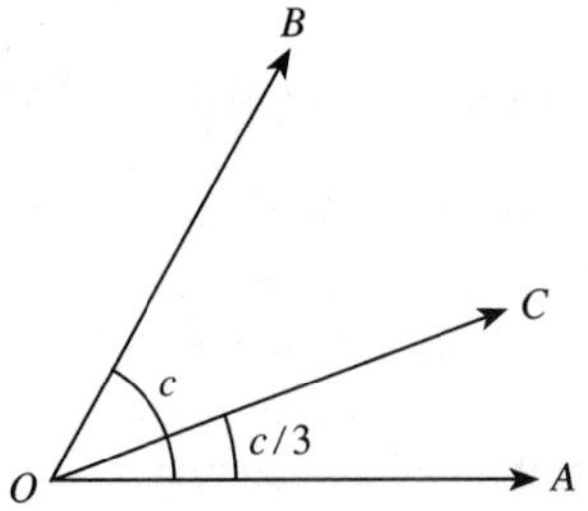

Figure 5.15

Ordinarily, this problem is approached by showing that a single example, such as trisecting an angle of 60 degrees, is impossible, and hence, a general case is impossible. In Figure 5.16, suppose that $m\ \angle BOA = 60$ and $m\ \angle COA = 20$. If the circle is a unit circle, then $OD = \cos \angle BOA$ and $OE = \cos \angle COA$. The problem of constructing the smaller angle is equivalent to that of finding OE, given OD. The trigonometric identity relating the cosines of an angle and a second angle with one-third the measure of the first angle is

$$\cos\theta = 4\cos^3\left(\frac{\theta}{3}\right) - 3\cos\left(\frac{\theta}{3}\right)$$

If $\cos(\theta/3) = x$, and $\cos\theta$ is $\frac{1}{2}$(since $\theta = 60°$), then

$$4x^3 - 3x - \frac{1}{2} = 0 \qquad \text{or} \qquad 8x^3 - 6x - 1 = 0$$

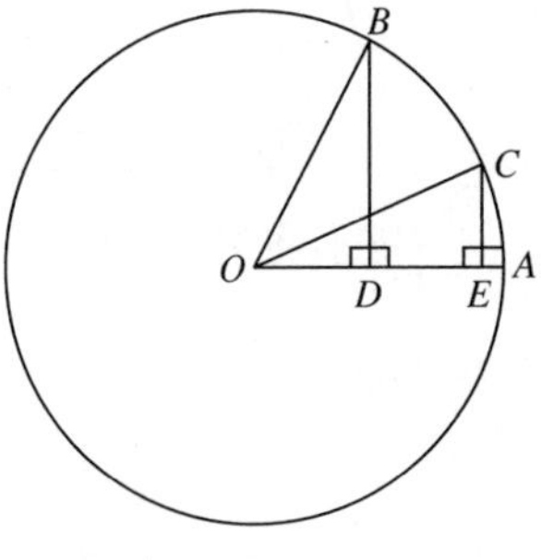

Figure 5.16

This equation has no rational solutions, because if a/b is a rational solution, then a is a positive or negative factor of 1, and b is a positive or negative factor of 8; all possibilities can be checked to see that none is a solution.

If the equation has a solution of the form $a+b\sqrt{c}$, it has $a-b\sqrt{c}$ as another solution. Suppose these represent the least inclusive extension field of which the solution is a member. The sum of the three roots must be zero, the coefficient of the x^2 term, so that if r is the third solution, (assumed from the relationship between roots and coefficients of an equation once studied in theory of equations)

$$\begin{aligned}(a+b\sqrt{c}) + (a-b\sqrt{c}) + r &= 0\\ 2a + r &= 0\\ r &= -2a\end{aligned}$$

This contradicts the assumption that $a+b\sqrt{c}$ represented a number in the least inclusive extension field of solutions. The conclusion is that the solutions of $8x^3 - 6x - 1 = 0$ are not constructible numbers, so the following theorem is established.

THEOREM 5.5 ➤ The construction of trisecting every angle cannot be performed by use of the straightedge and compass alone.

The third and most complex of the three famous Greek problems is that of squaring a circle. This means finding the length of one edge of a square that has the same measure of area as a circle whose radius is known. Algebraically, if x is the required length of the edge and if the length of the known radius is 1, then $x^2 = \pi$. The length $\sqrt{\pi}$ is the required measure of the edge. The problem is illustrated in Figure 5.17, in which the two figures have the same area. Proving that this number cannot be constructed depends on the two statements following the definition, both of which are true but not proved here.

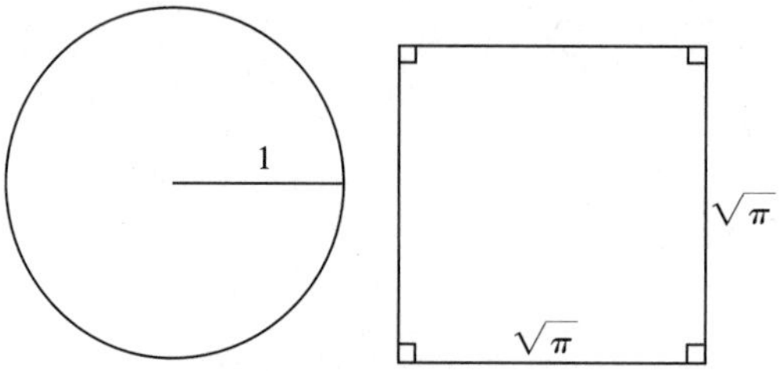

Figure 5.17

DEFINITION ➤ An algebraic number is a solution of an algebraic equation of the form

$$a_n x^n + a_{n-1} x^{n-1} + \cdots + a_0 = 0$$

with integral coefficients, and with $n \geq 1$ and $a_n \neq 0$.

1. All constructible numbers are algebraic.
2. The number π is not an algebraic number, so $\sqrt{\pi}$ is not algebraic either (J. Lindemann, 1882).

THEOREM 5.6 ➤ The construction of squaring a circle is impossible by means of a straightedge and compass alone.

As mentioned earlier, the Greeks could not perform these three constructions by the use of the straightedge and compass. But that did not keep them from finding solutions by other means. Many advances in mathematics, including much of the theory of conics, possibly arose as the result of attempts to provide solutions for construction problems. Two examples showing how the constructions can actually be performed are taken from the history of mathematics. One of these is the use of a device or curve called the *conchoid of Nicomedes*, named for the mathematician Nicomedes, who lived about 240 B.C. The actual method of trisecting an angle by using the conchoid is illustrated in Figure 5.18a. The curve through G is the conchoid. For a fixed line $\overleftrightarrow{CD}$ and a fixed point O not on $\overleftrightarrow{CD}$, the conchoid (which really consists of two branches, one on each side of $\overleftrightarrow{CD}$) is the set of points defined as follows: Consider the set of all lines through O and intersecting $\overleftrightarrow{CD}$. Take a fixed distance on the line beyond the point of intersection. For example, let $\overleftrightarrow{OB}$ be one of the lines, with EX the fixed distance. For a given point, line, and distance, the set of all points X the fixed distance from $\overleftrightarrow{CD}$ along the rays from O is a branch of a conchoid.

If the curve (or rather the mechanical instrument) is placed as in Figure 5.18, with $\overleftrightarrow{CD}$ perpendicular to $\overleftrightarrow{OA}$ and with EO equal to half the distance XE, then the given angle AOB can be trisected simply by locating point E on the line on $\overrightarrow{OB}$, drawing $\overline{EG}$ parallel to $\overleftrightarrow{OA}$, and connecting O to G, the point on the conchoid. $\overrightarrow{OG}$ is the trisector of the given angle.

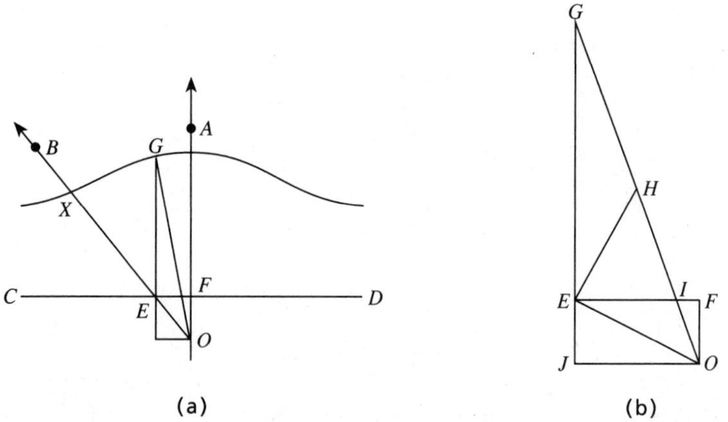

Figure 5.18

The reason why $\overrightarrow{OG}$ trisects the angle depends on the theory explained in connection with Figure 5.18b. $GI = 2EO$ by assumption. If H is the midpoint of $\overline{GI}$, then $\overline{EO}$, $\overline{HI}$, $\overline{GH}$, and $\overline{EH}$ are all congruent. Because of the isosceles triangles, $m\angle EOH = m\angle EHO = m\angle HGE + m\angle HEG = 2m\angle HGE$. But $\angle FOI \cong \angle HGE$ because of parallel lines, so $2\,m\angle HGE = 2m\angle FOI$. The result is that $\angle EOG$ has a measure twice that of $\angle FOI$, so $\overrightarrow{OI}$ is the trisector.

A second example of a curve from higher geometry used to construct a solution to one of the famous problems is the use of the *spiral of Archimedes* to solve the problem of squaring the circle. Archimedes (287–212 B.C.) is considered the greatest mathematician of antiquity. Besides his book *On Spirals*, which contains 28 propositions on the curve we call the spiral of Archimedes, Archimedes also wrote on many other mathematical topics including such geometric topics as computing π, finding the area of plane regions in a way that anticipated integral calculus, and surfaces in three dimensions. The spiral of Archimedes is the curve OCD shown in Figure 5.19, with polar equation $r = k\theta$, for k a given constant. If a circle has a radius k as shown, then the arc for right angle AOB has length $k\theta$, which is also the length of $r = OC$. The area of the circle is πk^2, which can be written as $(2k)(\pi k/2)$. But $\widehat{AB} = \pi k/2$, so that

$$\pi k^2 = 2k\left(\frac{\pi k}{2}\right)$$

If OA and OB are perpendicular, then

$$(2k)\left(\frac{\pi k}{2}\right) = 2k(OC)$$

If a square of side x is to have the same size as the circle, then

$$x^2 = 2k(OC)$$

and x can be constructed.

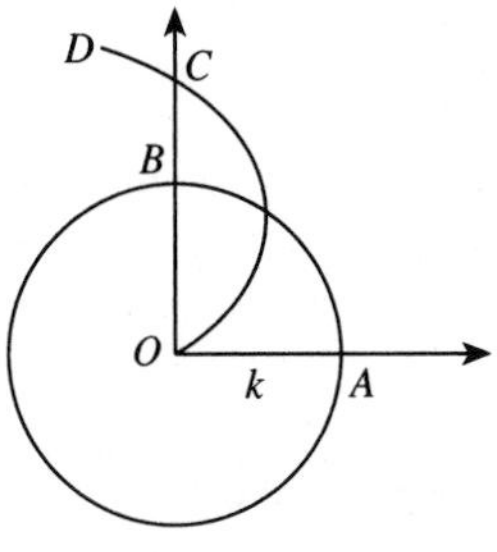

Figure 5.19

In addition to the three famous problems, the equally interesting problem of constructing a regular polygon inscribed in a given circle has concerned mathematicians since the time of the Greeks. In this case, the proof for the general problem was again algebraic and was provided by the great mathematician Karl Gauss. At the age of 18, he solved the previously unsolved problem of how to inscribe a regular polygon of 17 sides in a circle, using only a straightedge and compass. He also proved the theorem that tells which regular polygons can and cannot be inscribed. This theorem is stated here without proof.

THEOREM 5.7 ➤ A regular polygon can be inscribed in a circle by means of a straightedge and compass alone if and only if the number of sides, n, can be expressed as $2^x \cdot p_1 \cdot p_2 \cdot \cdots \cdot p_k$, for x a nonnegative integer and each p_i a distinct prime of the form $2^{2^y} + 1$, for $y \geq 0$.

Some of the regular polygons that are constructible, according to the theorem of Gauss, are those with 3, 4, 5, 6, 8, 10, 12, 15, 16, 17, 20, or 24 sides. The statement of the theorem gives no clue as to how to prove the possibility or impossibility in any specific case without reference to the general theorem. The specific cases for $n = 10$ and $n = 7$ are discussed briefly.

Figure 5.20 shows the analysis for the regular decagon. The central angle AOB has a measurement of 36°, and angles OAB and ABO each have measurements of 72°. If $\overline{BC}$ is the bisector of $\angle ABO$, then triangles ABC and BCO are both isosceles, so $\overline{AB} \cong \overline{BC} \cong \overline{CO}$. Also,

$$\triangle ABO \sim \triangle ACB$$

and

$$\frac{1}{x} = \frac{x}{1-x}$$

or $x^2 + x - 1 = 0$. The positive solution is

$$x = \frac{\sqrt{5}-1}{2}$$

which is a constructible number.

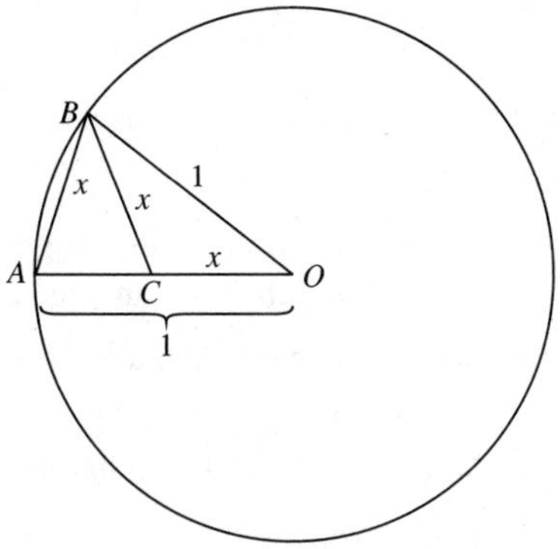

Figure 5.20

As Theorem 5.7 indicates, the regular decagon can be constructed in a circle, whereas the regular heptagon (seven sides) cannot. The proof that it is impossible to construct a heptagon in a circle is similar to the proof concerning the trisection of a general angle. The problem is equivalent to constructing a length $x = 2\cos 2\pi/7$. If $2\pi/7 = \theta$, then

$$3\theta + 4\theta = 360° \qquad \text{and} \qquad \cos 3\theta = \cos 4\theta$$

Trigonometric identities can be used to express these cosines in terms of x.

$$\begin{aligned} 2\cos 3\theta &= 2(4\cos^3\theta - 3\cos\theta) = x^3 - 3x \\ 2\cos 4\theta &= 2(2\cos^2 2\theta - 1) = 4(2\cos^2\theta - 1)^2 - 2 \\ &= (x^2 - 2)^2 - 2 \end{aligned}$$

Setting the two expressions equal yields a quartic equation in x:

$$x^4 - x^3 - 4x^2 + 3x + 2 = 0$$

which can be factored as $(x-2)(x^3+x^2-2x-1)=0$. It is left as an exercise to show why these factors do not give values for x that can all be constructed with straightedge and compass.

▲ EXERCISES 5.4 ▲

1. Which of the three famous Greek problems can be solved using compass and straightedge?

For Exercises 2–7, which of the numbers are algebraic numbers?

2. $3/4$
3. $\sqrt{5}$
4. $\log 2.4$
5. $\tan 19^\circ$
6. $\pi/2$
7. 3,748
8. Apply the definition to show that $\sqrt{2+\sqrt{2}}$ is an algebraic number.
9. Is $\sqrt[3]{2}$ an algebraic number? Why?
10. Find an approximate decimal solution for the length of an edge of a cube that has twice the volume of another cube with an edge 2 inches long.
11. Prove that if a number of the form $a+b\sqrt{c}$ is a solution of $x^3+dx^2+ex+f=0$, so is $a-b\sqrt{c}$.
12. Explain why the equation $8x^3-6x-1=0$ has no rational solutions.
13. Explain how a mechanical device might be made that would actually draw one branch of a conchoid.
14. Show that the spiral of Archimedes can also be used for trisecting an angle.
15. Use Theorem 5.7 to list those regular polygons with between 24 and 30 sides that can be constructed in a circle.
16. In the quartic equation for the regular heptagon, explain why the first factor cannot be used to find a solution.
17. In the quartic equation for the regular heptagon, explain why the second factor cannot be used to find a solution.
18. Show that $\sqrt[3]{7}$ is not constructible.
19. Use the theory presented to actually carry out the construction of a regular decagon.
20. ▷ Find and outline the proof of the theorem that all constructible numbers are algebraic.
21. ▷ Find and outline Lindemann's proof that the number π is not an algebraic number.

▷ **22.** Archimedes is given credit for devising one of the simplest methods for trisecting an angle, using a ruler with only two marked points on it. See if you can rediscover his method and prove that it does trisect the angle. Use a reference text if necessary.

▷ **23.** Find and outline the proof of Theorem 5.7.

▷ **24.** Read in a history of mathematics text about additional historical methods used to try to solve the three famous problems.

▷ **25.** Devise a physical model and use it to construct a conchoid.

▲ 5.5 Constructions by Paper Folding and by Use of Computer Software

Since the time of the ancient Greeks, other ways of accomplishing constructions besides using an ordinary compass and straightedge have interested many mathematicians. For example, you found in Section 5.4 that various instruments and equations for curves were used to solve problems that were impossible otherwise. In this section and the next, three additional ways of carrying out constructions will show some of the variety possible in this fascinating area of mathematics. One way of performing constructions that may seem elementary, yet has occupied the attention of mathematicians in recent years, is an approach through *paper folding*. In actual practice, waxed paper is ordinarily used so that the crease will remain visible. A few of the basic constructions are illustrated in this section so that you will catch the spirit of this work and thus be able to contrast it with other construction methods in Euclidean geometry. Paper folding has previously been used as an intuitive method of verifying the proofs of theorems in Chapter 4.

One construction carried out by means of paper folding is to *fold a straight line that is the perpendicular bisector of a given segment.* Suppose $\overline{AB}$ is given, as in Figure 5.21a. Fold the paper so that point A is superimposed on point B and crease, as in Figure 5.21b. $\overleftrightarrow{CD}$ is the required line.

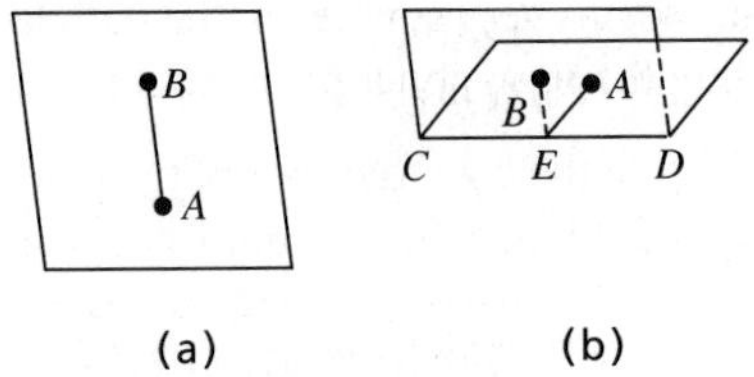

Figure 5.21

Another is to *fold the bisector of a given angle* (Figure 5.22). Fold the paper with the crease through the vertex and with one side of the angle superimposed on the other.

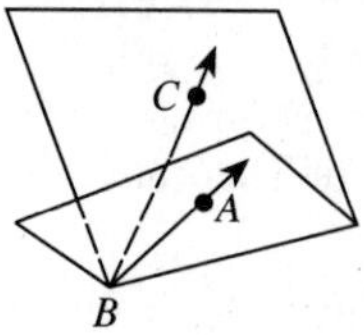

Figure 5.22

A third construction is to *fold a line through a given point and parallel to a given line*. First fold a crease for $\overleftrightarrow{CD}$ perpendicular to the given line $\overleftrightarrow{AB}$ (Figure 5.23), then fold a crease through the given point G for a line $\overleftrightarrow{EF}$ perpendicular to $\overleftrightarrow{CD}$. That line is parallel to $\overleftrightarrow{AB}$ and is the required line.

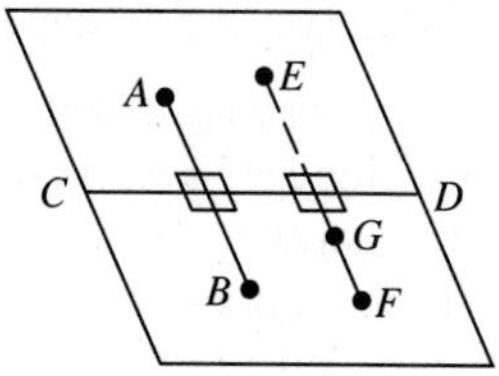

Figure 5.23

Still another is to *fold a perpendicular from a point to a line.* Simply fold a perpendicular to the line through the point by superimposing part of line $\overleftrightarrow{AB}$ on itself (Figure 5.24).

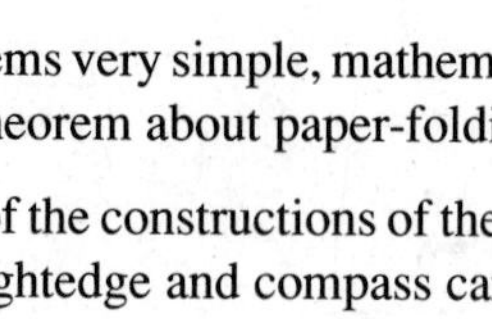

Figure 5.24

Although paper folding seems very simple, mathematicians have been able to prove the following rather startling theorem about paper-folding constructions.

THEOREM 5.8 ➤ All of the constructions of the plane Euclidean geometry that can be performed by straightedge and compass can also be performed by folding and creasing paper.

This remarkable theorem is based on several assumptions that can be found in the excellent reference on paper folding, *Paper Folding for the Mathematics Class*, by

Donovan A. Johnson. These include, for example, the assumption that paper can be folded in such a way that one line can be superimposed on another line on the same sheet of paper and that the crease formed is in fact a straight line. The theory of paper folding is just as mathematical and just as exact as the theory of constructions with straightedge and compass, but the methods differ widely. Also see *Geometric Exercises in Paper Folding*, by T.S. Row.

Paper folding can be used for a somewhat different type of construction not usually studied in ordinary college geometry: constructing tangents to a parabola. Figure 5.25a illustrates how a series of tangents to the parabola can be determined by folding the focus F onto the directrix d. The reason why this construction works is explained in Figure 5.25b. It can be proved that the tangent $\overleftrightarrow{BA}$ at a point on the parabola is the angle bisector of $\angle FBC$, F the focus and C the foot of the perpendicular from B to the directrix.

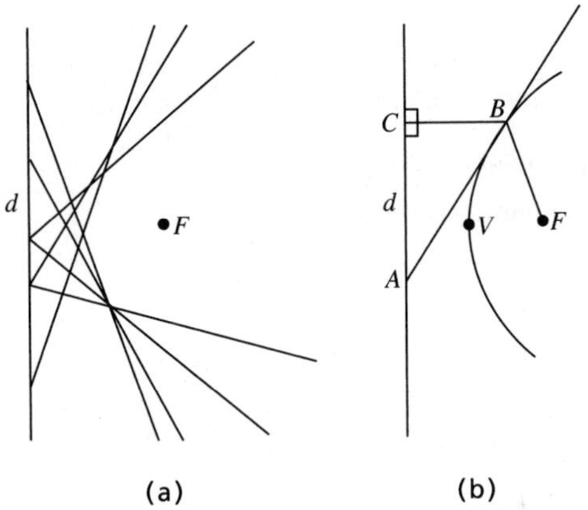

Figure 5.25

A second new way to perform constructions is by using computer software. Various software now available makes it possible to perform all the constructions of plane geometry as well as to use animation features to explore geometric ideas dynamically. The intuitive nature of this use of software, and the fact that it can be used with those with relatively weak geometric backgrounds, make it an effective means to radically change how some geometry is taught. On the other hand, some mathematicians would probably argue that computers technically can't "perform constructions" at all, in the classical sense.

▲ EXERCISES 5.5 ▲

Using waxed paper, perform the following constructions for Exercises 1–9.

1. Fold a straight line that is the perpendicular bisector of a given segment.

2. Construct the bisector of a given angle.
3. Construct a line through a given point and parallel to a given line not through the point.
4. Construct the perpendicular from a point to a line not through the point.
5. Construct ten tangents to a parabola, given the directrix and the focus.
6. Construct the tangent at a point on a circle.
7. Construct the incenter of a given triangle.
8. Construct the circumcenter of a given triangle.
9. Construct the orthocenter of a given triangle.
10. Prove that the paper-folding method of constructing lines tangent to a parabola is valid.

For Exercises 11I–14I, perform the following constructions with waxed paper. Consult the reference in the section, or another reference, if necessary.

▷ 11. Construct an equilateral triangle.

▷ 12. Construct a regular hexagon.

▷ 13. Construct tangents to an ellipse.

▷ 14. Construct tangents to a hyperbola.

▷▷ 15. Study the user's guide for *CABRI, The Interactive Geometry Notebook* (Brooks/Cole Publishing Co.) and use the software to carry out some of the constructions in this chapter.

▲ 5.6 Constructions with Only One Instrument

In Sections 5.1–5.3, constructions with straightedge and compass, the traditional instruments specified by Plato, were investigated. In the study of impossibility proofs in Section 5.4, we also encountered the use of various geometric curves to extend the list of constructions. Section 5.5 considered an alternate approach to constructions by means of paper folding. This section considers one additional important idea from the history of constructions—attempts to perform constructions by using only one of the two instruments specified by Plato.

The first alternative is to limit the instrument to the compass alone. Obviously, it is impossible to draw a straight line with the compass alone, so it must be understood that a line is completely determined if two points on it are found or given. Constructions with the compass alone are called *Mohr–Mascheroni constructions*. C. Mohr published the first known account of these constructions in 1672, although his book was not well known to mathematicians until 1928, when it was rediscovered. Meanwhile, the Italian mathematician Mascheroni (about 1797) had independently discovered the following theorem.

THEOREM 5.9 ➤ All constructions possible by use of the straightedge and compass can also be made by use of the compass alone.

For a discussion of this theorem, see H. Rademacher and O. Toeplitz, *The Enjoyment of Mathematics*.

The following example illustrates a construction performed entirely with the compass.

Example ➤ *Find the midpoint of a given segment by use of a compass alone.*

1. Consider $\overline{AB}$, as in Figure 5.26. Construct the circle with B as center and BA as radius.
2. With AB as radius, mark off three arcs $\overset{\frown}{AC}$, $\overset{\frown}{CD}$, $\overset{\frown}{DA'}$, locating A' so that AA' is a diameter of the circle.
3. Construct the circle with center A and radius AB.
4. Construct the circle with center A' and radius $A'A$, intersecting the circle of step 3 at the points E and F.
5. Construct the circles with centers E and F and radius EA. These meet at A and the required point G.

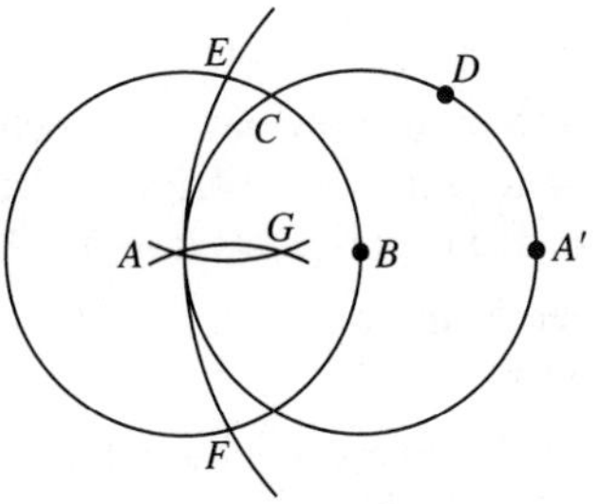

Figure 5.26

The proof that this rather elaborate construction results in the correct point is based on the fact that A' and G are *inverse points* with respect to the circle with center A and radius AB. The geometry of inverse points is discussed in Chapter 6, but the proof used here does not require a knowledge of that concept.

Triangles $A'EA$ and EGA are similar, so

$$\frac{AA'}{AE} = \frac{AE}{AG} \quad \text{or} \quad AA' \cdot AG = (AE)^2$$

$$\text{Because} \quad AA' = 2AB = 2AE$$

$$2AE \cdot AG = (AE)^2$$

$$2AG = AE = AB$$

and G is the midpoint of $\overline{AB}$, as was to be proved.

The proof of Theorem 5.9 consists of showing how to find the points of intersection of a straight line and a circle and the points of intersection of two straight lines with the compass alone, since it is obviously simple enough to draw a circle and to find the points of intersection of two circles with the compass alone. The method of *constructing the points of intersection of a circle and a line not through the center* is shown in Figure 5.27. Let B and C be the given points, with A the center of the given circle.

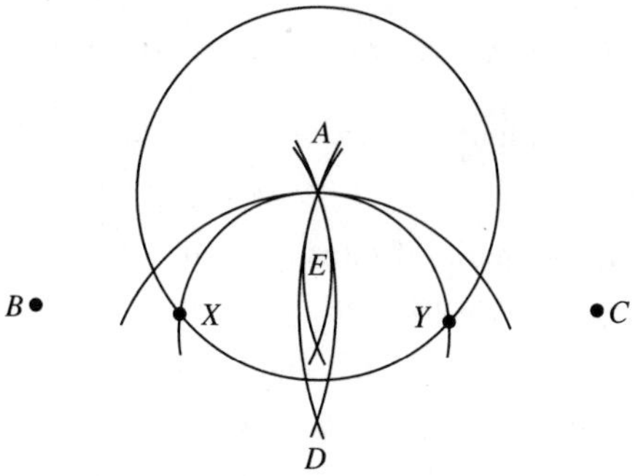

Figure 5.27

1. Construct circles with centers B and C, the given points on the line, passing through A and intersecting again in D.
2. Following steps 4 and 5 of the previous construction, find a point E such that $AD \cdot AE = r^2$, for r the radius of the original circle. (This construction must be modified if D is inside the original circle.)
3. The circle with center E and radius EA intersects the original circle in the required points X and Y.

The proof that X and Y are the points of intersection of the given line and circle is left as Exercise 1 in Exercises 5.6. For a discussion of the case in which the line passes through the center of the circle, see Howard Eves, *A Survey of Geometry*.

Not all the Euclidean constructions can be done with the straightedge alone. The Arabian mathematician Abul-Wefa suggested using the straightedge and *rusty compasses*. The use of the compass with a fixed opening is equivalent to being given one circle and the center. The basic result is stated in what is known as the Poncelet-Steiner construction theorem, which follows.

THEOREM 5.10 ➤ All constructions that can be performed with the straightedge and compass can be performed with the straightedge alone, given a single circle and its center.

The straightedge alone does suffice for making all the constructions in projective geometry, Chapter 7. In that geometry, neither a circle nor length of a segment is invariant, but the property of being a line is an invariant property. Constructions with the straightedge and compass will be encountered again in inversive geometry.

▲ EXERCISES 5.6 ▲

1. Complete the proof of the Mascheroni construction of the point of intersection of a circle and a line not through the center.

With the compass alone, perform the following Mohr–Mascheroni constructions and briefly describe each step. It is assumed that a line is given by two distinct points on the line.

2. Given a circle and a point inside the circle but distinct from the center, find a point collinear with the center and the given point such that the product of the distances of the two points from the center is equal to the square of the radius of the given circle.
3. Find a distinct point C on $\overleftrightarrow{AB}$ such that $AB = BC$, if A and B are given points.
4. Find the midpoint of a given arc of a circle.
5. Construct the perpendicular to a given line at a given point on the line.
6. Construct the parallel to a given line through a point in the plane but not on the line.
7. Given two points on each line, construct the point of intersection of the two lines.

For Exercises 8–13, give the proof of the constructions performed previously in the indicated exercises.

8. Exercise 2
9. Exercise 3
10. Exercise 4
11. Exercise 5
12. Exercise 6
13. Exercise 7

For Exercises 14 and 15, perform the construction with a straightedge and rusty compass (compass with a fixed opening).

14. Construct the midpoint of a segment.
15. Construct the perpendicular to a given line at a given point on the line.
16. ▷ Investigate the use of Mascheroni constructions for other problems than those given for Exercises 2–7.
17. ▷▷ Demonstrate three additional constructions using a straightedge and rusty compass.

▲ CHAPTER 5 REVIEW EXERCISES ▲

1. Using a straightedge and a collapsing compass, construct a perpendicular from a point to a given line.
2. Carry out the construction of partitioning a segment into nine congruent segments.
3. Carry out the construction for partitioning a segment internally and externally in the ratio of 3 to 5.
4. Construct a segment whose length is c divided by d, if c and d have the lengths given. The unit segment is also given:

c ________________ d __________ 1 _____

5. Construct a segment whose length is the square root of the segment given. The unit segment is also given:

a __________________________ 1 _______

6. Given a unit segment, construct a segment for the number $4 + \sqrt{3}$.
7. Given a unit segment, construct a segment for the number $7 + \sqrt{4 + \sqrt{3}}$.
8. What is the purpose of the analysis step in a construction problem?
9. What does it mean to say that a triangle is determined in species?
10. As used in a construction problem, what is an auxiliary triangle?

For Exercises 11 and 12, sketch the analysis figure and write the analysis step for the construction.

11. Construct a triangle, given the length of one side, the length of the internal bisector of the opposite angle, and the ratio of lengths of the other two sides.
12. Construct a triangle, given the length of two medians and the length of the altitude to the third side.
13. Are all constructible numbers also algebraic numbers?
14. Apply the definition to show that $\sqrt{3 + \sqrt{5}}$ is an algebraic number.
15. Is the number $\sqrt[5]{13}$ in an extension field having the rational numbers as a subset?
16. List those regular polygons with from 33 to 40 sides that can be constructed.
17. When using paper folding to construct tangents to a parabola, which point is folded over which line?
18. Use wax paper to construct the Euler line for a given triangle.
19. List the steps in the construction, with compass only, of the perpendicular from a given point to a line given by two points on it.

Additional chapter review exercises may be selected from any exercises not already assigned from previous exercise sets in this chapter.

6 The Transformation of Inversion

6.1 Basic Concepts

The transformation of inversion is one of the topics in modern geometry developed during the first half of the nineteenth century. The invention of the transformation of inversion is sometimes credited to L.J. Magnus (1790–1861), who published his work on the subject in 1831. But prior to this time, François Vieta in the sixteenth century and Robert Simson in the eighteenth century were aware of elements of the theory. Other mathematicians give the major credit to J. Steiner (1796–1863) and Lord Kelvin. Regardless of who deserves credit for the invention of inversion, many mathematicians worked during the 1830s and 1840s to further develop the general theory.

There are several reasons for studying the material in this chapter. These include: to understand a type of transformation different from those of Euclidean geometry, to discover some significant invariant properties under this type of transformation, and to explore several practical applications of the new theory.

The study of the transformation of inversion begins with the definition of inverse points.

DEFINITION ➤ Two points P and Q are *inverse points* with respect to a point O and a given positive real number c if $OP \cdot OQ = c$, and O, P, Q are collinear.

Both lengths are measured in the same direction from O. The circle with center O and radius $\sqrt{c}$, shown in Figure 6.1, is the *circle of inversion*. The point O is called the *center of inversion*. For inversion with respect to a circle, the product of the distances of two inverse points from the center of inversion is equal to the square of the radius of the circle of inversion, called the *radius of inversion*. In this chapter, lengths are always considered positive.

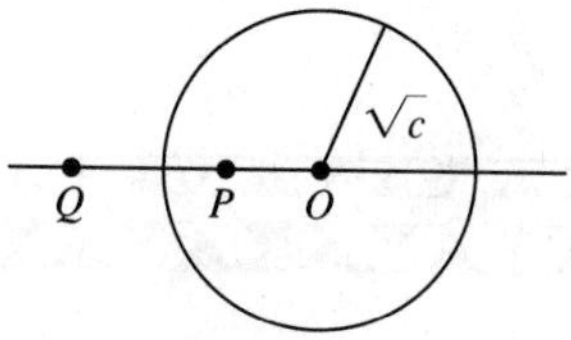

Figure 6.1

In its simplest interpretation, the concept of inversion is numerical in nature, consisting merely of two distances that always have the same product. From this intuitive statement, it is easy to see that if one of the points moves away from the center, the second must move toward it.

Example ➤ Suppose that point P (in Figure 6.1) is at a distance of 2 units from the point O and that the constant product, called the *constant of inversion*, is 9. Then

$$2 \cdot OQ = 9$$

and

$$OQ = \tfrac{9}{2}$$

Example ➤ Suppose that the radius of the circle of inversion is 1 and the distance from the center to point P is 3. Find the distance from the center to the inverse point Q.

$$OP \cdot OQ = 1$$
$$3 \cdot OQ = 1$$
$$OQ = \frac{1}{3}$$

If the radius of inversion is 1, then the numerical values of the distances are reciprocals, and one can be found from the other by inverting the number.

Inversion makes it possible to pair the points on a line in reference to one fixed point on the line with a constant that represents the square of the radius of the circle of inversion. In this pairing of points, two special cases become apparent. A point on the circle of inversion is its own inverse point. In other words, these points are *self-inverse*. The second special case is that no point on the line is the inverse of the center of inversion, since division by zero does not result in a real number. This last statement shows that we need to consider inversion from a more general point of view.

A review of the definition of transformation given in Chapter 2 shows that inversion for the Euclidean plane is not a transformation, since there is an exception to the pairing. But if the *inversive plane* is created by including on each line an inverse point for the center of inversion, then the definition of transformation is satisfied. A line in the inversive plane is formed by including, in addition to the ordinary points, a special point called an *ideal point*, defined to be the image of the center of inversion. With the addition of the ideal point, each point on the line has a unique image under inversion. The line is not an ordinary Euclidean line. The ideal point in the inversive plane is not a concept from Euclidean geometry but a new concept created to guarantee a desired one-to-one correspondence of points.

Because the center of inversion in the inversive plane is to have a unique image in the inversive plane, then each line through the center must have the same inverse point for it, and the new point created must lie on each line through the original point. The inversive plane is illustrated symbolically in Figure 6.2, with O the center of inversion and I (the ideal point on the inversive plane) the inverse of the center on any line through O. The image of the ideal point I is the center of the circle of inversion. It should be understood thoroughly that the inversive plane, because it includes one ideal point that lies on each line, has some properties that differ from the Euclidean plane with which you are familiar. The transformation of inversion in the inversive plane is a new transformation that can be studied in its own right. Before reading ahead, can you discover any invariant properties under inversion?

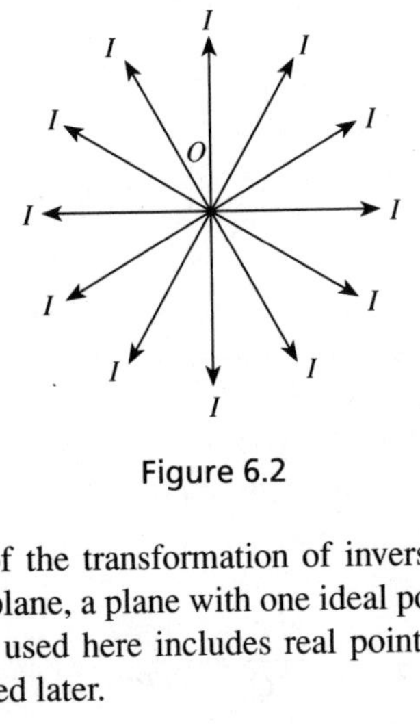

Figure 6.2

The basic properties of the transformation of inversion depend on the peculiar properties of the inversive plane, a plane with one ideal point. It should be understood that the inversive plane as used here includes real points only and not the complex points that will be introduced later.

THEOREM 6.1 ➤ The circle of inversion is an invariant under the transformation of inversion.

PROOF: This theorem follows because each point of the circle of inversion is its own inverse under inversion with respect to that circle. The circle is pointwise invariant, which means that every point is its own image.

THEOREM 6.2 ➤ The inverse of a line through the center of inversion is the same line.

This line is not pointwise invariant because a second point on the line is the image of a given point on the line, with the exception of the points on the circle of inversion. That the inverse points and the center of inversion are collinear is actually an immediate consequence of the definition of inverse points.

A more significant theorem, and possibly one that may not have been expected, concerns the inverse curve for a straight line not passing through the center of inversion in the inversive plane.

THEOREM 6.3 ➤ The image of a straight line not through the center of inversion, under the transformation of inversion, is a circle passing through the center of inversion.

PROOF: In Figure 6.3, let O be the center of inversion and P the foot of the perpendicular from O to the line. Let Q be any other point on the line. Let P and P' and Q and Q' be pairs of inverse points with respect to the circle of inversion with center O.

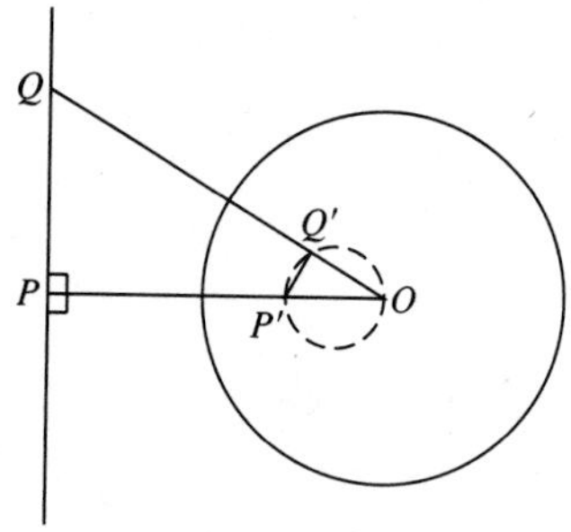

Figure 6.3

By the definition of inverse points,

$$OP \cdot OP' = OQ \cdot OQ'$$

or

$$\frac{OP}{OQ} = \frac{OQ'}{OP'}$$

This proportion implies that $\triangle OPQ \sim \triangle OQ'P'$ because they have an angle in common. Thus $\triangle OQ'P'$ is a right triangle, with the vertex of the right angle at Q'. If Q is considered a point on the given line, then Q' is a vertex of a right triangle inscribed in a semicircle, with $\overline{OP'}$ the diameter of the circle.

The proof of the converse of Theorem 6.3 consists of reversing the steps in the previous argument and is left as an exercise (Exercise 15, Exercises 6.1).

The image of a line under inversion is sometimes a circle rather than a line, so the property of being a straight line is not always an invariant property under the transformation of inversion. This is a major difference between this transformation and the ones we have previously examined. The relationship between circles and lines under inversion can be further extended by investigating the inverse of a circle not passing through the center of inversion. Before reading ahead, you should predict what this image will be.

In Figure 6.4, let O be the center of inversion, with circle O' the given circle. For any line $\overleftrightarrow{OP}$ intersecting the circle in two distinct points and passing through the center of inversion, if P' and Q' are the inverse points for P and Q, then

$$OP \cdot OP' = OQ \cdot OQ' = c$$

which is the constant of inversion. Also, from Theorem 4.9,

$$OP \cdot OQ = (OA)^2 = k$$

a constant equal to the square of the length of the tangent from O to the given circle. Then

$$\frac{OP \cdot OP'}{OQ \cdot OP} = \frac{OQ \cdot OQ'}{OQ \cdot OP} = \frac{c}{OQ \cdot OP}$$

which can be simplified to

$$\frac{OP'}{OQ} = \frac{OQ'}{OP} = \frac{c}{k}$$

which is a constant.

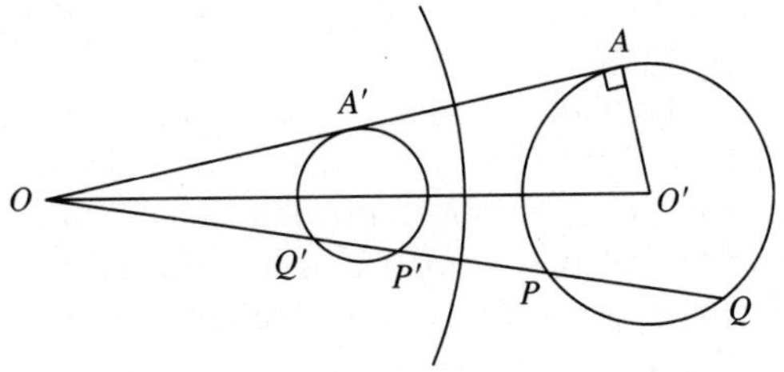

Figure 6.4

The situation reduces to a dilation (see Section 2.8) with center O. This means that, when P is a point on the given circle, the image P' is a point on a circle that is the image of the given circle; thus c/k is the ratio of similarity. This completes the proof of the next theorem.

THEOREM 6.4 ➤ The image under inversion of a circle not passing through the center of inversion is a circle not passing through the center of inversion.

Example ➤ Describe the image of Figure 6.5a under inversion with respect to the circle with center O.

The result is shown as Figure 6.5b. Actually, this figure should be superimposed on Figure 6.5a, but is moved to a different location so it will be easier to examine. Circle O (set 1) is an invariant set of points (labeled $1'$ in Figure 6.5b). Line AB, set 2, is changed into a circle through O, set $2'$. Circle 3 is changed into circle $3'$, and passes through invariant points A and B.

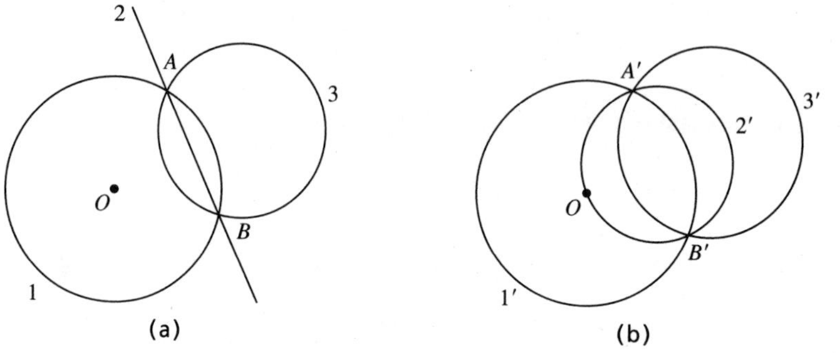

Figure 6.5

In the example, the original figure contains two circles and a line, while the new figure contains three circles. The technique of inverting a figure is one of the skills needed in Section 6.4 in proving new theorems by inversion. For example, the original theorem might refer to two circles and a line, while the new theorem would refer to the three circles of the new figure.

This section concludes with one other important property of the transformation of inversion. Suppose, as in Figure 6.6, that P and P' are inverse points with respect to a given circle with center O. The transformation takes P to its inverse point P'. If the same inversive transformation is applied again, the image of P' is P. The result is that the product of an inversion followed by the same inversion is the identity. The inversion transformation is called an *involutory transformation*, or a *transformation of period two*, because two successive applications result in the identity. It should be observed that an inversion transformation is its own inverse, which provides another good reason for the name given to the correspondence.

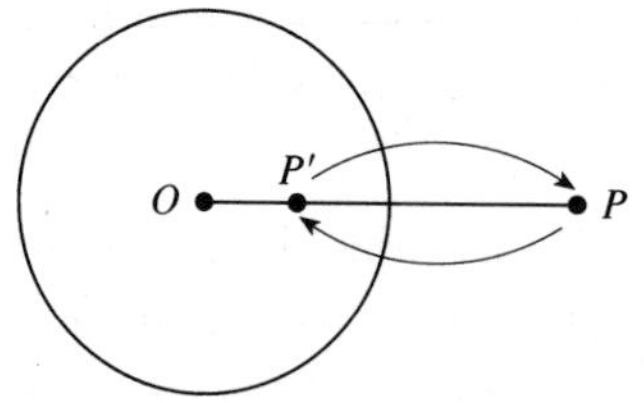

Figure 6.6

▲ EXERCISES 6.1 ▲

Complete the following table:

	Distance from center of inversion to original point	Radius of circle of inversion	Distance of inverse point from center of inversion
1.	3	4	?
2.	3	3	?
3.	3	2	?
4.	3	1	?
5.	$\frac{9}{8}$	$\frac{3}{7}$	?
6.	$\frac{2}{3}$	?	$\frac{5}{6}$
7.	?	$\frac{9}{5}$	$\frac{6}{7}$

8. If P and Q are inverse points with respect to S, and if $SP = m$ and the radius of the circle of inversion is n, find the length SQ.

9. Do each two points in the inversive plane determine a unique line?

10. Describe the location of the inverse of any point inside the circle of inversion.

11. State and prove a theorem about the inverse of the set of all points in the inversive plane outside the circle of inversion.

12. When is the image of a circle under inversion a circle, and when is it not?

13. Why must each line, not just lines through the center of inversion, contain the ideal point?

14. Suppose you know that two points are inverse points with respect to some circle. Will this information alone make it possible to find: (a) the radius of the circle? (b) the position of the center of the circle?

15. Prove the converse of Theorem 6.3.

16. What is the product of inversive transformations if the same transformation is used as a factor an even number of times?

17. What are possible images under inversion for a set of points consisting of two distinct intersecting lines?

18. Describe the inverse of a circle outside the circle of inversion and concentric to it.

19. Suppose a curve intersects its inverse curve. Where are the points of intersection?

20. In Section 5.6, the construction was given to find the midpoint of a given segment by use of a compass alone. Use the concept of inversion to prove that this construction results in the required point.

21. State and prove a theorem about the inverse of a line tangent to the circle of inversion.
22. State and prove a theorem about the inverse of the set of all lines passing through point P, if P is a point outside the circle of inversion.
23. Before reading ahead, decide whether you think the set of all inversive transformations is a group. Give supporting statements for your position.
24. Study more about the history of the geometry of inversion.

6.2 Additional Properties and Invariants under Inversion

In Section 6.1, we found that the circle of inversion is pointwise invariant under inversion and that a line through the center of inversion is invariant, but not pointwise invariant. We further found that coincidence of points and curves is preserved under inversion. These facts provide the basis for the continued investigation of invariant properties under the inversion transformation. One of the most useful of the invariants under inversion is indicated in the next theorem.

THEOREM 6.5 ➤ The measure of the angle between two intersecting curves is an invariant under the transformation of inversion.

The meaning of this theorem is shown symbolically in Figure 6.7. The angle between two curves C_1 and C_2 is defined to be the angle between their tangents at the point of intersection. One of the two supplementary angles must be designated. The images of the two intersecting curves are two intersecting curves C'_1 and C'_2. The angle between their two tangents at the point of intersection has the same measure as between the two original tangents.

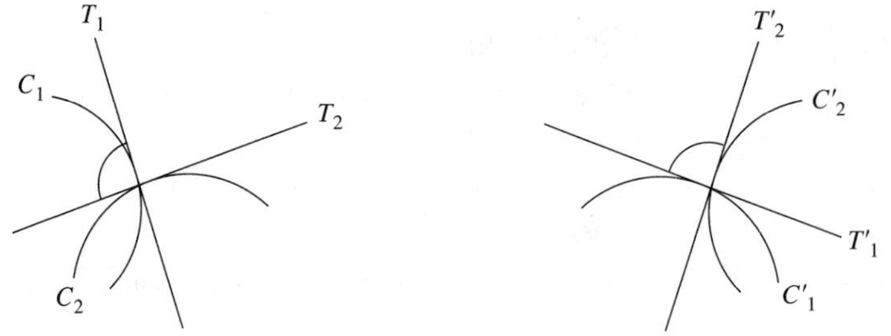

Figure 6.7

PROOF: The proof of Theorem 6.5 depends on proving that the angle that a curve makes with a line through the center of inversion is congruent to the angle the inverse curve makes with the same line. In Figure 6.8, let curves PQ and $P'Q'$

be inverse curves, with O the center of inversion and with P, P' and Q, Q' pairs of inverse points. The four points P, P', Q', Q all lie on a circle (the proof of this statement is left as an exercise), so opposite angles are supplementary. Thus,

$$\angle OPQ \cong \angle P'Q'O$$

The limiting position of $\overrightarrow{OQ}$ is $\overrightarrow{OP}$, as Q approaches P along the original curve. The tangents $\overleftrightarrow{PA}$ and $\overleftrightarrow{P'A}$ are the limiting positions of the secants $\overleftrightarrow{PQ}$ and $\overleftrightarrow{P'Q'}$. Thus, the limiting position of the angle from $\overrightarrow{OP}$ to the original curve is $\angle OPA$, whose supplement, $\angle OPB$, is congruent to $\angle OP'A$, the limiting position of the angle from $\overrightarrow{OP}$ to the image curve.

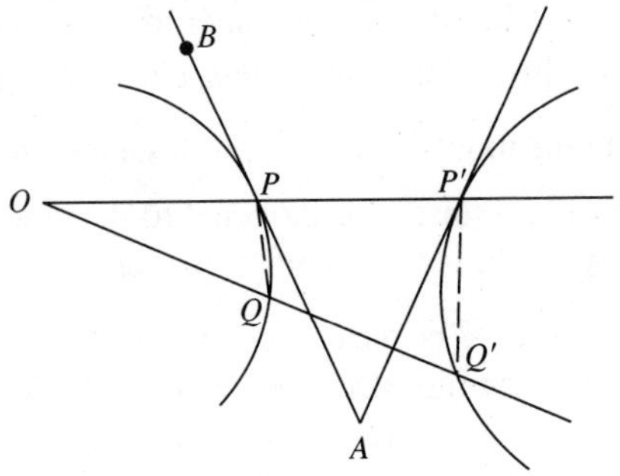

Figure 6.8

It is important to observe that, although the measure of the angle is preserved under inversion, the direction of the angle is reversed. In Figure 6.8, the angle from $\overrightarrow{OP}$ to $\overrightarrow{PB}$ is measured in a clockwise direction, whereas the angle from $\overrightarrow{OP'}$ to $\overrightarrow{P'A}$ is measured in a counterclockwise direction. For this reason, inversion (like reflection) is another example of an opposite transformation. The set of all inversive transformations in a plane does not form a group of transformations, since the product of two inversions is not an inversion. One inversion and the identity do constitute a rather trivial finite group with two elements, however.

As a corollary to Theorem 6.5, two curves that intersect at right angles have images that also meet at right angles. *Orthogonal circles*, two circles intersecting at right angles, play a very important role in the theory of inversion. A fundamental concept is included in the next theorem.

THEOREM 6.6 ➤ A circle orthogonal to the circle of inversion is an invariant set of points (but not pointwise invariant) under the inversion transformation.

PROOF: In Figure 6.9, circle O is the circle of inversion, and circle O' is any circle orthogonal to it. If a line through O intersects circle O' at C and D, then $OC \cdot OD = (OE)^2$, for E a point of intersection of the two circles. But since

$\overline{OE}$ is also the radius of the circle of inversion, C and D are by definition inverse points with respect to the circle of inversion.

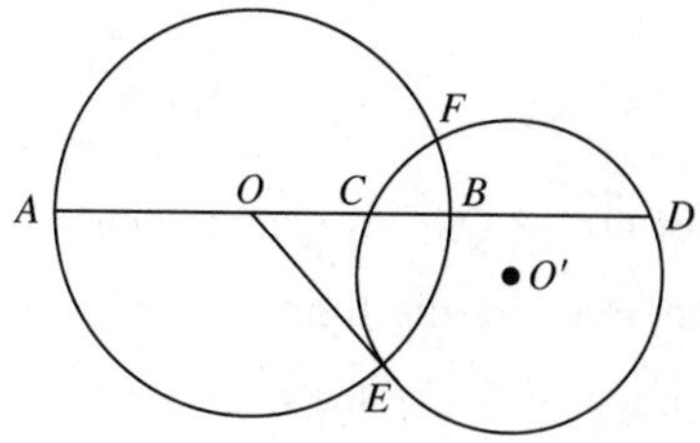

Figure 6.9

For each point of $\overset{\frown}{FDE}$, the image under inversion is a point of $\overset{\frown}{FCE}$, with the endpoints F and E invariant points. Thus, circle O' is a set of invariant points.

A second useful theorem about orthogonal circles under inversion is the following.

THEOREM 6.7 ➤ If two circles orthogonal to the circle of inversion intersect each other in two points, these two points are inverse points.

PROOF: In Figure 6.10, let O be the circle of inversion with O' and O'' two circles orthogonal to it and intersecting each other at A and B. It can be proved that point O is collinear with A and B because it is a point from which congruent tangents can be drawn to circles O' and O''. (The proof of this statement is left as an exercise.) Then, from the proof of Theorem 6.6, A and B are inverse points.

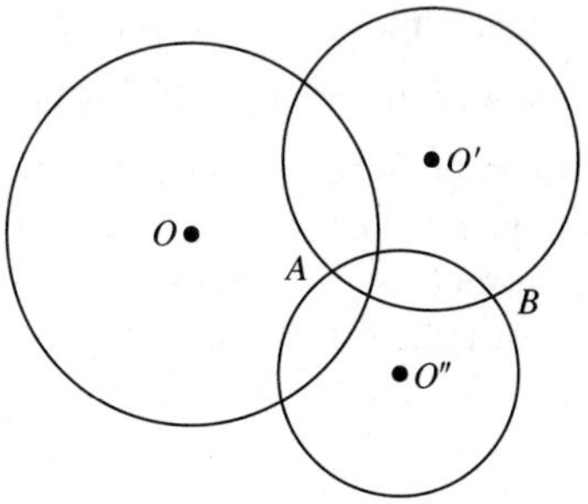

Figure 6.10

Recall that the concept of inverse point was encountered briefly in the last section of Chapter 5 on construction. Speculation about how a point inverse to a given point might actually be constructed could lead to discovery of an additional fundamental property of inverse points that is implied in the next theorem.

THEOREM 6.8 ➤ The inverse point, with respect to a circle, for a point outside that circle lies on the line joining the points of intersection of the tangents from the point to the circle.

Proof: In Figure 6.11, P is the external point. From similar right triangles OAP and $OP'A$,

$$\frac{OA}{OP'} = \frac{OP}{OA}, \qquad (OA)^2 = OP \cdot OP'$$

so P and P' are inverse points.

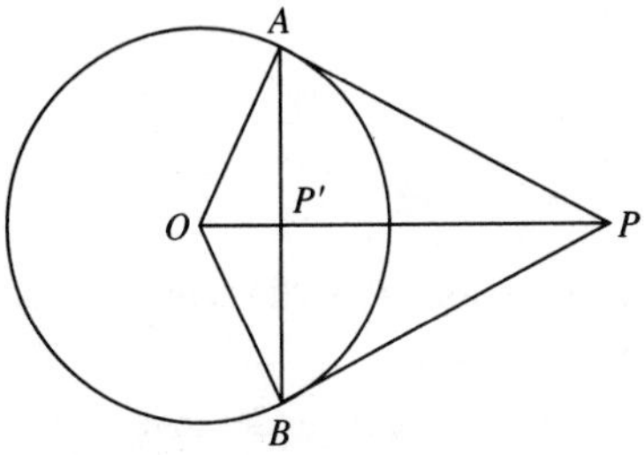

Figure 6.11

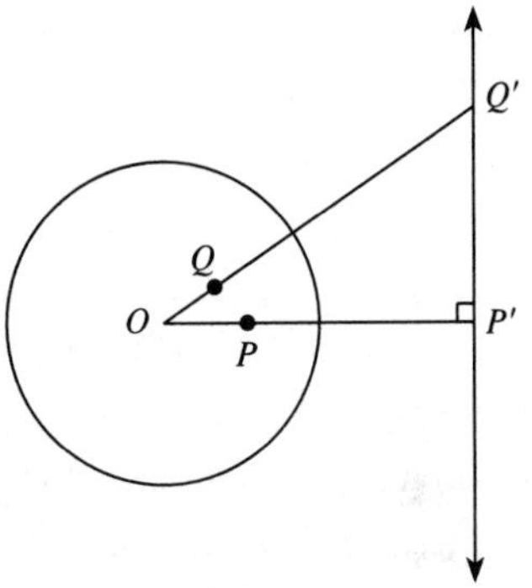

Figure 6.12

From Theorem 6.8 and Figure 6.11, it is possible to devise a method of constructing the point inverse to a given point, whether the given point is inside or outside the circle. If the point P' is inside the circle, for example, you will need to construct the perpendicular $\overleftrightarrow{AB}$ to $\overleftrightarrow{OP'}$ at P'. Then you must construct a perpendicular to $\overleftrightarrow{OA}$ at A. This line is a tangent to the circle, since it is perpendicular to the radius. The inverse point P is the intersection of the tangent and $\overleftrightarrow{OP'}$. If the point P is outside the circle, you will need to construct the tangents to the circle. The line joining the points of tangency, A and B, intersects $\overleftrightarrow{OP}$ at the required point P'. These construction problems are included in the exercises.

DEFINITION ➤ The line through an inverse point and perpendicular to the line joining the original point to the center of the circle of inversion is called the *polar* of the original point, whereas the point itself is called the *pole* of the line.

In Figure 6.11, $\overleftrightarrow{AB}$ is the polar of P, and P is the pole of $\overleftrightarrow{AB}$. The words *pole* and *polar* have different meanings in Chapters 1 and 7.

In Figure 6.12, P and P' are inverse points, then $\overleftrightarrow{P'Q'}$ is the polar of P, and P is the pole of $\overleftrightarrow{P'Q'}$. From Figure 6.12, you might surmise that the reciprocal relation between poles and polars exists as stated formally in the next theorem.

THEOREM 6.9 ➤ If a second point is on the polar of a first, with respect to a given circle of inversion, then the first is on the polar of the second with respect to the same circle.

Proof: In Figure 6.12, if $\overleftrightarrow{PQ'}$ is the polar of P, then it must be shown that the polar of Q' passes through P. If Q is the inverse point for Q', then all that is needed is to show that $\overleftrightarrow{QP}$ is perpendicular to $\overleftrightarrow{OQ'}$. The two pairs of inverse points, P, P' and Q, Q', lie on a circle. Because $\angle PP'Q'$ is a right angle, $\angle Q'QP$, the opposite angle in the inscribed quadrilateral, is also a right angle.

Polar reciprocation with respect to a circle is a special case of polar reciprocation with respect to conics, to be introduced in Chapter 7. In polar reciprocation, the image of a set of points is a set of lines, and the image of a set of lines is a set of points.

▲ EXERCISES 6.2 ▲

1. List the possibilities for the images under inversion of two lines meeting at right angles at a point other than the center of inversion.

2. List all the possibilities for the image of a right triangle under inversion.

3. In the proof of Theorem 6.5, prove that P, P', Q', Q all lie on a circle.

4. In the proof of Theorem 6.7, prove that point O is collinear with points A and B.

5. Given a circle, construct another circle orthogonal to it.

6. Given a point outside a given circle, construct its inverse point with respect to the circle.

7. Given a point inside a given circle, construct its inverse point with respect to the circle.

8. Prove that the polars of every point on a line are concurrent at the pole of the line.

9. Construct the pole of a line not intersecting the circle of inversion.

10. Construct the pole of a line intersecting the circle of inversion in two distinct points.

11. A triangle is *self-polar* with respect to the circle of inversion if each side is the polar of the opposite vertex. Explain how to construct a triangle self-polar to a given circle. (In inversive geometry, a triangle consists of three lines, not three segments.)

12. Describe the inverse of a circle with respect to a point on the circle.

13. Through two points inside a given circle but not collinear with the center, construct a circle orthogonal to the given circle.

14. Sketch and describe the inverse of a square with respect to its center, if the four sides are tangent to the circle of inversion. (Think of the sides of the square as lines, not segments.)

15. Sketch and describe the inverse of a square with respect to the center, if the four vertices are on the circle of inversion. (Think of the sides of the square as lines, not segments.)

16. Prove that any three circles can be inverted into three circles whose centers are collinear.

▷ 17. Investigate how it would be possible for a point to have the same inverse point with respect to two different circles of inversion.

▷ 18. The pole-polar relationship with reference to the circle of inversion provides a specific case of polar reciprocation with respect to a conic (explained in Chapter 7). Investigate properties of this relationship.

▷ 19. Investigate what happens to a checkerboard pattern inside the circle of inversion (see Figure 6.13) under inversion.

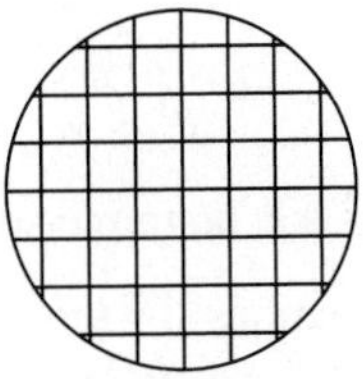

Figure 6.13

▲ 6.3 The Analytic Geometry of Inversion

The equations for inversion can be developed using Figure 6.14. The circle with center at the origin and with radius r can be considered the circle of inversion. Inverse points P and P' have coordinates (x, y) and (x', y'), respectively.

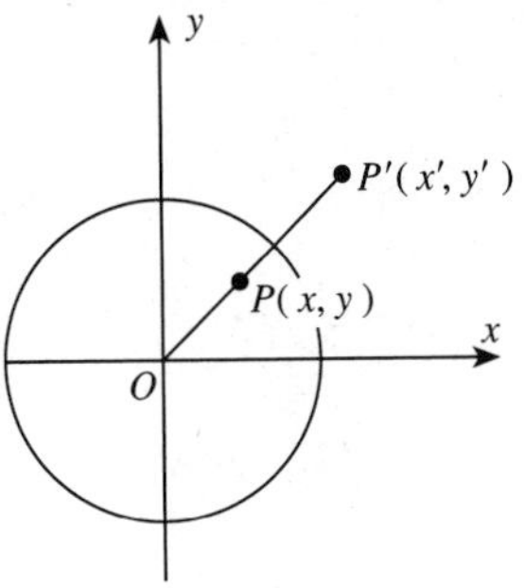

Figure 6.14

Since $OP \cdot OP' = r^2$,

$$\left(\sqrt{x^2+y^2}\right)\left(\sqrt{x'^2+y'^2}\right) = r^2 \qquad \text{or} \qquad \left(x^2+y^2\right)\left(x'^2+y'^2\right) = r^4$$

Because P and P' are collinear with 0, $x'/x = y'/y, x'y = xy'$, or $x'^2y^2 = x^2y'^2$; therefore,

$$\left(x^2 + y^2\right)\left(x'^2 + \frac{x'^2y'^2}{x^2}\right) = r^4$$

$$\frac{1}{x^2}\left(x^2 + y^2\right)\left(x'^2x^2 + x'^2y^2\right) = r^4$$

$$\frac{x'^2}{x^2}\left(x^2 + y^2\right)\left(x^2 + y^2\right) = r^4$$

Taking the square root of both sides, and choosing the square root so that x and x' have the same sign, shows that

$$x'\left(x^2 + y^2\right) = xr^2$$

or

$$x' = \frac{xr^2}{x^2 + y^2}$$

Similarly, the expression for y' can be found so that the next theorem is proved.

THEOREM 6.10 ➤ The equations for the inversion transformation of a point $P(x, y)$ into $P'(x', y')$ relative to the circle $x^2 + y^2 = r^2$ are

$$x' = \frac{xr^2}{x^2 + y^2}, \qquad y' = \frac{yr^2}{x^2 + y^2}$$

The equations in Theorem 6.10 show clearly that the image of the center of inversion is not a real point since $x^2 + y^2 = 0$, and x' and y' do not have real values. They also show that for points on the circle of inversion, $x = x'$ and $y = y'$.

Example ➤ Find the inverse of (2, 3) with respect to the circle of radius 3 and center at the origin.

$$x' = \frac{9(2)}{4+9} = \frac{18}{13}, \qquad y' = \frac{9(3)}{13} = \frac{27}{13},$$

The inverse point is

$$\left(\frac{18}{13}, \frac{27}{13}\right)$$

Since the inversion transformation is its own inverse, x and x' can be interchanged and y and y' can be interchanged so that

$$x = \frac{x'r^2}{x'^2 + y'^2}, \qquad y = \frac{y'r^2}{x'^2 + y'^2}$$

The requirement that the circle of inversion have its center at the origin is of course not mandatory. The substitution

$$x'' = x + h, \qquad y'' = y + k$$

shows that more general forms of the equations for inversion are

$$x' - h = \frac{(x-h)r^2}{(x-h)^2 + (y-k)^2}, \qquad y' - k = \frac{(y-k)r^2}{(x-h)^2 + (y-k)^2}$$

It is instructive to use the analytic formulas to verify the inversion theorems of the last two sections. For example, what is the image of the line $x - 4 = 0$ under inversion with respect to a circle with center at the origin and radius three? (See Figure 6.15.)

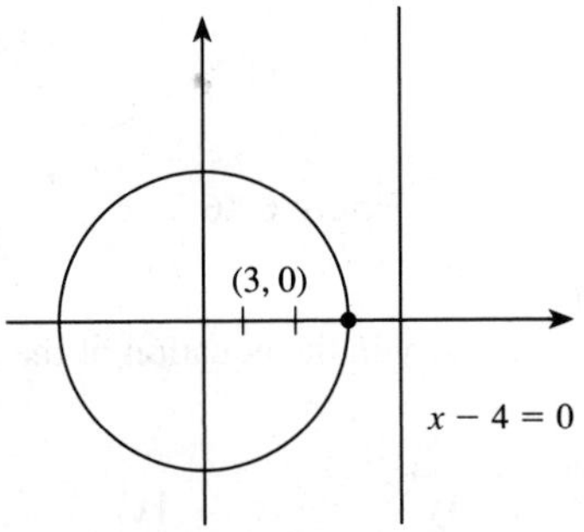

Figure 6.15

The equation of the image is obtained by substituting

$$\frac{9x'}{x'^2 + y'^2}$$

for x in the equation of the line. Thus,

$$\frac{9x'}{x'^2 + y'^2} - 4 = 0$$
$$\frac{9x'}{x'^2 + y'^2} = 4$$
$$9x' = 4(x'^2 + y'^2)$$

or

$$4x'^2 - 9x + 4y'^2 = 0$$

This is a circle through the origin, as expected.

The equations of the transformation of inversion and its inverse transformation also make it possible to find the equations for the inverse of curves not covered in previous theorems. An example illustrates the wide variety of possibilities. Find the image of the parabola $y^2 = 4x$ under inversion with respect to the circle $x^2 + y^2 = 4$. (See Figure 6.16.)

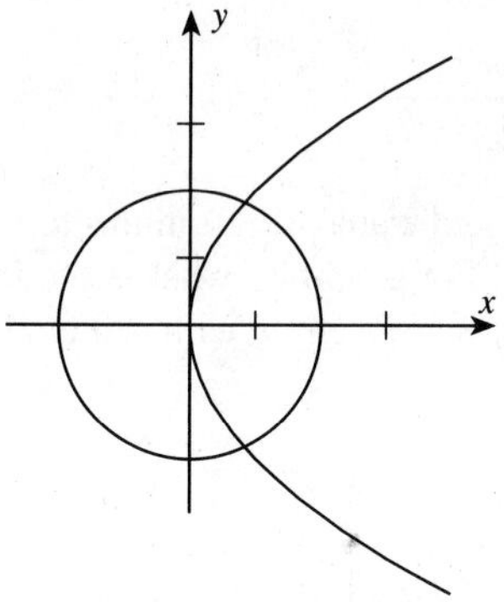

Figure 6.16

The substitution for x and y in the equation of the parabola results in

$$\frac{(4y')^2}{(x'^2 + y'^2)^2} = 4\frac{4x'}{x'^2 + y'^2}$$

This equation can be somewhat simplified, as follows:

$$\begin{aligned} 16y'^2 &= 16x'(x'^2 + y'^2) \\ y'^2 &= x'^3 + x'y'^2 \\ y'^2 - x'y'^2 &= x'^3 \\ y'^2 &= \frac{x'^3}{1 - x'}. \end{aligned}$$

The inverse of the parabola in this case is not a parabola, or even a conic, so the property of being a conic is not an invariant.

When the study of inversion is extended to complex numbers, a rather unexpected simplification appears in the equation of the transformation.

In Figure 6.17, let the real and imaginary axes be given. If the circle of inversion is a unit circle, and if $z' = c + di$ is the inverse of $z = a + bi$ with respect to this circle, then the following theorem shows the relationship between z and z'.

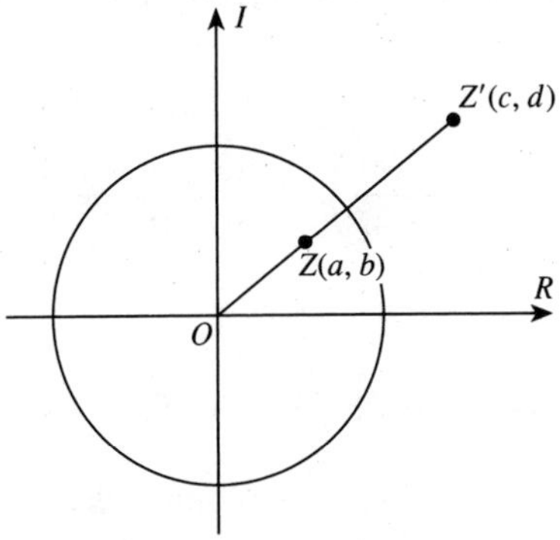

Figure 6.17

THEOREM 6.11 ➤ The transformation of complex points with respect to the unit circle with center at the origin has the equation $z' = 1/\overline{z}$, where $\overline{z}$ is the conjugate of z. If $z = a + bi$, then

$$\begin{aligned} z' &= \frac{1}{\overline{z}} = \frac{1}{a - bi} \cdot \frac{a + bi}{a + bi} \\ &= \frac{a + bi}{a^2 + b^2} \end{aligned}$$

To show that z and z' are inverse points with respect to the unit circle, it is necessary to:

1. Show that the product of their distances from the origin is one.
2. Show that they are collinear with the origin.

Demonstrating that z and z' are inverse points is left as an exercise.

Example ➤ Find the inverse of the point $2 + 3i$ with respect to a circle with radius 1 and center at the origin.

$$\begin{aligned} z' &= \frac{1}{2 - 3i} \\ &= \frac{1(2 + 3i)}{(2 - 3i)(2 + 3i)} \\ &= \frac{2}{13} + \frac{3}{13}i \end{aligned}$$

Inversion with respect to the circle with center at the origin and radius for complex numbers is accomplished by the transformation

$$z' = \frac{r^2}{\overline{z}}$$

Example ➤ The inverse of $3 + 4i$, with respect to the circle with center at the origin and radius two, is $4/(3 - 4i)$, which can be written in the form $a + bi$ as follows:

$$\left(\frac{4}{3-4i}\right)\left(\frac{3+4i}{3+4i}\right) = \frac{4(3+4i)}{9+16} = \frac{12}{25} + \frac{16}{25}i$$

▲ EXERCISES 6.3 ▲

For Exercises 1–8, use the equations for inversion with respect to a circle with center at the origin to find the inverse of the following real points.

Point	Radius of inversion	Radius of inversion
(3, 4)	**1.** 1	**2.** 2
(1, 1)	**3.** 2	**4.** 3
(0, 3)	**5.** 3	**6.** 4
(7, 1)	**7.** 4	**8.** 5

For Exercises 9–12, find the inverse of the given point with respect to a circle with center (2, 3).

Point	Radius of inversion	Radius of inversion
(6, 2)	**9.** 2	**10.** 3
(5, 5)	**11.** 3	**12.** 2

13. Find the equations for the inverse of the inversive transformation by actually solving for x and y in the original equations.

For each set of points given in Exercises 14–19, find the image under inversion with respect to the circle $x^2 + y^2 = 9$.

14. $x - 2 = 0$

15. $x - 5 = 0$

16. $x + y + 2 = 0$

17. $y^2 = 4x$

18. $x^2 = y$

19. $x^2 + y^2 = 16$

20. Complete the proof of Theorem 6.11.

For complex numbers, find the inverse of the points given in Exercises 21–28, with reference to a circle with center at the origin and with the given radius.

Point	Radius of inversion	Radius of inversion
$2 - 5i$	**21.** 1	**22.** 2
$3 + 7i$	**23.** 1	**24.** 2
$1 - 2i$	**25.** 2	**26.** 3
$3 + 11i$	**27.** 3	**28.** 4

For Exercises 29–31, derive general equations for the inverse of each set of points with respect to the circle $x^2 + y^2 = r^2$.

29. $x^2 + y^2 = k^2$

30. $ax + by + c = 0$

31. $y^2 = ax$

For Exercises 32 and 33, use equations to give an analytic proof of these theorems from previous sections.

32. Prove that the image of a line through the center of inversion, under inversion, is the same line.

33. Prove that the image of a circle not through the center of inversion, under the transformation of inversion, is a circle not passing through the center of inversion.

▷ **34.** Investigate whether distance is an invariant property under inversion. Take pairs of points and compare the distance between them with the distance between the image points.

▷ **35.** Investigate the application of inversion in a text on complex variables.

▲ 6.4 Some Applications of Inversion

A new fact about inverse points shows an application to an earlier construction problem.

THEOREM 6.12 ➤ Two inverse points with respect to a circle divide internally and externally in the same ratio the diameter on which they lie.

PROOF: In Figure 6.18, $OC \cdot OD = (OB)^2$, or

$$\frac{OD}{OB} = \frac{OB}{OC}$$

By a theorem on proportions, if $a/b = c/d$, then

$$\frac{a+b}{a-b} = \frac{c+d}{c-d}$$

so that

$$\frac{OD+OB}{OD-OB} = \frac{OB+OC}{OB-OC}$$

or

$$\frac{AD}{BD} = \frac{AC}{CB}$$

so that the ratios of division of $\overline{AB}$ by C and D are equal (directed segments have not been considered here).

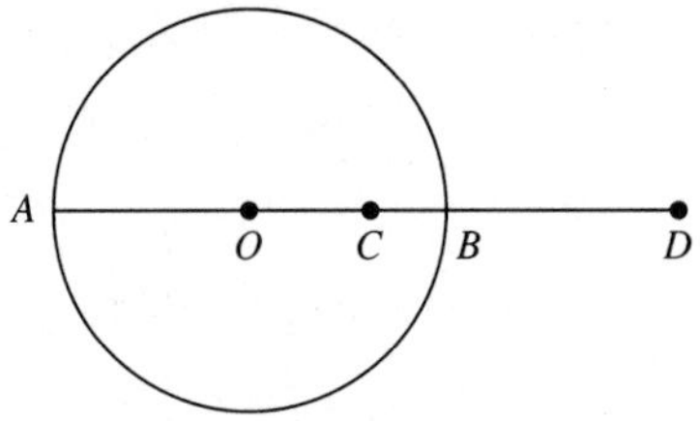

Figure 6.18

The fact that an inversion transformation sometimes transforms a straight line into a circle, or a circle into a straight line, is the basis for physical applications of inversion in linkages that change linear motion into curvilinear motion or vice-versa. The harnessing of the tides, the driving of a locomotive, and the production of electricity are all illustrations of one type of motion changed into another type of motion. The mechanical problem of devising linkages to do this sort of thing has concerned engineers for the past hundred years. One such device is *Peucellier's cell*, whose construction is based on the theory of inversion. The device is illustrated in Figure 6.19.

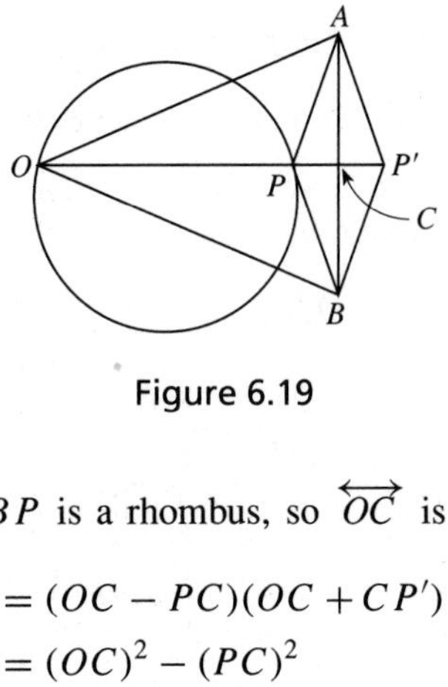

Figure 6.19

The quadrilateral $AP'BP$ is a rhombus, so $\overleftrightarrow{OC}$ is the perpendicular bisector of $\overline{AB}$.

$$\begin{aligned} OP \cdot OP' &= (OC - PC)(OC + CP') \\ &= (OC)^2 - (PC)^2 \\ &= (OA)^2 - (AC)^2 - [(PA)^2 - (AC)^2] \\ &= (OA)^2 - (PA)^2 \end{aligned}$$

But in the mechanical device, which consists of the six rods forming the rhombus as well as $\overline{OA}$ and $\overline{OB}$, the distances OA and PA are constant, and thus P and P' are inverse points. If point P moves around the circle, P' traces a straight line.

Quite a different application of inversion is that of finding new theorems by inverting familiar ones. More precisely, the figure for a familiar Euclidean theorem, when inverted, may yield a new theorem by suggesting the corresponding properties. Even further, the new theorem does not have to be proved from the beginning; it is true because of the proof of the original theorem and the accepted properties of inversion. Examples will illustrate the technique.

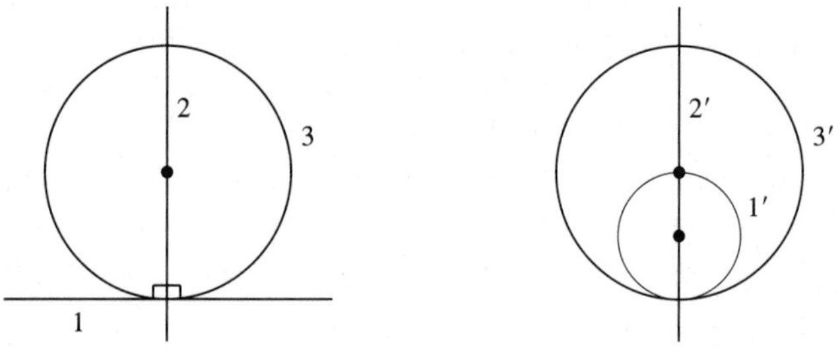

Figure 6.20

Example ➤ One of the elementary theorems from geometry is the statement that if a line is perpendicular to a radius at its endpoint on the circle, then it is tangent to the circle. This theorem is illustrated by Figure 6.20a. The sets of points in the theorem can be numbered for easy identification with the sets of points in the figure as follows:

If a line ① is perpendicular to a radius ② at its endpoint on the circle ③, then it ① is tangent to the circle ③.

Now the figure can be inverted to find a new theorem. For this example, assume that the given circle is chosen as the circle of inversion. The inverted figure is shown in Figure 6.20b, with the sets of points numbered with primes to show corresponding sets.

Line 2 (a line through the center of inversion) is invariant. Circle 3 (the circle of inversion) is invariant. Line 1 (a line not through the center of inversion) becomes a circle through the center of inversion. The three sets of points still have a common point of intersection, and the angles of intersection remain the same size. The conclusion of the new theorem is that $1'$ and $3'$ are tangent. The new theorem may be stated as follows.

THEOREM 6.13 ➤ A circle whose diameter is a radius of a second circle is internally tangent to the second circle.

When finding new theorems by inversion, it is very important to realize that the new figure (Figure 6.20b) is concerned only with the relationships among the image sets of points—it is not concerned with the relationship between the original figure and the new, or with areas and other extraneous factors, but only with the relationship within the new figure that makes it possible to state a new theorem.

Example ➤ Figure 6.21a shows the figure for the theorem that states that an angle inscribed in a semicircle is a right angle. Figure 6.21b shows the result of inverting the sets of points in Figure 6.21a with respect to O as the center of inversion, with the given circle the circle of inversion. In Figure 6.21b, circle 1 and line 2 are invariant sets of points. $\overleftrightarrow{AB}$ has circle 3 as its image, and $\overleftrightarrow{BC}$ has circle 4 as its image. The new theorem involves the conclusion that circles 3 and 4 will be orthogonal. The formulation of the wording of the theorem is left as an exercise.

The choice of the center of inversion was arbitrary, and various different theorems can be written if other centers of inversion are chosen. As might be expected, choosing a key point of the figure itself as the center of inversion results in the greatest simplification, as in the previous example.

Example ➤ Figure 6.21c shows the result of inversion with respect to a point not on any of the given lines or circles. Verify the following statements with respect to Figure 6.21c:

1. The image of circle O, a circle not through the center of inversion, is a circle (circle 1).
2. The image of the line $\overleftrightarrow{AC}$ is a circle (circle 2) that goes through the center of inversion and is orthogonal to the image of circle O.
3. The image of the line $\overleftrightarrow{AB}$ is a circle (circle 3) that passes through the intersection of circles 1 and 2 and also through the center of inversion.
4. The image of the line $\overleftrightarrow{BC}$ is a circle (circle 4) that passes through the intersections of 1, 2 and 1, 3 and also through the center of inversion.

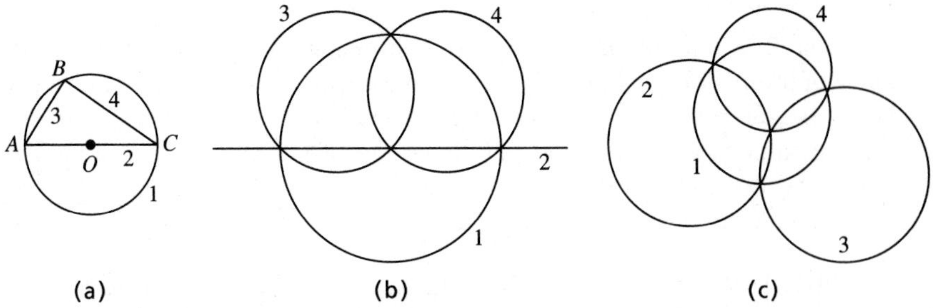

Figure 6.21

The new theorem will involve four circles, and the conclusion will be that two of them are orthogonal.

THEOREM 6.14 ➤ For 4 circles in a plane, if a) Circles 1 and 2 are orthogonal, and if b) Circle 3 passes through one intersection of circles 1 and 2 and also intersects circle 1 in a distinct point, and if c) Circle 4 passes through the second intersection of circles 1 and 2 and the second intersection of circles 1 and 3, and if d) Circles 2, 3, 4 have another distinct point in common, then circles 3 and 4 are orthogonal.

In the discussion thus far, inversion has been used to get new theorems from old. In general, the new theorems have been more complicated because some of the straight lines have been inverted into circles. In the application of inversion called *proof by inversion*, the procedure is somewhat reversed. Here the idea is to prove a more complicated theorem by inverting the figure and then using a simpler theorem already established. The classic example of this is in the proof of Feuerbach's theorem.

THEOREM 6.15 ➤ *Feuerbach's theorem.* The nine-point circle of a triangle is tangent to the incircle and to each of the three excircles of the triangle.

PROOF: In Figure 6.22, let I and I' be the centers of the inscribed circle and any one excircle, with A' the center of the side $\overline{BC}$. If ID and $I'E$ are radii perpendicular to $\overline{BC}$, then A' is also the midpoint of $\overline{DE}$. This can be shown because $\overline{BD} \cong \overline{CE}$, since both have measures equal to half the perimeter of the triangle minus the length of side $\overline{AC}$.

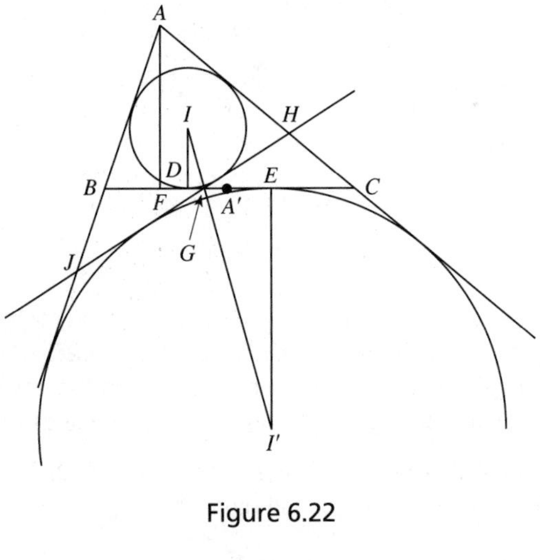

Figure 6.22

Now if A' is taken as the center of inversion and if $A'D = A'E$ is taken as the radius of inversion, then the incircle and the excircle are both invariant, since they are orthogonal to the circle of inversion. The nine-point circle passes through A' and F, the foot of the altitude from A. But A' is the center of inversion and G is the inverse of F with respect to the circle of inversion. Thus, the image of the nine-point circle is a line through G. Because angles are preserved under inversion, the angle between the image of the nine-point circle and $\overline{BC}$ is congruent to the angle that the tangent to the nine-point circle at A' makes with $\overline{BC}$. This tangent is parallel to the opposite side of the orthic triangle (the triangle through the feet of the altitudes). This line is antiparallel to $\overline{BC}$. Two antiparallel lines make congruent angles with the bisector of the angle between the other two sides of the inscribed quadrangle. But it is the other internal tangent of circles I and I' that makes the same angle with BC that $\overline{II'}$ does, so it is the tangent $\overleftrightarrow{HJ}$ that is the image of the nine-point circle. This means that the nine-point circle is tangent to both the incircle and the excircle. In a similar way, the nine-point circle can be shown to be tangent to the other excircles.

Another example of proof by inversion appears in Chapter 9 on non-Euclidean geometry. In this example, inversion is used in the Poincaré model for hyperbolic geometry to find the sum of the measures of the angles of a triangle.

This section concludes with a look at inversion in three dimensions. Analogous to the circle of inversion is the sphere of inversion, as shown in Figure 6.23. If P and P' are inverse points with respect to a sphere of radius r, then

$$OP \cdot OP' = r^2$$

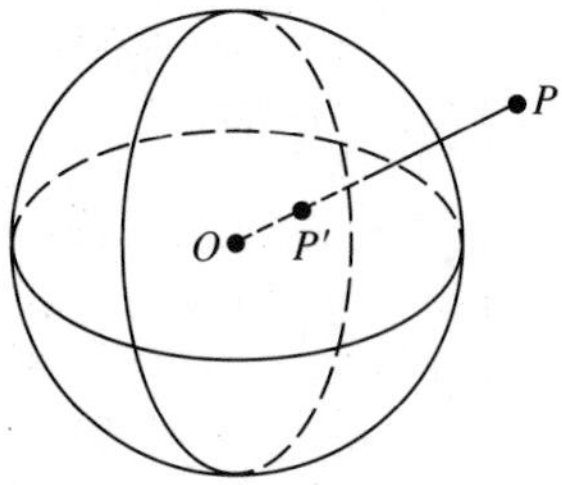

Figure 6.23

Inversion in two dimensions can be seen to be a cross section of the three-dimensional drawing, with the plane passing through the center of the sphere. Again, it is necessary to add an ideal point so that inversion is a transformation of inversive space into itself. This ideal point lies on all planes, is the image of O, and maps to O.

Example ➤ Find the inverse point P' for a point $P(2, 3, 4)$ with respect to a sphere with center O at the origin and radius 5.

$$OP \cdot OP' = 25$$

$$\sqrt{2^2 + 3^2 + 4^2}\,OP' = 25$$

$$OP' = \frac{25}{\sqrt{29}}$$

The direction cosines for P and P' are $2/\sqrt{29}$, $3/\sqrt{29}$, and $4/\sqrt{29}$, so that the coordinates of P' are

$$x = \frac{2}{\sqrt{29}} \cdot \frac{25}{\sqrt{29}} = \frac{50}{29}$$

$$y = \frac{3}{\sqrt{29}} \cdot \frac{25}{\sqrt{29}} = \frac{75}{29}$$

$$z = \frac{4}{\sqrt{29}} \cdot \frac{25}{\sqrt{29}} = \frac{100}{29}$$

Corresponding to inversion relationships between the line and the circle in two dimensions are those between the plane and the sphere in three dimensions. An example is provided in the next theorem. (See Exercise 14, Exercises 6.4)

THEOREM 6.16 ➤ The image under inversion with respect to a sphere of a plane not through the center of inversion is a sphere passing through the center of inversion, and conversely.

A special case of inversion with respect to a sphere is known as *stereographic projection* and is very useful in mapmaking. This method of projection is illustrated in Figure 6.24. Plane α is tangent to both the sphere with O as center and the sphere with B as the center at the common point A. Plane α and the sphere with O as center are inverses with respect to the second sphere. The effect of this inversion is to establish a one-to-one correspondence between the points on the sphere with O as center and on the inversive plane. One of the important features of a map prepared in this way is that all angle measures are preserved under inversion. Stereographic projection has been suggested as being particularly suited to produce a map that will show the patterns of plate tectonics. See Athelstan Spilhaus, "New Look in Maps Brings Out Patterns of Plate Tectonics," *Smithsonian*, August 1976. Stereographic projection is also used in obtaining the Poincaré model for hyperbolic geometry, to be introduced in Chapter 9.

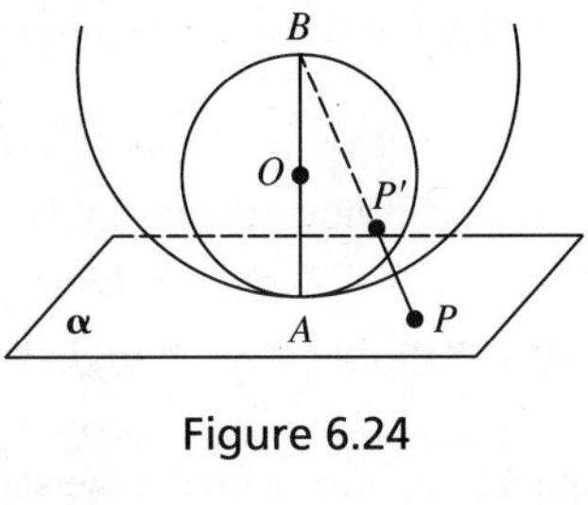

Figure 6.24

Inversion can also be generalized in ways other than increasing the dimensions. One possibility is to invert with respect to an ellipse or hyperbola instead of a circle. For example, in Figure 6.25, A and B could be defined as inverse points with respect to an ellipse if $AO \cdot BO = CO^2$, for C a point on the ellipse collinear with A, B, and the center O.

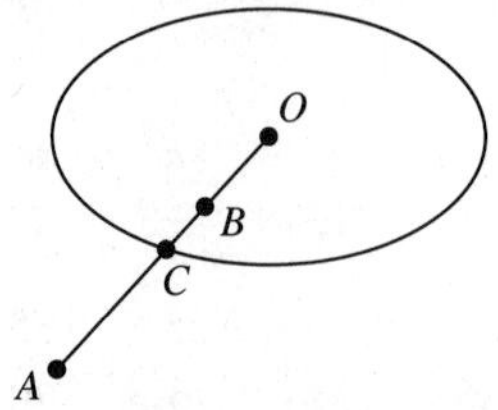

Figure 6.25

The transformation of inversion has given us an important new transformation of historical interest that has significant applications. Unlike many previous sets of transformations, the set of inversions does not constitute a group. The equations are not linear, and the results depend on whether the center of inversion is a point of the set of points being inverted. Though neither distance nor ratios of distances are invariant, the size of angles is an invariant. The next chapter introduces a different kind of transformation that does form a group—a more general group that contains some of the groups of Chapter 2 as subgroups.

▲ EXERCISES 6.4 ▲

1. Explain the special case of Theorem 6.12 if one of the points is the center of inversion.
2. Use cardboard and paper fasteners to prepare a working model of Peucellier's cell.
3. Invert the figure of a line perpendicular to a radius of a circle at its endpoint on the circle, with respect to a point on the tangent, and state the resulting theorem.
4. Invert the figure of a triangle inscribed in a semicircle with respect to one of the endpoints of the diameter and state the resulting theorem.

For Exercises 5–8, write new theorems by inverting the figure.

5. If the opposite angles of a quadrilateral are supplementary, the quadrilateral can be inscribed in a circle.
6. The line joining the centers of two intersecting circles is perpendicular to their common chord.
7. If two circles are tangent to the same line at the same point, the line joining the centers passes through the common point of tangency.
8. The altitudes of a triangle are concurrent.

Exercises 9–12 concern the proof of Theorem 6.15.

9. Verify that $\overline{BD} \cong \overline{CE}$.

10. Establish that G divides $\overline{DE}$ internally in the same ratio that F divides it externally.

11. Show that the tangent to the nine-point circle at a midpoint of a side is parallel to the opposite side of the orthic triangle.

12. Prove that two antiparallel lines make congruent angles with the bisector of the angle bisector between the other two sides of the inscribed quadrangle.

13. Use inversion to prove the theorem that a circle passing through two inverse points on two mutually inverse circles intersects the two circles at the same angle.

14. Outline the proof of Theorem 6.16, using ideas from the proof of Theorem 6.3.

15. Under inversion, with respect to a sphere, what is the image of a plane through the center of the sphere?

For Exercises 16 and 17, find the inverse point for the given point with respect to a sphere with center at the origin and with the given radius.

16. (3, 1, 2); radius 7

17. (2, 5, 3); radius 4

For Exercises 18I–21I, for the exercise listed, write the entire set of possible new theorems by taking the center of inversion in each different location involving intersection of sets or location on one of the sets of points.

▷ **18.** Theorem in Exercise 5

▷ **19.** Theorem in Exercise 6

▷ **20.** Theorem in Exercise 7

▷ **21.** Theorem in Exercise 8

▷ **22.** Show how the construction problem of drawing a circle tangent to two given circles and passing through a given point may be solved by using the transformation of inversion.

▷ **23.** Investigate more about stereographic projection. Give the images of various figures, and state some advantages and disadvantages of this type of map projection.

▷▷ **24.** Construct a model from wood or plastic for Peucellier's cell, then demonstrate how to use it.

▲ CHAPTER 6 REVIEW EXERCISES ▲

1. Which points in the inversive plane are the only ones that are their own image?
2. When is a circle its own image in inversive geometry?
3. If the distance from the center of inversion to the original point is 4, and if the radius of the circle of inversion is 5, find the distance of the inverse point from the center of inversion.
4. If a point is 3 units from the center of inversion and if its image is 2 units from the center of inversion, find the radius of the circle of inversion.
5. In general, is the product of two inversions an inversion?
6. What other transformation must be included along with a given inversion to form a trivial example of a group of transformations?
7. Describe the image under inversion of two intersecting circles, orthogonal to the circle of inversion.
8. Describe all possible images under inversion for a right triangle.
9. Would it be possible for the image of a triangle with three ordinary vertices to be a triangle with three ordinary vertices, under the transformation of inversion?
10. Under what special circumstances would the two points of intersection of two circles be inverse points?
11. Construct the inverse point for point P in Figure 6.26.
12. Explain how Theorem 6.7 can be used to provide an alternate method for constructing the inverse point for a given point.
13. Find the inverse for the point (2, 3) with respect to a circle with center at the origin and with radius 2.

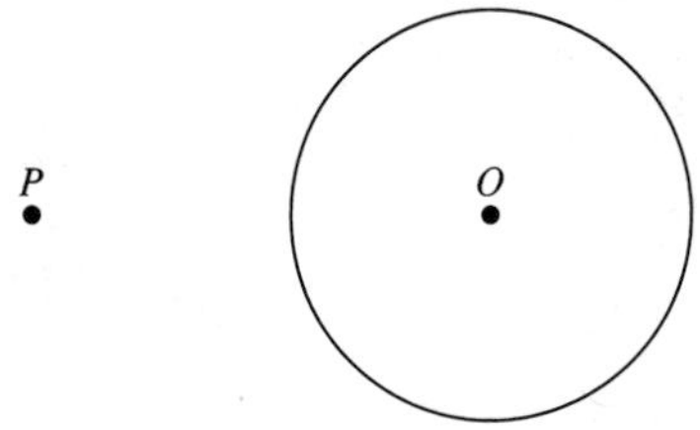

Figure 6.26

14. Find the image of the line $y + 3 = 0$ under inversion with respect to a circle with center at the origin and radius 3.
15. Find the image of the circle $x^2 + y^2 = 9$ under inversion with respect to the circle $x^2 + y^2 = 1$.
16. In the complex plane, find the inverse of the point $3 + 3i$ with reference to a circle with center at the origin and with a radius of 4.

For Exercises 17 and 18, write a new theorem by inverting the figure for the given theorem. Take the center of inversion as a point not on any of the given sets of points.

17. If two perpendicular lines intersect at a point on a circle, and if one of these lines is tangent to the circle, then the second line passes through the center of the circle.

18. If a line passes through the points of intersection of two circles, then it is perpendicular to the line through the centers of the two circles.

19. Find the inverse point for $(-1, 3, 2)$ with respect to the sphere with center at the origin and radius 9.

Additional chapter review exercises may be selected from any exercises not already assigned from previous exercise sets in this chapter.

7 Projective Geometry

7.1 Fundamental Concepts

Projective geometry is one of the modern geometries introduced in the early nineteenth century. Euclidean geometry is a special case of projective geometry in the sense that the group of Euclidean motions is a subgroup of the group of projective transformations.

This chapter uses both groups of transformations and a set of axioms to present what might be described as a minicourse in projective geometry. The emphasis is on development of a more general geometry than Euclidean geometry or the geometry of similarities from Chapter 2. The concept of a group of projective transformations and the study of invariant properties under this set of transformations are central ideas. Projective geometry includes many mathematical concepts not met in earlier geometries, and has more practical applications than you might expect.

In ordinary analytic geometry, the projection of a segment onto an axis is used. As illustrated in Figure 7.1, $\overline{AB}$ is projected onto the x-axis by dropping perpendiculars $\overline{AA'}$ and $\overline{BB'}$. Although AB and $A'B'$ are not usually equal, a one-to-one correspondence is established by the projection between points on the two segments.

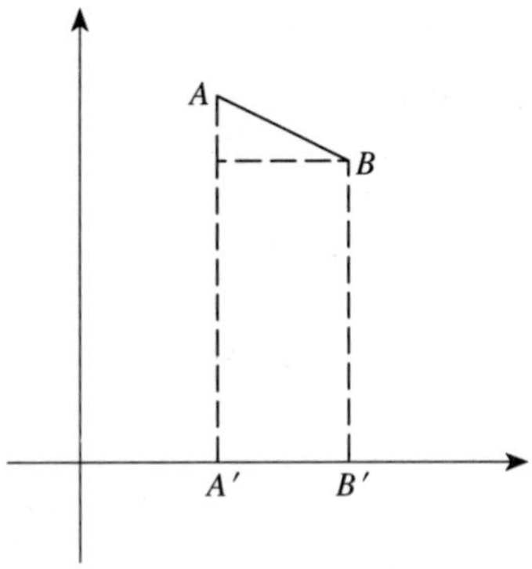

Figure 7.1

Another familiar use of the word *projection* is in connection with a motion picture projector. Here, the picture on the film is projected onto the screen, as suggested by Figure 7.2a. It should be apparent that pictures on the film and on the screen will be similar to each other, since the shape remains invariant while the size changes uniformly. For Figure 7.2b, however, the picture on the screen will be distorted because the film and the screen are not parallel. The figures are no longer similar, and it is no longer apparent what properties of figures remain unchanged. Even the ratios of distances are no longer invariant.

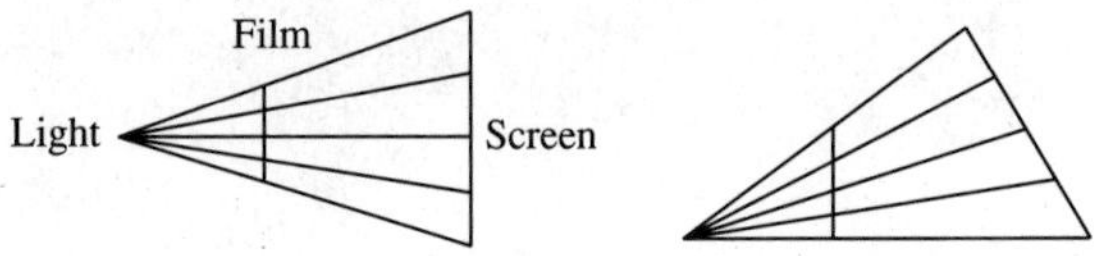

(a) (b)

Figure 7.2

Figure 7.2 also may suggest the relationship between projection and the concept of perspective as used in art. Many famous painters, particularly during the Renaissance, attempted to use mathematical ideas to help create the illusion of depth in a painting. One idea used was to think of the position of objects in the painting as being determined by the point of intersection of the canvas and the line from the eye to the object, as illustrated in Figure 7.3.

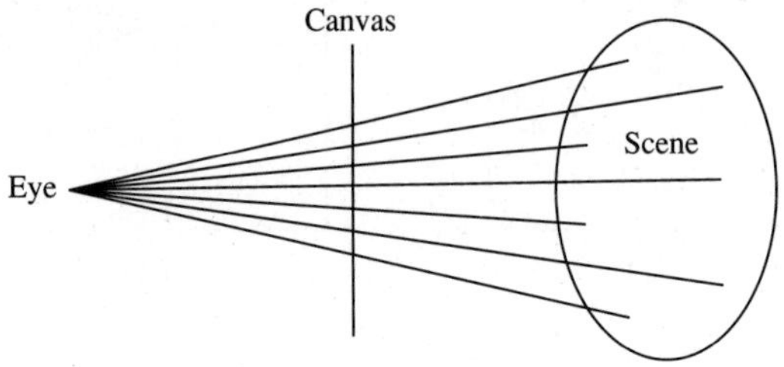

Figure 7.3

A second use of geometry to give an illusion of depth in a painting is called *linear perspective*. This concept includes the principle that parallel lines seem to converge as they recede from the viewer. For example, railroad tracks appear to meet on the horizon line. Artists use this idea by having parallel lines (in the real scene) converge at one or more vanishing points in the painting. A famous example of a painting with one vanishing point is the Last Supper by Leonardo da Vinci, shown in Figure 7.4.

Figure 7.4

Figures 7.5a and 7.5b show sketches in which the vanishing point is not at the center. You should be able to locate this point in each case. Additional ideas about vanishing points will be found in Section 7.7.

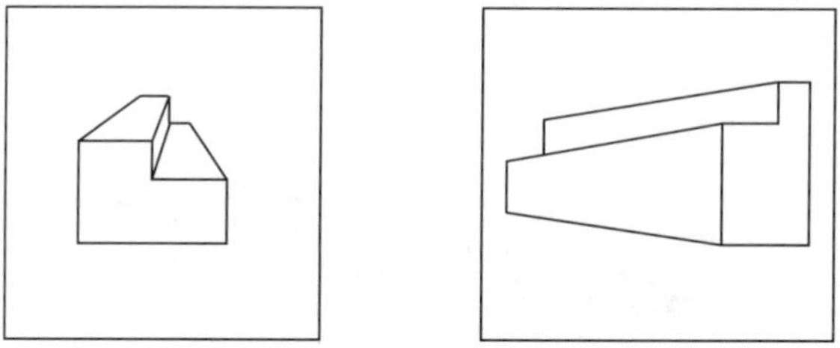

(a) (b)

Figure 7.5

DEFINITION ➤ The sets of points on two lines are *perspective* if pairs of corresponding points are collinear with a fixed point not on either line. The correspondence between the sets of points is a *perspectivity*. The fixed point is called the *center of perspectivity*. The sets of points on the two lines are called *ranges* of points and the set of lines through the fixed point is called a *pencil* of lines.

An example of a perspectivity is shown in Figure 7.6a. The notation for this perspectivity is $ABCD \overset{P}{\barwedge} A'B'C'D'$. This notation shows that P is the center of perspectivity and A, A'; B, B'; C, C'; D, D' are pairs of corresponding points.

You may have noticed that there are exceptions in Figure 7.6a, so there is no one-to-one correspondence of points on the two lines. See Figure 7.6b. If $\overleftrightarrow{PE}$ is parallel to l', then there is no corresponding point E' on l'. If $\overleftrightarrow{PF'}$ is parallel to l, there is no corresponding point on l. These exceptions show the need to use a line in projective geometry that is not the same as the ordinary Euclidean line.

Figure 7.6c shows several perspectivities in the same drawing. The points on l_1 and l_2 are perspective, with A the center of perspectivity. The points on l_2 and l_3 are perspective, with point B the center of perspectivity, and the points on l_3 and l_4 are perspective, with point C the center of perspectivity. The points on l_1 and l_4 are not perspective. Instead, they are defined to be projective. From this point of view, a projectivity is considered a chain of perspectivities.

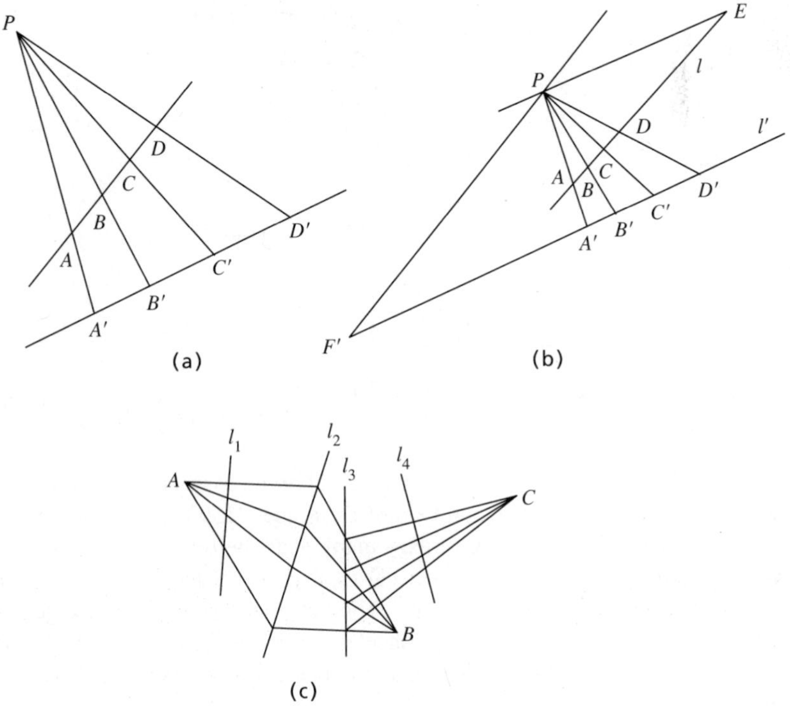

Figure 7.6

DEFINITION ➤ A projectivity is the product of a finite sequence of perspectivities.

In projective geometry, the invariant properties of projectivities are studied. While it is not obvious which properties may be preserved, it should be evident that there is a one-to-one correspondence and that collinear points remain collinear even though distances and relative distances may change.

The concept of perspectivity can also be extended to three dimensions.

DEFINITION ➤ The sets of points on two planes are perspective if pairs of corresponding points are collinear with a fixed point not on either plane. The fixed point is called the *center of perspectivity*.

Figure 7.7 shows a perspectivity involving polygons.

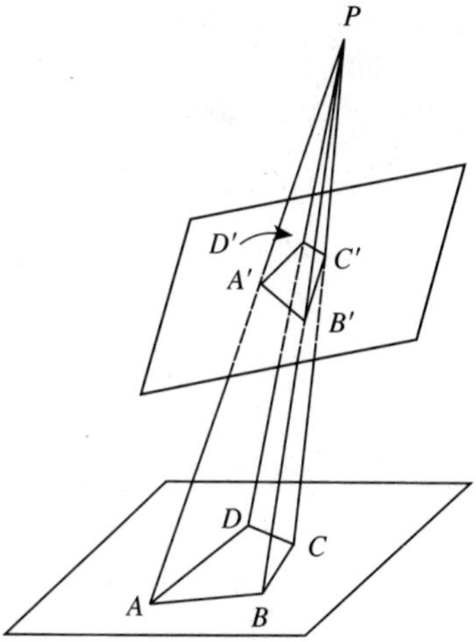

Figure 7.7

This intuitive discussion has introduced several of the basic concepts of projective geometry. The French mathematician Gerard Desargues (1593–1662) anticipated the formal development of projective geometry many years before it actually was developed. His 1639 treatise on conic sections, a classic in the early development of synthetic geometry, was neglected for two centuries. The first real text in projective geometry, *Traité des propriétés des figures* by J. V. Poncelet (1788–1867), was published in 1822. Most of the work on this book was done while Poncelet was a prisoner in Russia, after the wars of Napoleon.

Other advances in projective geometry were made during the nineteenth century. Karl von Staudt (1798–1867), in his *Geometrie der Lage*, showed how projective geometry could be developed without the use of any metrical basis for measurement. Felix Klein gave projective geometry the prominent place it deserves in the classification of geometries in his Erlanger program of 1872. The Erlanger program is the name given to Klein's inaugural address upon appointment to the faculty and senate of the University of Erlangen, and to the program of geometric study advocated in that address. Since that time, postulational developments of projective geometry have been written, and finite projective geometries have been introduced.

▲ EXERCISES 7.1 ▲

1. Write an equation relating the lengths of $\overline{AB}$ and $\overline{A'B'}$ in Figure 7.1.

For Exercises 2 and 3, describe the location of the vanishing point in:

2. Figure 7.5a
3. Figure 7.5b
4. In general, is the product of two perspectivities a perspectivity?
5. Make a rough sketch of a picture looking down a road, with sets of utility poles on either side. Make the lines in the drawing converge at a point on the horizon.

For Exercises 6–16, state which of these properties, which are invariant in Euclidean geometry, also seem to be invariant in projective geometry.

6. Measures of angles
7. Collinearity
8. Measures of area
9. Ratios of distances
10. Property of being a circle
11. Property of being a square
12. Property of being a triangle
13. Property of being a rectangle
14. Property of being a parallelogram
15. Property of three points being the vertices of an equilateral triangle
16. Property of one point being the midpoint of the segment determined by two other points
17. Make a drawing similar to Figure 7.6, in which several perspectivities result in the points on a line being projected back into points (but not the same points) on the same line.
18. Make a drawing similar to Figure 7.7, in which a square is projected into a nonsquare quadrilateral.
19. ▷ Read more about the history of projective geometry in a history of mathematics text.
20. ▷▷ Investigate more about how famous Renaissance artists used perspectivity in their painting. Study some paintings of this period to find vanishing points.

▲ 7.2 Postulational Basis for Projective Geometry

A postulational basis for projective geometry is simpler than is the basis for Euclidean geometry. The following set of axioms (the first five of which are adapted from a set appearing in Coxeter, *Projective Geometry*) is based on the undefined terms *point*, *line*, and *incidence*. The intuitive meaning of *incidence* is simply the idea of "lying on" or "containing." For example, Axiom 1 means that any two distinct points lie on just one line. Axiom 2 means that any two lines have at least one point lying on both of them. The new terms used in Axiom 4 will be explained after the statement of the axioms.

Axioms for Projective Geometry

1. Each two distinct points are incident with exactly one line.
2. Any two coplanar lines are incident with at least one point.
3. There exist four points of which no three are collinear.
4. The three diagonal points of a complete quadrangle cannot be collinear.
5. If a projectivity leaves invariant three distinct points on a line, it leaves invariant each point on the line.
6. Not all points are coplanar.
7. Any two distinct planes have at least two points in common.

A plane can be defined using the three undefined terms. If a point P and line p are not incident, then the plane incident with the point and line consists of all points incident with lines joining P to points of p, and all the lines incident with pairs of distinct points chosen in this manner.

DEFINITION ➤ A *complete quadrangle* is a set of four points (vertices) in a plane, no three collinear, and the lines joining these vertices in pairs.

A complete quadrangle is shown in Figure 7.8. Complete quadrangle $ABCD$ has three pairs of *opposite sides*, $\overleftrightarrow{AB}$ and $\overleftrightarrow{CD}$, $\overleftrightarrow{AD}$ and $\overleftrightarrow{BC}$, and $\overleftrightarrow{AC}$ and $\overleftrightarrow{BD}$. Opposite sides will intersect. Opposite sides are two lines, one determined by any two vertices and the other determined by the remaining vertices of a complete quadrangle. For example, vertices A and C determine one side, so the opposite side is the line determined by the two remaining vertices, B and D.

The opposite sides of a complete quadrangle meet by twos in three points other than vertices (points E, F, G of Figure 7.8). These three points are the *diagonal points* of the complete quadrangle, as used in Axiom 4. The dotted lines in Figure 7.8 are the sides of the *diagonal triangle*, whose vertices are the diagonal points.

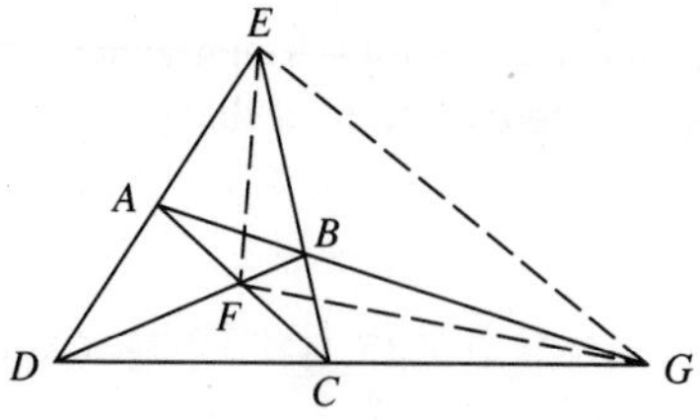

Figure 7.8

Many of the axioms for projective geometry seem the same as those for Euclidean geometry, and this is correct. One that is not the same, however, and that is responsible for giving projective geometry its characteristic structure is Axiom 2. The significance of this axiom is illustrated in Figure 7.9.

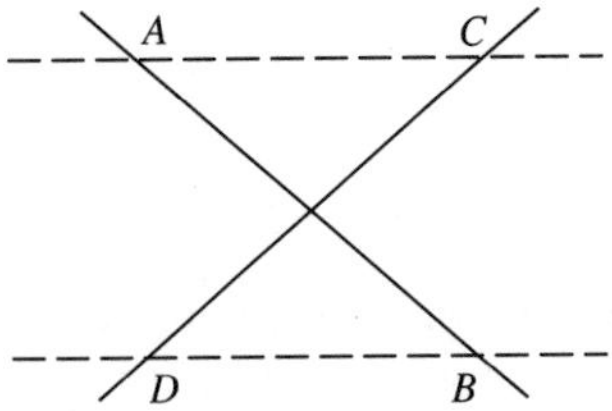

Figure 7.9

$\overleftrightarrow{AB}$ and $\overleftrightarrow{CD}$ have an obvious point of intersection, but $\overleftrightarrow{AC}$ and $\overleftrightarrow{DB}$ have no point of intersection shown. They seem to be parallel. The consequence of the axiom, however, is that *any* two lines of a plane, even those such as $\overleftrightarrow{AC}$ and $\overleftrightarrow{DB}$, meet in a point. That is, there are no parallel lines in projective geometry.

In Euclidean geometry, every point of the Euclidean plane is an ordinary point. Two parallel lines have no point in common. In the extended Euclidean plane, parallel lines are said to meet at *ideal points*. Saying that two lines meet at an ideal point in the extended Euclidean plane is just another way of saying that the lines are parallel in the ordinary plane. The extended Euclidean plane consists of the union of the ordinary and ideal points in the plane. The introduction of ideal points eliminates the exceptions indicated in the perspectivity in Figure 7.6b. Now we can say that the corresponding points for E and F' are ideal, and that there is a one-to-one correspondence between the sets of points on the two lines.

In the planes and spaces of projective geometry, the ideal points lose their special nature, and parallelism cannot occur. We may think intuitively of projective geometry as being derived from the geometry of the extended Euclidean plane through the elimination of any distinction between real and ideal points. The points A, B, C, in Figure 7.6 could just as easily be ideal points as real points. In Figure 7.8, the opposite sides could intersect in ideal points or in real points. Parallelism no longer exists, and projective geometry is displayed as a general geometry with ordinary geometry as a very special case.

Although the study of projective geometry can be approached intuitively by extending the Euclidean plane, the more common mathematical approach is just the opposite. If we begin with the general projective plane, we can then think of specializing one line—the one containing all the ideal points (points at infinity or intersections of parallel lines). If this line is removed from the projective plane, the resulting geometry is called *affine*, a term first used by Euler about 1750. The transformations of affine geometry form a proper subgroup of the group of all projective transformations. The group of affine transformations has the group of similarities and the group of motions as proper subgroups. Thus, affine geometry, in which parallelism plays a significant role, is a geometry between Euclidean and projective geometry in the sense that it is more general than Euclidean geometry, but only a subgeometry of the more general projective geometry.

▲ EXERCISES 7.2 ▲

1. Give at least one reason why we should expect the postulational basis for projective geometry to be simpler than that for Euclidean geometry.
2. Which of the five axioms for projective geometry are also axioms for ordinary Euclidean plane geometry?

For Exercises 3–9, which of the axioms for projective geometry are also axioms for the finite geometry named?

3. Geometry of three points and three lines
4. Geometry of four points
5. Geometry of four lines
6. Geometry of Fano
7. Geometry of Pappus
8. Geometry of Desargues
9. Geometry of Young
10. Prove that, in projective geometry, two distinct lines cannot have more than one point in common.

In Exercises 11 and 12, use the set of points shown in Figure 7.8.

11. If $AEGC$ is considered the basic complete quadrangle, how many of its diagonal points are labeled?
12. If $AEBF$ is considered the basic complete quadrangle, how many of its diagonal points are labeled?
13. Suppose the four vertices of a complete quadrangle are the vertices of a parallelogram in Euclidean geometry. What effect would this have on the diagonal points?
14. Could a triangle be a diagonal triangle for more than one complete quadrangle? Make a drawing to support your answer.

For Exercises 15–19, use the axioms for projective geometry.

15. Prove that for a given line there exists a point not on that line.
16. Prove that every point lies on at least three lines.

17. Prove that there exist four coplanar lines of which no three are concurrent.
18. Prove the existence of a diagonal triangle for a given quadrangle.
19. Prove that a perspectivity establishes a one-to-one correspondence between the points on two lines.
▷ 20. Investigate the consequences of the axioms in terms of the minimum number of points that must lie on each projective line.
▷ 21. Find a set of axioms for affine geometry, and compare that geometry with both Euclidean and projective geometry.

▲ 7.3 Duality and Some Consequences

The concept of duality, a basic idea in projective geometry, does not appear in Euclidean geometry. In Euclidean geometry, two points always determine exactly one line, but two lines in the same plane do not always determine a point (they may be parallel). In projective geometry, on the other hand, this exception has been eliminated. A comparison of the statements "two points determine a line" and "two lines determine a point," with respect to a plane, shows that one may be changed to the other by interchanging the words *point* and *line*. This is an example of *plane duality*. For sets of points in the projective plane (the plane of projective geometry), every statement remains true when the words *point* and *line* are interchanged if certain other pairs of words, such as *collinear* and *concurrent*, are changed accordingly. For example, the plane dual of the expression "three collinear points" is "three concurrent lines."

The set of plane axioms for projective geometry have plane duals that are axioms or theorems. Axioms 1 and 2 involve dual concepts. The plane dual of Axiom 3 is a theorem that can be proved.

THEOREM 7.1 ➤ There exist four lines of which no three are concurrent.

PROOF: Let A, B, C, and D in Figure 7.10 be four points, no three of which are collinear. By Axiom 1, these four points determine the four lines $\overleftrightarrow{AB}$, $\overleftrightarrow{AD}$, $\overleftrightarrow{CD}$, and $\overleftrightarrow{BC}$. No three of these lines can be concurrent without violating Axioms 1 or 2.

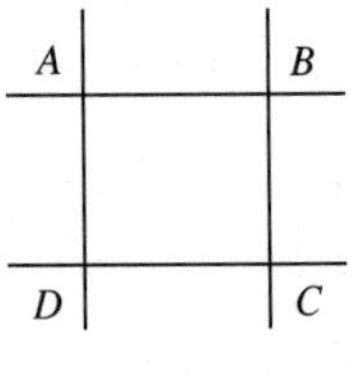

Figure 7.10

Axioms 3 and 4 also involve dual concepts. The plane dual of Axiom 4 introduces new terminology that must be explained.

DEFINITION ➤ A *complete quadrilateral* is a set of four lines, no three concurrent, and the points of intersection of these lines in pairs (see Figure 7.11).

The dual of Axiom 4 states that the three diagonal lines of a complete quadrilateral are never concurrent. The definition of a complete quadrilateral was written by taking the plane dual of the definition of a complete quadrangle.

The properties of the complete quadrilateral can be studied as the plane dual of the corresponding properties of the complete quadrangle. The six points of intersection of the four given lines are in three sets of opposite vertices. In Figure 7.11, A and B are opposite vertices, as are C and D and E and F. The lines joining opposite vertices are diagonal lines, and the figure formed by the three diagonal lines is the *diagonal trilateral*. In projective geometry, it is customary to speak of a triangle as being self-dual, since it includes both the vertices and the sides, and to use diagonal triangle for both the complete quadrangle and the complete quadrilateral.

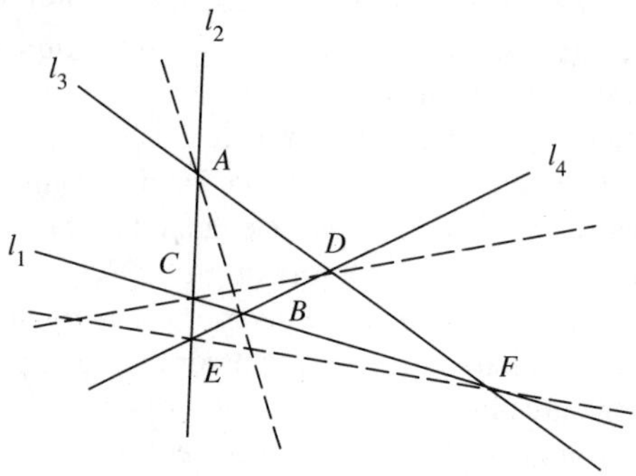

Figure 7.11

The dual of Axiom 4 implies that the dotted lines in Figure 7.11 cannot be concurrent. This should be easy to see, because if they were concurrent, then the three diagonal points of quadrangle $ABCD$ would be collinear, contrary to Axiom 4.

Finally, the dual of Axiom 5 states that if a projectivity leaves invariant each of three distinct lines on a point, it leaves invariant every line on the point. The proof that this dual is a true statement (see Exercises 7.3) results in the completion of the argument that all the duals of the axioms of projective geometry are true and that the concept of plane duality may be used freely. It is assumed that the plane dual of a definition or proved theorem is a definition or another theorem whose proof has been established without further work.

Recall that the concept of plane duality has an important application in helping to classify various finite geometries. In Chapter 1, the axioms for the four-point and four-line geometries are plane duals of each other. The finite projective geometries $PG(n, q)$ of Fano are self-dual, which means that for them, as in the projective geometry of Chapter 7, the plane dual of each true statement is a true statement for the geometry.

The concept of *space duality* is based on an interchange of the words *point* and *plane*, with the word *line* being self-dual in space.

Example ➤ In the following statements, the second gives the space dual for the first.

Any two distinct planes have at least two common points. Space dual: Any two distinct points have at least two common planes.

One of the most fundamental, yet nontrivial, applications of the concept of plane duality is involved in the proof of Desargues's theorem and its converse. Recall that this theorem was first encountered in a finite geometry in Chapter 1. Desargues's definition of triangles perspective from a point and line is made clear by Figure 7.12. Two triangles such as ABC and $A'B'C'$ are perspective from point O, since corresponding vertices are collinear with O. Two triangles such as DEF and $D'E'F'$ are perspective from line l, since corresponding sides meet at points X, Y, and Z on line l.

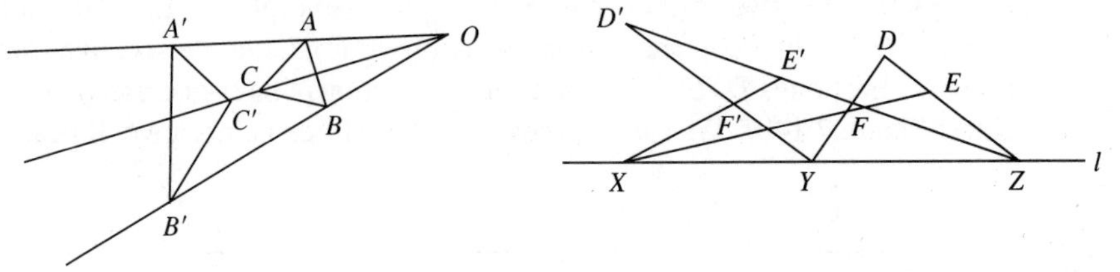

Figure 7.12

THEOREM 7.2 ➤ *Desargues's theorem.* If two triangles are perspective from a point, they are perspective from a line.

Unlike most theorems of geometry, Desargues's theorem is easier to prove for two triangles in different planes than for two triangles in the same plane.

PROOF:

a. Assume that the two given triangles are in different planes, as in Figure 7.13. Since O, B', C', B, C lie in a plane, $\overleftrightarrow{BC}$ and $\overleftrightarrow{B'C'}$ must meet, and this point must be a point Z on the line of intersection of the two planes π and π'. Similarly, O, A', C', A, C determine a plane, and $\overleftrightarrow{AC}$ and $\overleftrightarrow{A'C'}$ meet at a point Y on the line of intersection of π and π'. Lines $\overleftrightarrow{AB}$ and $\overleftrightarrow{A'B'}$, likewise, must meet on the same line of intersection of π and π', so $\triangle ABC$ and $\triangle A'B'C'$ are perspective from a line.

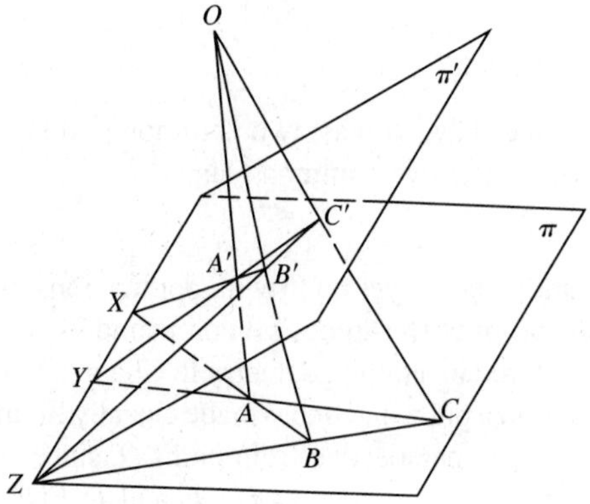

Figure 7.13

b. Assume that the two given triangles are in the same plane. Let ABC and $A'B'C'$ be any two coplanar triangles, as in Figure 7.14a, perspective from point O. Let points D, E, F be points of intersection of pairs of corresponding sides (Figure 7.14b). We need to show that these three points are collinear.

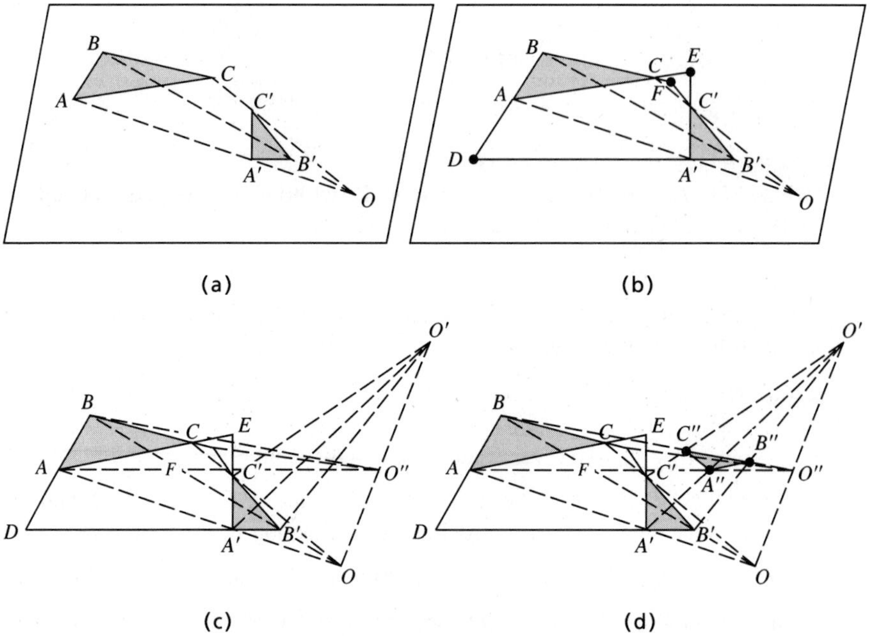

Figure 7.14

Let O' and O'' be any two points collinear with O, but not on the plane of the given triangles (Figure 7.14c). Connect O' and the vertices of triangle $A'B'C'$ and O'' and the vertices of triangle ABC. Since the lines $O'O$ and OA determine a plane, AO'' and $O'A'$ lie in this plane and must meet at a point A''. Similarly, $O''B$ and $O'B'$ meet at B'', and $O''C$ and $O'C'$ meet at C''. Triangles ABC and $A''B''C''$ are in different planes and are perspective from O''. By Desargues's theorem for noncoplanar triangles, they are perspective from a line common to their two planes. Also, triangles $A'B'C'$ and $A''B''C''$ are noncoplanar and perspective from O', so they are also perspective from the same common line. The result is that $\overleftrightarrow{AB}$ and $\overleftrightarrow{A'B'}$ meet at a point on the common line of the plane containing the given triangles and the plane of triangle $A''B''C''$. Similarly, $\overleftrightarrow{BC}$ and $\overleftrightarrow{B'C'}$ and $\overleftrightarrow{AC}$ and $\overleftrightarrow{A'C'}$ meet on this same line and the two given triangles are perspective from this line, $\overleftrightarrow{DE}$.

From the principle of planar duality, the plane dual of Desargues's theorem has also been established.

THEOREM 7.3 ➤ If two triangles are perspective from a line, they are perspective from a point.

The plane dual of Desargues's theorem is also its converse. This is not usual, but neither is it a unique case.

Several other remarks should be made about Figure 7.14. The set of points and lines in the plane of the two given triangles is called the *Desargues's configuration.* It consists of ten points with three lines on each point and ten lines with three points on each line. Observe that this has already been used as the basis for the finite geometry of Desargues in Chapter 1. This is an example of a finite projective geometry. In Desargues's configuration are not only one pair of perspective triangles but a total of ten pairs, each perspective from a different point and line.

▲ EXERCISES 7.3 ▲

1. A complete quadrilateral has how many pairs of opposite vertices?

2. A complete quadrangle has how many diagonal points?

Write the plane dual of each statement.

3. If a and b are distinct lines of a plane, there is at least one point on both lines.

4. On a point are at least four lines.

5. A triangle consists of three noncollinear points and the lines joining them in pairs.

6. The three diagonal lines of a complete quadrilateral cannot be concurrent.

Write the space dual of each of the next three statements.

7. Three planes not on the same line determine a point.

8. A plane is determined by two intersecting lines.

9. A point and a line in the same plane might not be incident.

10. Explain why the word *line* is self-dual in space.

11. Prove the plane dual of Axiom 5.

12. Prove the converse of Desargues's theorem directly for triangles in different planes, without assuming the original theorem or using duality.

13. Draw a figure to represent Desargues's configuration, then indicate the ten pairs of perspective triangles, naming the center and axis of perspectivity for each pair.

14. Prove that if three triangles have a common center of perspectivity, their three axes of perspectivity have a common point.

15. Prove that the diagonal triangle of a quadrangle is perspective with each of the four triangles whose vertices are three of the vertices of the quadrangle.

16. For two coplanar triangles to be perspective from a point is a very special case. Given one triangle, explain the restrictions that must apply to the selection of each vertex of the second triangle so that the two triangles will be perspective from a given point in the plane.

17. Given one triangle, explain the restrictions that must apply to the selection of a second coplanar triangle so that the two triangles will be perspective from a given line in the plane.

▷ **18.** For Theorem 7.2, would it be possible for the line of perspectivity to pass through the point of perspectivity?

▷ **19.** Prove that your answer for Exercise 18I is correct.

▶▷ **20.** Prepare a physical model for Figure 7.13. Use two sheets of cardboard, taped together, for the planes. Let point O be inside the angle formed by the planes, and use rubber bands for $\overline{CC'}$, $\overline{BB'}$, and $\overline{AA'}$.

▶▷ **21.** For the drawing in Exercise 13, prepare separate overlays to show each of the ten pairs of perspective triangles.

▲ 7.4 Harmonic Sets

Harmonic sets are special sets of four points or lines that have great significance in projective geometry. Since a complete quadrangle consists of four points and six lines, an arbitrary line of the plane will meet the six sides in six distinct points. This set of six points, not necessarily all distinct, is called a *quadrangular set of points*. Figure 7.15 shows complete quadrangle $ABCD$ and the quadrangular set of points E, F, G, H, I, J. Desargues's theorem can be used to show that each point of the quadrangular set can be determined uniquely if the other five are known, where G and H, E and J, F and I are on opposite sides of the new quadrangle.

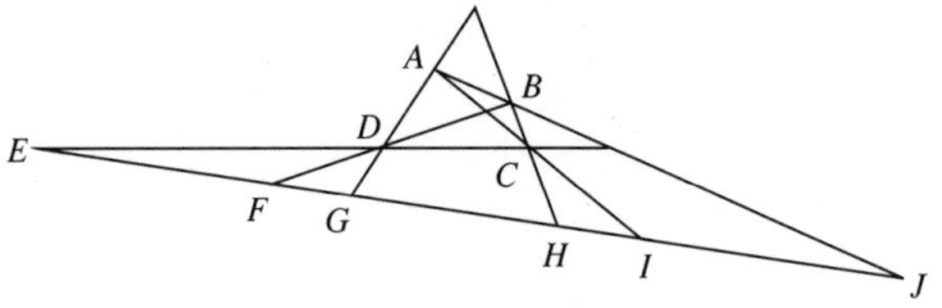

Figure 7.15

DEFINITION ➤ A harmonic set of points, or a *harmonic range*, is a quadrangular set of points consisting of the four points of intersection of the sides of a complete quadrangle with a line passing through two diagonal points.

Figure 7.16 shows a harmonic set of points, $EFXY$, consisting of the four points of intersection of the sides of complete quadrangle $ABCD$, with a line passing through diagonal points X and Y. Theorem 7.4 will show that three of the points (when properly paired) uniquely determine the fourth point of the harmonic set.

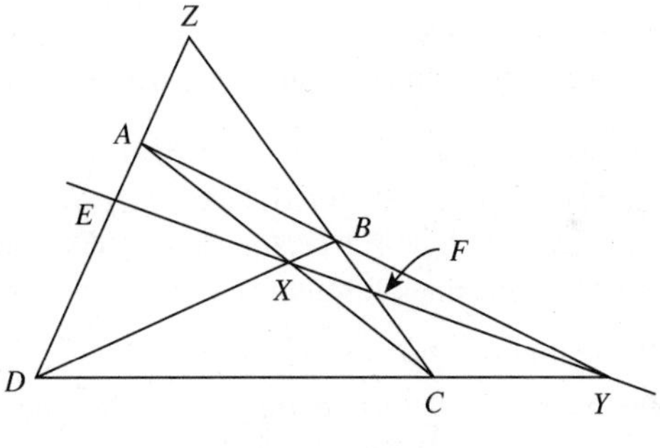

Figure 7.16

Figure 7.17 shows three harmonic sets of points determined by complete quadrangle $ABCD$. The four points $NFIG$ are a harmonic set of points on the line passing through the diagonal points F and G. Similarly, the four points $LEMG$ are a harmonic set on the line through diagonal points E and G, whereas the points $EJFK$ are a harmonic set on the line through diagonal points E and F. In each case, two of the four points of the harmonic set are diagonal points, and the other two points are on the sides passing through the third diagonal point. A special notation for harmonic sets emphasizes this distinction between the two pairs of points in the set. For example, $H(FG, NI)$ indicates the harmonic set composed of the four points F, G and N, I. One pair, either FG or NI, indicates the two diagonal points, whereas the other two points are the points of intersection of the sides of the quadrangle through the third diagonal point. The same harmonic set could be indicated by $H(FG, IN)$, $H(GF, NI)$, or $H(GF, IN)$.

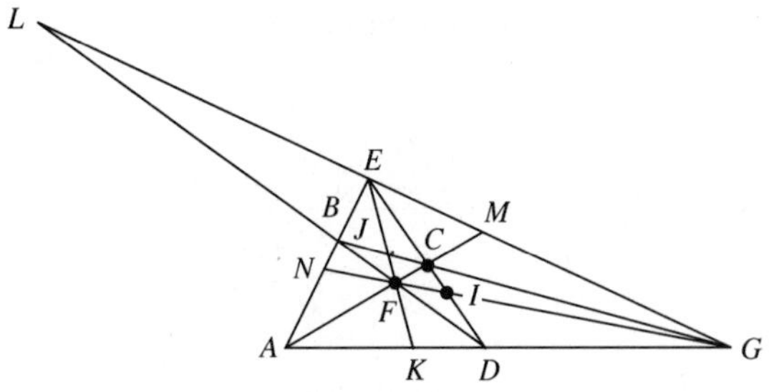

Figure 7.17

In the notation $H(FG, NI)$, N is the *harmonic conjugate* of I with respect to F and G. Each point of the harmonic set is the harmonic conjugate of the other member of its pair with respect to the other pair of points. Since Axiom 4 specifies that the three diagonal points of a complete quadrangle are not collinear, a point and its harmonic conjugate must be distinct points, so there are at least four points on each line in a projective geometry.

The four points of a harmonic set cannot all be located independently. A statement of dependence is given in the following theorem.

THEOREM 7.4 ➤ The harmonic conjugate of a point A with respect to two other given collinear points B and C is uniquely determined.

Proof: In Figure 7.18, let B, A, C be any three given collinear points. A harmonic conjugate D of A with respect to B and C, $H(BC, AD)$, can be found by constructing complete quadrangle $EFGH$ so that B and C are two of the diagonal points, and A and D lie on lines through the third diagonal point. This can be done by choosing any point F not on $\overleftrightarrow{AB}$ and connecting F and points A, B, C. A line can be drawn through B intersecting $\overleftrightarrow{FA}$ at H and $\overleftrightarrow{FC}$ at G. Let $\overleftrightarrow{CH}$ intersect $\overleftrightarrow{BF}$ at E. Then $\overleftrightarrow{EG}$ intersects $\overleftrightarrow{AB}$ at the required point D.

The fact that D is unique is somewhat more difficult to prove. Assume, as in Figure 7.18, that a second quadrangle $E'F'G'H'$ has been constructed as before, beginning with the choice of F' as a point not on $\overleftrightarrow{AB}$ and also distinct from F. It is required to show that $\overleftrightarrow{E'G'}$ intersects $\overleftrightarrow{AB}$ at D.

Triangles FGH and $F'G'H'$ are perspective from line $\overleftrightarrow{AB}$, and hence, are perspective from a point J by the converse of Desargues's theorem. Triangles FEH and $F'E'H'$ are also perspective from line $\overleftrightarrow{AB}$ and are then perspective from the same point J, so $EE'J$ are collinear. This means that triangles FEG and $F'E'G'$ are perspective from J and, by Desargues's theorem, are perspective from a line. This line is $\overleftrightarrow{AB}$, and $\overleftrightarrow{EG}$ and $\overleftrightarrow{E'G'}$ must meet at a point D on $\overleftrightarrow{AB}$; thus, the harmonic conjugate of A with respect to B and C is a unique point.

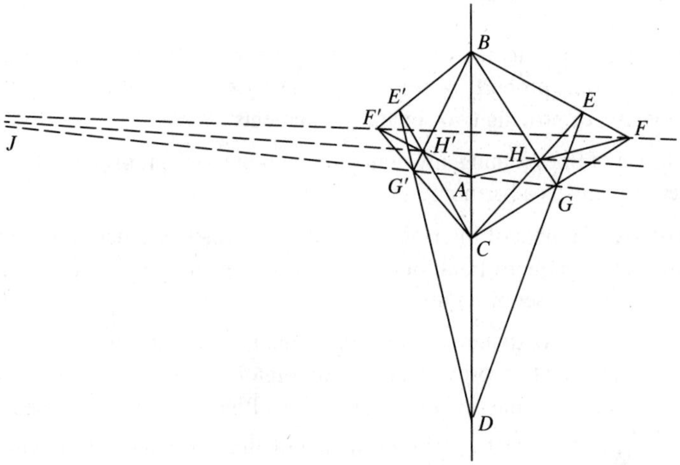

Figure 7.18

The proof of Theorem 7.4 includes a method of constructing the harmonic conjugate of a point with respect to two other given collinear points. This construction, as anticipated in Section 5.6, uses only a straightedge. You should be able to write down the steps in this construction to use in the exercises.

The dual of the definition of a harmonic set of points is the definition of a harmonic set of lines.

DEFINITION ➤ A *harmonic set of lines*, or a *harmonic pencil*, is a set of four concurrent lines such that two of them are diagonal lines of a complete quadrilateral and the other two pass through the two vertices lying on the third diagonal line.

In Figure 7.19, let a, b, c, and d be the sides of the quadrilateral, with e, f, g the diagonal lines. One set of harmonic lines is $H(eg, hi)$.

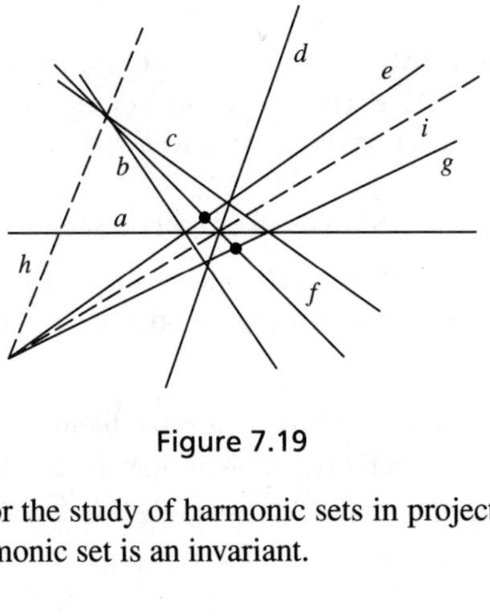

Figure 7.19

The main reason for the study of harmonic sets in projective geometry is that the property of being a harmonic set is an invariant.

THEOREM 7.5 ➤ The harmonic property is preserved under projectivities.

Proof: A projectivity is a finite sequence of perspectivities, so it suffices to show that the harmonic property is preserved in a single perspectivity. To prove that this is so consists of proving two separate statements:

a. The set of lines joining any noncollinear point to the four points of a harmonic set of points is a harmonic set of lines.

b. The set of points of intersection of the four lines of a harmonic set with any line not through the point of concurrency of the lines (the center of the pencil) is a harmonic set of points.

Since the two statements are plane duals, proving one will be sufficient. The first statement is proved here. In Figure 7.20, let $H(AB, CD)$ be any harmonic set of points, with O any noncollinear point. We need to show $H(\overleftrightarrow{OA}, \overleftrightarrow{OB}, \overleftrightarrow{OC}, \overleftrightarrow{OD})$. The existence of the harmonic set of points implies the existence of a complete quadrangle $OEFG$, with A and B two diagonal points and C and D points on lines through the third diagonal point. But $\overleftrightarrow{GF}$, $\overleftrightarrow{AE}$, $\overleftrightarrow{AB}$, and $\overleftrightarrow{GE}$ are the four sides of a complete quadrilateral. $\overleftrightarrow{AO}$ and $\overleftrightarrow{OB}$ are diagonal lines, whereas $\overleftrightarrow{OC}$ and $\overleftrightarrow{OD}$ are lines through the other two vertices, F and D, which lie on the third diagonal line. This implies $H(\overleftrightarrow{OB}, \overleftrightarrow{OA}, \overleftrightarrow{OC}, \overleftrightarrow{OD})$.

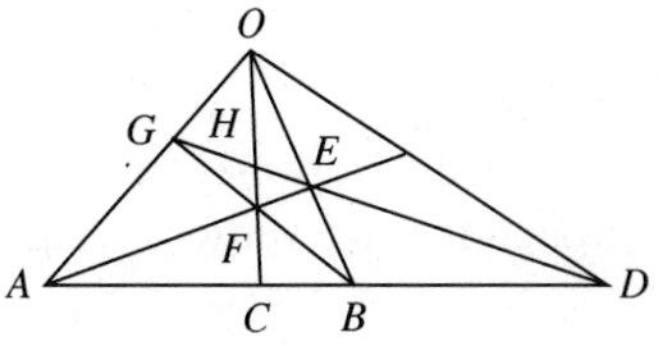

Figure 7.20

A symbol for a perspectivity, $ABCD \overset{S}{\barwedge} A'B'C'D'$, indicates that S is the center of perspectivity and A and A', B and B', ... are corresponding points. The symbol $ABCD \barwedge A'B'C'D'$ indicates a projectivity in which AA', BB', ... are corresponding points.

A third type of notation, used in the next section shows a perspectivity of a set of lines and points. The notation $(\overleftrightarrow{AA'}, \overleftrightarrow{AB'}, \overleftrightarrow{AC'} -) \barwedge (A'B'C'-)$ shows a perspective relationship between the set of points $A'B'C'-$ on a line and the set of lines through a point A not on the same line.

Example ➤ Use the given information about harmonic sets of points in Figure 7.17, along with a perspectivity, to show that $H(EI, CD)$ is a harmonic set. $H(EF, JK)$ is a given harmonic set. Since $EFJK \overset{G}{\barwedge} EICD$, $H(EI, CD)$ is also a harmonic set.

Beginning with three distinct points on a line, the harmonic conjugate of one of these points with respect to the other two can be uniquely determined. For the set of four points, sets of three determine other harmonic conjugates. All the harmonic conjugates on a line are said to be *harmonically related* to the three original points.

DEFINITION ➤ The set of all points harmonically related to three distinct collinear points is a *harmonic net* of points.

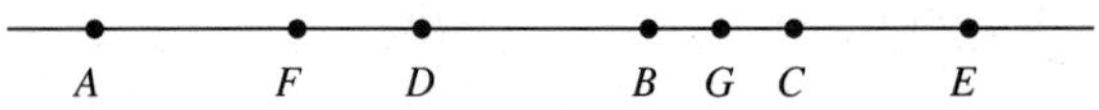

Figure 7.21

Figure 7.21 shows just a few of the points in the harmonic net determined by points A, B, C. The harmonic sets are $H(AB, CD)$, $H(AC, BE)$, $H(AD, BF)$, and $H(BC, EG)$.

The concept of harmonic net is fundamental in the development of a system of coordinates for points in projective geometry (see Section 7.6).

▲ EXERCISES 7.4 ▲

For Exercises 1 and 2, given three points A, B, C in this arrangement on a line:

1. Locate D by construction with a straightedge so that $H(AC, BD)$ is a harmonic set.

2. Locate D by construction with a straightedge so that $H(AB, CD)$ is a harmonic set.

3. Suppose, in the Euclidean sense, that B is the midpoint of $\overline{AC}$. Describe the location of the harmonic conjugate of B with respect to A and C.

For Exercises 4–8, which of these are harmonic sets in Figure 7.17? Why?

4. (AD, KG) **5.** (AC, FM) **6.** (AB, NE) **7.** (LF, BD) **8.** (BC, JG)

9. Prove statement b under Theorem 7.5 directly without using duality.

10. Prove that the sixth point of a quadrangular set of points can be determined uniquely if the other five are known.

11. Prove that a line in projective geometry has more than four distinct points.

12. Suppose that $ABCD$ is a quadrangle and PQR is its diagonal triangle. Show that the other three vertices of the quadrangle can be found if triangle PQR and point A are given.

13. Prove that a harmonic net is determined by any three distinct points on it, not just the three original beginning points.

14. Construct a moving model for Figure 7.20, so that C and D move along the line, while A, B, and O stay fixed.

15. If $H(AB, CD)$ and A, B are diagonal points of a complete quadrangle, show that this implies $H(CD, AB)$ with C, D the diagonal points of a complete quadrangle.

16. Construct ten more points of a harmonic net, given the three collinear points determining the net.

7.5 Projectivities

A projectivity, defined as a finite sequence of perspectivities, is a transformation, since it is a one-to-one onto mapping of points. The inverse of a projectivity is a projectivity, and the product of two projectivities is a projectivity; thus, the following theorem is true (Exercise 16, Exercises 7.5).

THEOREM 7.6 ➤ The set of all projectivities for which the product is defined in a plane constitutes a group of transformations.

Projective geometry is the study of properties of sets of points that are invariant under the group of projective transformations. The algebraic approach to the study of projective geometry must wait until the introduction of coordinates for the projective plane in the next section. Before this, however, it is possible to study projectivities from a different viewpoint. A question that may have occurred to you is: What is the minimum information needed to determine a projectivity? Since projectivities involve points and their images, how many pairs of points and images must be given so that all pairs can be determined? The answers to these questions appear in what is called the *fundamental theorem of projective geometry.*

THEOREM 7.7 ➤ *Fundamental theorem.* A projectivity between the sets of points on two lines in a plane is uniquely determined by three collinear points and their images.

PROOF: "Uniquely determined" means that there cannot be two different transformations mapping the three given points to their given images. The image of each point X is fixed as one certain point by the given information. The proof is indirect. Assume, as illustrated in Figure 7.22, that $ABCX \,\overline{\wedge}\, A'B'C'X'$, and $ABCX \,\overline{\wedge}\, A'B'C'X''$, for X' and X'' distinct points.

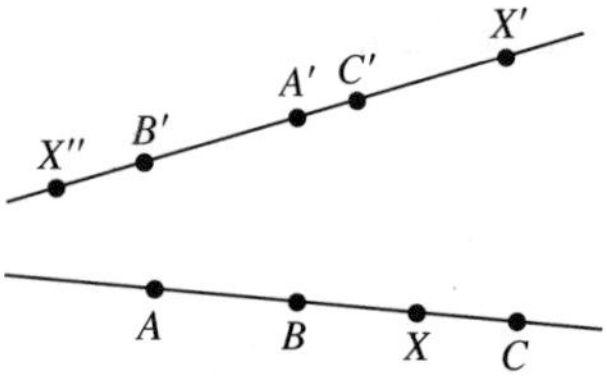

Figure 7.22

Then $A'B'C'X' \,\overline{\wedge}\, ABCX \,\overline{\wedge}\, A'B'C'X''$, and $A'B'C'X' \,\overline{\wedge}\, A'B'C'X''$. But in this last projectivity, points A', B', and C' are fixed. By Axiom 5 of Section 7.2, all the other points on the line must also be fixed. The assumption that X' and X'' are distinct contradicts the axiom, so X' and X'' must be identical.

While the fundamental theorem of projective geometry is very important from a theoretical point of view, it does not yet provide a constructive method for actually determining additional pairs of corresponding points in a projectivity. One such method is based on establishing a minimum sequence of perspectivities for the projectivity.

THEOREM 7.8 ➤ Three distinct points A, B, C on one line may be projected into any three distinct points A', B', C' on a second line by means of a sequence of at most two perspectivities.

PROOF: In Figure 7.23, let A, B, C, A', B', C' represent the six given points. Choose a center of perspectivity other than A or A' on $\overleftrightarrow{AA'}$ and a line l (not containing B', C') through A'. Then points B'' and C'' are determined on l so that $ABC \stackrel{S}{\overline{\wedge}} A'B''C''$. The intersection of $\overleftrightarrow{C'C''}$ and $\overleftrightarrow{B'B''}$ determines a point S' such that $A'B''C'' \stackrel{S'}{\overline{\wedge}} A'B'C'$, so $ABC \,\overline{\wedge}\, A'B'C'$.

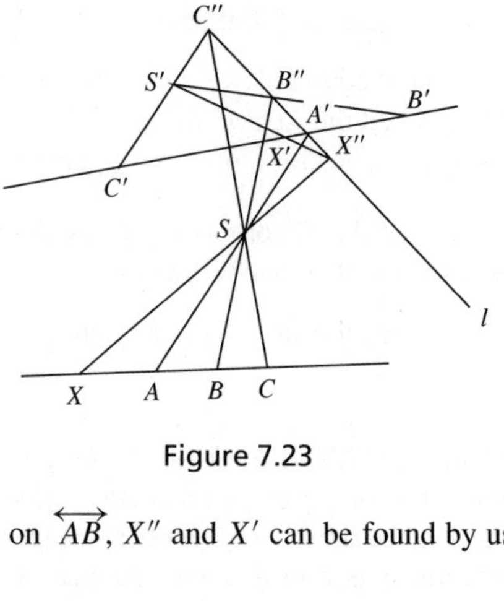

Figure 7.23

Given any point X on $\overleftrightarrow{AB}$, X'' and X' can be found by using the perspectivities already defined.

Example ➤ Find the image of point D in the projectivity established between A, B, C and $A'B'C'$ in Figure 7.24a.

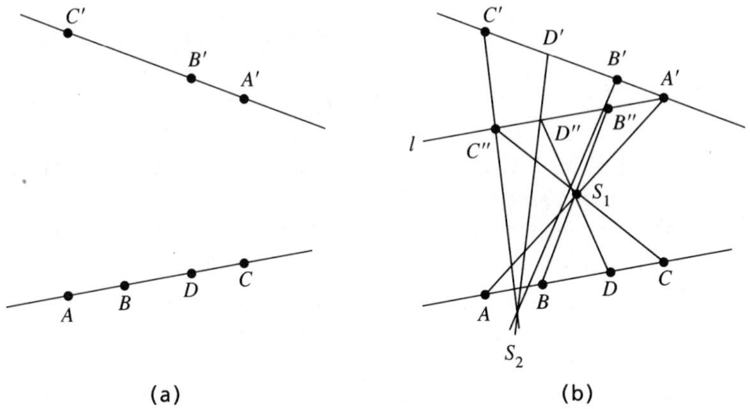

Figure 7.24

Figure 7.24b shows the solution determined by setting up the two perspectivities $ABC \overset{S_1}{\barwedge} A'B''C''$ and $A'B''C'' \overset{S_2}{\barwedge} A'B'C'$.

If the six given points all lie on the same line, then one extra perspectivity is required. See Exercise 7 of Exercises 7.5. If one of the given points is its own image in the projectivity, then the projectivity is a perspectivity. See Exercise 11.

The corresponding points in a projectivity are not, in general, perspective from a point, nor are the corresponding lines in a projectivity perspective from a line. But projectivities do have points or lines that are somewhat analogous to centers and axes of perspectivity. These concepts are useful in Section 7.10.

THEOREM 7.9 ➤ A projectivity between two sets of points on two distinct lines determines a third line called the *axis of projectivity*, or *axis of homology*, which contains the intersections of the cross joins of all pairs of corresponding points.

This theorem is often called the Theorem of Pappus, after Pappus of Alexandria, who proved the Euclidean version in about 300 A.D.

DEFINITION ➤ The *cross joins* of two pairs of points, such as A, A' and B, B', are $\overleftrightarrow{AB'}$ and $\overleftrightarrow{BA'}$.

Proof: To prove Theorem 7.9, it must be shown that the axis is unique and that all of the intersections of the cross joins are collinear. This can be accomplished by showing that the axis is independent of the choices of points. See Figure 7.25a. Since there is a projectivity between the points on l_1, and l_2,

$(\overleftrightarrow{A'A}, \overleftrightarrow{A'B}, \overleftrightarrow{A'C} -) \overline{\overline{\wedge}} (ABC-) \overline{\wedge} (A'B'C'-) \overline{\overline{\wedge}} (\overleftrightarrow{AA'}, \overleftrightarrow{AB'}, \overleftrightarrow{AC'} -)$, so that $(\overleftrightarrow{A'A}, \overleftrightarrow{A'B}, \overleftrightarrow{A'C} -) \overline{\wedge} (\overleftrightarrow{AA'}, \overleftrightarrow{AB'}, \overleftrightarrow{AC'} -)$.

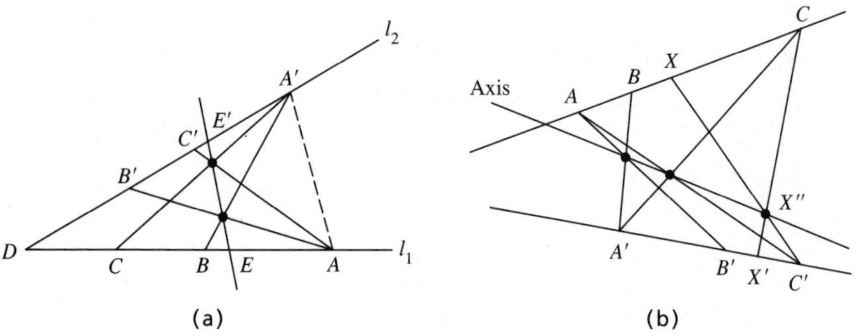

Figure 7.25

In this projectivity, $A'A$ is a self-corresponding line, so the projectivity is equivalent to a perspectivity whose line passes through the intersection of $\overleftrightarrow{A'B}$ and $\overleftrightarrow{AB'}$, $\overleftrightarrow{A'C}$ and $\overleftrightarrow{AC'}$, and so on.

If another pair of points, such as B, B' for example, had been taken as the centers of the projective pencils, rather than A, A', the same axis of projectivity would have been determined. This is true because in any event, the axis always passes through the two images of the common point of the two lines. In Figure 7.25a, if E' is the image of D considered as a point on l_1, and if E is the image of D considered as a point on l_2, then for the two pairs of points B, B' and D, E', the cross joins meet at E' on the axis of projectivity. For the two pairs of points B, B' and D, E, the cross joins meet at E on the axis of projectivity.

The axis of projectivity may be used to construct additional pairs of corresponding points in a projectivity, as shown in the following example. In Figure 7.25b, assume that a projectivity is determined by the given pairs of points A, A', B, B', C, C'. The axis of projectivity can be determined by the intersection of the cross joins $\overleftrightarrow{AC'}$, $\overleftrightarrow{A'C}$ and $\overleftrightarrow{AB'}$, $\overleftrightarrow{A'B}$. Let X represent any fourth point on $\overleftrightarrow{AB}$. Now $\overleftrightarrow{XC'}$ intersects the axis of projectivity at a point X'', and $\overleftrightarrow{CX''}$ meets $\overleftrightarrow{A'C'}$ at the required point X' on $\overleftrightarrow{A'C'}$. Note that this construction uses only a straightedge.

The duals of the theorems already presented in this section are interesting in their own right. The proofs given depend on the concept of duality, even though the theorems could be proved directly.

THEOREM 7.10 ➤ *Dual of the fundamental theorem.* A projectivity between the sets of lines on two points in a plane is determined by three concurrent lines and their images.

THEOREM 7.11 ➤ *Dual of Theorem 7.9.* A projectivity between two sets of lines on two distinct points determines a third point, called the *center of projectivity*, or *center of homology*, which lies on the joins of the cross intersections of corresponding lines.

Theorem 7.11 is illustrated in Figure 7.26. Lines a, b, c and a', b', c' are given. Point X is the intersection of a and b', and Y is the intersection of a' and b, so $\overleftrightarrow{XY}$ passes through S, the center of projectivity. Likewise, c and b' intersect at W, and c' and b meet at Z, so S lies on $\overleftrightarrow{WZ}$.

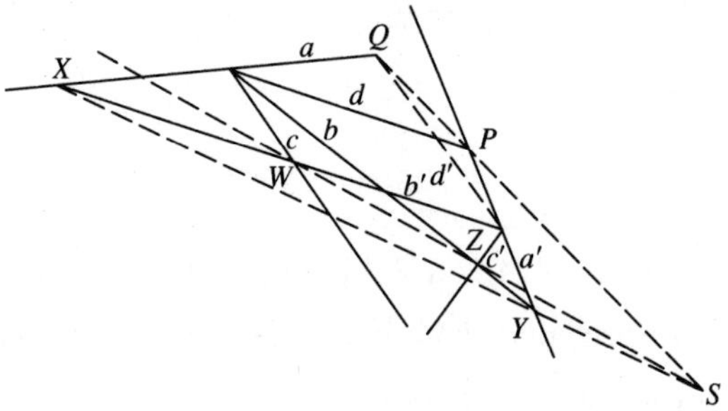

Figure 7.26

The center of projectivity can be used to construct additional pairs of corresponding lines in a projectivity. In Figure 7.26, suppose that d is given. Lines d and a' meet at P. Line a meets d' at point Q on $\overleftrightarrow{PS}$. The line connecting Q and the point of concurrency of a', b', c' is the desired line d'.

▲ EXERCISES 7.5 ▲

For Exercises 1–3, use Figure 7.27a.

1. Use the method in the proof of Theorem 7.8 to establish two perspectivities so that a projectivity is established between A, B, C and A', B', C'.

2. Construct the axis of projectivity for the projectivity with A, A', B, B', and C, C' as corresponding points.

3. Find the image of point X in the projectivity of Exercise 2.

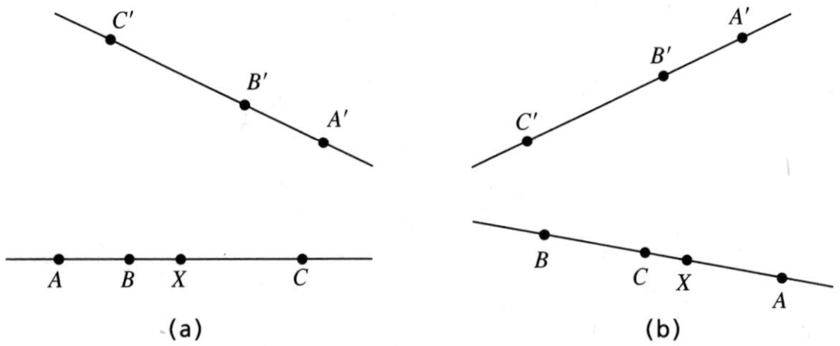

Figure 7.27

For Exercises 4–6, use Figure 7.27b.

4. Follow the instructions in Exercise 1.

5. Follow the instructions in Exercise 2.

6. Find the image of point X in the projectivity of Exercise 5.

7. Prove that three distinct points A, B, C on a line can be projected into any three distinct points A', B', C' on the same line by means of a chain of three perspectivities.

8. Copy Figure 7.28. Set up a chain of three perspectivities so that A and A', B and B', and C and C' correspond.

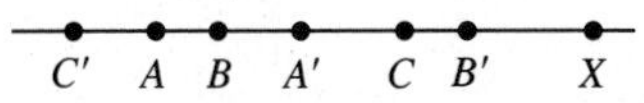

Figure 7.28

9. For the projectivity in Exercise 8, find the image of X.

10. Show how three collinear points A, B, C can be projected into the corresponding points B, C, A by a chain of three perspectivities.

11. For the projectivity of Theorems 7.7 and 7.8, prove that if one of the given points is its own image, the projectivity is a perspectivity.

12. Copy Figure 7.29a. Find the center of projectivity for the projectivity with aa', bb', and cc' as corresponding lines.

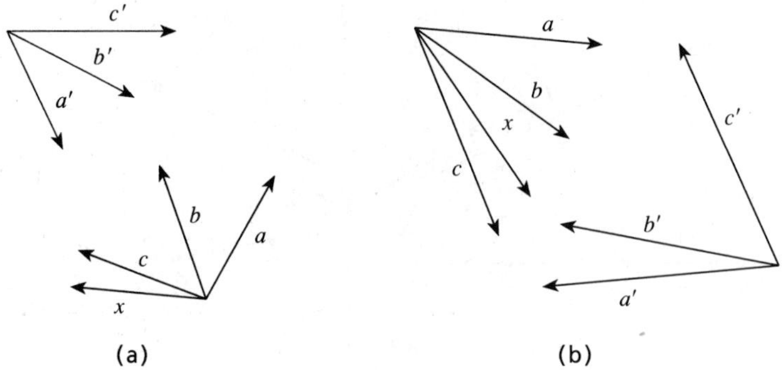

Figure 7.29

13. Find the image of line x in the projectivity of Exercise 12.
14. Copy Figure 7.29b. Find the center of projectivity for the projectivity with aa', bb', and cc' as corresponding lines.
15. Find the image of line x in the projectivity of Exercise 14.
16. Complete the details of the proof of Theorem 7.6.
17. Prove the plane dual of the fundamental theorem, without using duality.
18. Prove that two harmonic sets of points are related by a unique projectivity.
19. Prove Theorem 7.11 by writing the dual of the proof of Theorem 7.9.
20. State and prove the Theorem of Pappus for a plane hexagon. Use a reference text.
21. Make a list of fundamental theorems from other branches of mathematics (such as the fundamental theorems of arithmetic, algebra, and calculus). Explain why Theorem 7.7 is significant enough to be called the fundamental theorem of projective geometry.

7.6 Homogeneous Coordinates

The Cartesian coordinate system of ordinary analytic geometry will not suffice for projective geometry, since ideal points are included; hence, a more general system of coordinates is needed.

Prior to a more formal presentation of the coordinate system of projective geometry, called *homogeneous coordinates*, it is important to see intuitively how the two systems of coordinates are related. In one dimension, a number line for the geometry of the extended Euclidean plane is shown in Figure 7.30a and number lines for projective geometry are shown in Figures 7.30b–d. In Figure 7.30a, the point with coordinate infinity is indicated as an extra point with special properties. But in projective geometry, the point with coordinate infinity may be any place on the line, and is considered an ordinary point. Thus, a new coordinate system must be established so that something other than infinity can be used.

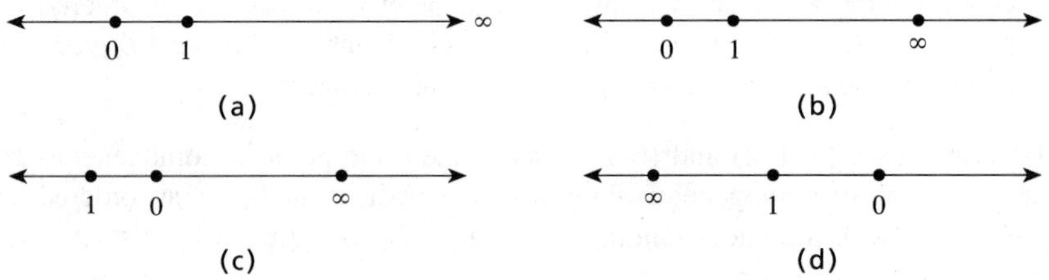

Figure 7.30

Figure 7.31a shows the framework of the ordinary Cartesian coordinate system, with points E and F indicated as "points at infinity." Figure 7.31a may be projected into Figure 7.31b, with the corresponding elements indicated by primes. The ideal points E' and F' are no longer special points, and $\overleftrightarrow{E'F'}$ is the ideal line. The points in the first quadrant of Figure 7.31a are projected into the points in the interior of triangle $A'E'F'$.

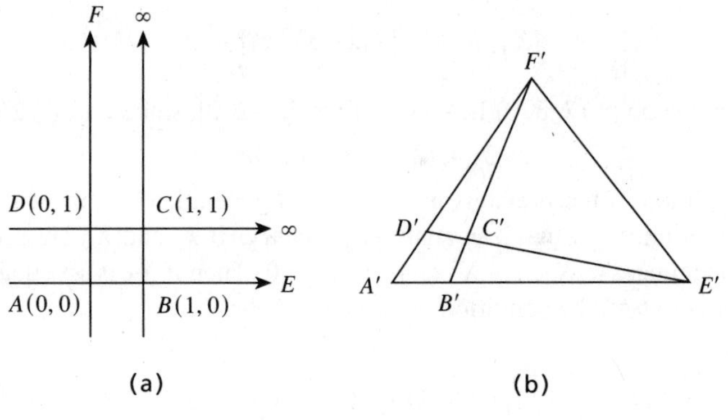

Figure 7.31

The homogeneous coordinates must be defined in such a way that they will distinguish among an infinite number of ideal points and yet not require any restrictions in the way these points are handled. The substitution

$$x = \frac{x_1}{x_3}, \qquad y = \frac{x_2}{x_3} \qquad \text{for } x_3 \neq 0$$

relates the ordinary Euclidean point with coordinates (x, y) to the corresponding point in the projective plane with homogeneous coordinates (x_1, x_2, x_3). For example, (2, 3) becomes (2, 3, 1); (3, 4, 5) is the same point as $\left(\frac{3}{5}, \frac{4}{5}\right)$. For ideal points, the restriction $x_3 \neq 0$ must be removed, and the stage is set for a more formal approach. Homogeneous coordinates of points in the plane were introduced about 1830 by Julius Plücker (1801–1868).

DEFINITION ➤ In analytic projective geometry, a *point* is an ordered triple (x_1, x_2, x_3), not all zero, with the understanding that the notation (ax_1, ax_2, ax_3) represents the same point as (x_1, x_2, x_3) for any nonzero a.

For example, (3, 1, 2) and (6, 2, 4) name the same point in homogeneous coordinates. A line in homogeneous coordinates is defined dually as an ordered triple $[X_1, X_2, X_3]$, with the understanding that $[aX_1, aX_2, aX_3]$ names the same line for any nonzero a. For example,

$$[3, 2, 7] \qquad \text{and} \qquad \left[1, \frac{2}{3}, \frac{7}{3}\right]$$

are two names for the same line. A point x lies on a line X if and only if

$$X_1x_1 + X_2x_2 + X_3x_3 = 0$$

This condition could be written as the vector dot product

$$[X_1, X_2, X_3] \cdot [x_1, x_2, x_3] = 0.$$

For example, the point (2, 3, 0) lies on the line [3, −2, 0], since

$$(2 \cdot 3) + (3 \cdot -2) + (0 \cdot 0) = 0.$$

The condition in the previous paragraph is a generalization of the condition for points lying on lines in Euclidean analytic geometry. If x_3 and X_3 are set equal to 1, the condition reduces to $X_1x_1 + X_2x_2 + (1 \cdot 1) = 0$. Then if X_1 is set equal to a/c and X_2 is set equal to b/c, the condition reduces to

$$\left(\frac{a}{c}\right)x_1 + \left(\frac{b}{c}\right)y_1 + (1 \cdot 1) = 0 \qquad \text{or} \qquad ax_1 + by_1 + c = 0$$

But this is the condition for (x_1, y_1) to lie on the Euclidean line

$$ax + by + c = 0$$

For example, the equation of the line [2, −5, 7] is $2x_1 - 5x_2 + 7x_3 = 0$. Dually, the equation of the point (3, 1, 2) is $3X_1 + X_2 + 2X_3 = 0$. Points do not have equations in Euclidean analytic geometry, but the duality in the projective plane should have prepared you for their appearance in the study of homogeneous coordinates.

It is now possible to relate the definition of homogeneous coordinates to Figure 7.31, recalling the intuitive development in terms of ratios. Study Figure 7.32 carefully, noting the homogeneous coordinates of each point as related to the Cartesian coordinates of the corresponding points in Figure 7.31a.

From Figure 7.32, it is not difficult to determine the equations of the various lines and to relate them to the corresponding lines in Euclidean geometry. The relationships are as follows:

$$\overleftrightarrow{A'E'} \qquad x_2 = 0$$

$$\overleftrightarrow{A'F'} \qquad x_1 = 0$$

$$\overleftrightarrow{E'F'} \qquad x_3 = 0$$

$$\overleftrightarrow{B'C'} \qquad x_1 - x_3 = 0$$

$$\overleftrightarrow{D'C'} \qquad x_2 - x_3 = 0$$

These equations can be derived formally from the condition

$$X_1x_1 + X_2x_2 + X_3x_3 = 0$$

for a point to lie on a line, by substituting the coordinates of the two points on a line and then solving the set of simultaneous equations. For example, for $\overleftrightarrow{C'D'}$,

$$X_2 + X_3 = 0$$
$$X_1 + X_2 + X_3 = 0$$

Let $X_3 = 1$; therefore,

$$X_2 = -1$$
$$X_1 = 0$$

and the equation is $-x_2 + x_3 = 0$ or $x_2 - x_3 = 0$ For the lines given, the equations can also be derived intuitively simply by examining the coordinates to notice a pattern.

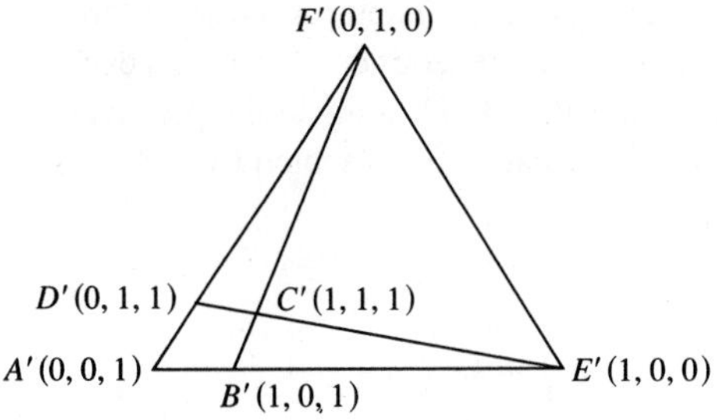

Figure 7.32

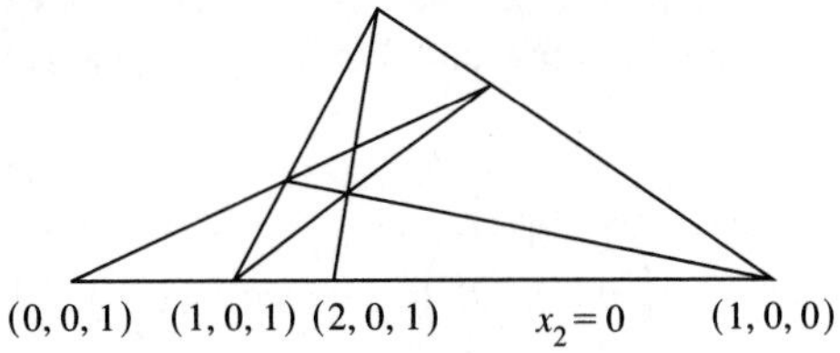

Figure 7.33

Example ➤ Give the homogeneous coordinates of the point of intersection of

$$x_1 + x_2 - x_3 = 0$$
$$x_1 - x_2 + 2x_3 = 0.$$

Because the lines are not parallel in the Euclidean sense, $x_3 \neq 0$. Let $x_3 = 1$; then $x_1 + x_2 = 1$, $z_1 - x_2 = -2$ and $x_1 = -\frac{1}{2}, x_2 = \frac{3}{2}$. The point of intersection is $(-\frac{1}{2}, \frac{3}{2}, 1)$.

It should be understood that the triangle in Figure 7.32 includes not the entire projective plane but only that part corresponding to the first quadrant. The points $A'E'F'$ are vertices of what is called the *fundamental triangle*, whereas C', with coordinates $(1, 1, 1)$, is called the *unit point*.

The procedure by which coordinates are found for additional points in the projective plane is based on the concept of a harmonic net. This procedure is illustrated for the line $x_2 = 0$ in Figure 7.33. The point $(2, 0, 1)$ is defined as the harmonic conjugate of $(0, 0, 1)$ with respect to $(1, 0, 1)$ and $(1, 0, 0)$. The point can be constructed with a straightedge, as explained in Section 7.4. The series of constructions could be carried on to show that $(3, 0, 1)$ is the harmonic conjugate of $(1, 0, 1)$ with respect to $(2, 0, 1)$ and $(1, 0, 0)$. Also, $(\frac{1}{2}, 0, 1)$ is the harmonic conjugate of $(1, 0, 0)$ with respect to $(0, 0, 1)$ and $(1, 0, 1)$. The procedure can be continued for points with negative coordinates. For example, $(-1, 0, 1)$ is the harmonic conjugate of $(1, 0, 1)$ with respect to $(0, 0, 1)$ and $(1, 0, 0)$. Not all the points on the line, but all those in the harmonic net determined by the three beginning points, can be constructed.

Desarguesian and Pappian planes, with an algebraic structure based on the homogeneous coordinates for points, are important particular classes of projective planes. Homogeneous coordinates for points in a finite projective plane also provide a different approach for the study of the finite projective geometries of Chapter 1. For example, the four-line geometry and the finite geometry of Desargues arise from figures now familiar to the reader following the study of projective geometry to this point. See Examples 27I–29I of Exercises 7.6.

Homogeneous coordinates play an important role in computer graphics, which gives projective geometry a new major practical application. Remember that computer graphics conventions are used for the matrices in this chapter and that they may differ from those you have studied in linear algebra. As explained in Chapter 2, a translation cannot be represented as a 2×2 matrix. But, by using homogeneous coordinates, it can be represented as a 3×3 matrix. For example, a translation of 2 units in the positive direction for x and 3 units in the positive direction for y can be represented by

$$\begin{pmatrix} 1 & 0 & 0 \\ 0 & 1 & 0 \\ 2 & 3 & 1 \end{pmatrix}, \quad \text{since} \quad (x \quad y \quad 1\,) \begin{pmatrix} 1 & 0 & 0 \\ 0 & 1 & 0 \\ 2 & 3 & 1 \end{pmatrix} = (x+2 \quad y+3 \quad 1)$$

Each of the previous 2×2 matrices from Chapter 2 corresponds to a 3×3 matrix with the same effect for homogeneous coordinates, with a 1 in the lower right-hand corner. For example, the transformation $\begin{pmatrix} 1 & 0 \\ 0 & 3 \end{pmatrix}$ representing a scale change in the y direction is given by

$$\begin{pmatrix} 1 & 0 & 0 \\ 0 & 3 & 0 \\ 0 & 0 & 1 \end{pmatrix}$$

for homogeneous coordinates.

In addition to making possible the representation of translations as matrices, there are two other reasons why homogeneous coordinates are commonly used as a tool in computer graphics. The first has to do with the value of using ratios when manipulating points with very large coordinates. Since a typical microcomputer cannot store an integer larger than 32767, a point with coordinates larger than this could not be stored. But with homogeneous coordinates, a number with larger coordinates, such as (75000, 60000) could be represented in homogeneous coordinates as (7500, 6000, .1), so that the microcomputer could handle it easily.

A third reason why homogeneous coordinates are used in computer graphics is that they make it possible to show points at infinity and to transform three-dimensional drawings into perspective drawings to create an illusion of depth. You may be interested in trying to discover for yourself how homogeneous coordinates can be used in this way. This subject is discussed again in Section 7.7.

▲ EXERCISES 7.6 ▲

1. How can one tell an ideal point by looking at its coordinates?
2. Write homogeneous coordinates for these points:
   a. (3, 8) **b.** $\left(2, \frac{3}{4}\right)$ **c.** (1, −4)
3. Write Cartesian coordinates for these points:
   a. (2, 5, 1) **b.** (−1, 3, 2) **c.** (−2, 5, −3)
4. Write three other names, using homogeneous coordinates, for the point (5, 2, 3).
5. Does the point (1, 1, 1) lie on the line [1, 1, 1]?
6. Find the coordinates for the ideal point on the line [1, 1, 1].
7. Write a general formula for the ideal point on the line $[a, b, c]$.
8. Write the equation of the line [3, 1, −2].
9. Write the equation of the point (3, 1, −2).
10. Develop the equation for $\overleftrightarrow{B'C'}$, Figure 7.32.
11. Find the point of intersection of the lines [1, 1, 1] and [2, 1, 2].

12. Find the equation of the line through (1, 1, 1) and (2, 1, 2).

13. Find the point of intersection of the lines $ax_1 + bx_2 + cx_3 = 0$ and $dx_1 + ex_2 + fx_3 = 0$, assuming the point of intersection is not an ideal point.

14. Give the coordinates of the harmonic conjugate of $(0, 1, 0)$ with respect to $(0, 1, 1)$ and $(0, 3, 1)$.

For Exercises 15–18, construct with a straightedge the required points on the line $x_2 = 0$, given $(0, 0, 1)$, $(1, 0, 1)$ and $(1, 0, 0)$.

15. $(2, 0, 1)$
16. $(3, 0, 1)$
17. $(-1, 0, 1)$
18. $\left(\frac{1}{2}, 0, 1\right)$

For Exercises 19 and 20, give homogeneous coordinates for each point so that no coordinate is larger than 32,767.

19. (84632, 50415)
20. (245316, 169208)

For Exercises 21 and 22, write the matrix for the translation and use matrix multiplication to find the image of the point $\left(-2, \frac{3}{2}\right)$, in homogeneous coordinates.

21. Translation of 3 units in the positive x direction and 7 units in the negative y direction.

22. Translation of 2 units in the negative x direction and 6 units in the positive y direction.

23. Prove that the product of two translations is a translation.

▷ 24. For the number lines in Figures 7.30b–d, shade the portion for the negative numbers. Sketch additional arrangements of the three basic points and shade the portion for the negative numbers for each.

▷ 25. The triangle in Figure 7.32 includes only the first quadrant of the projective plane. Make an extended drawing and show all the quadrants.

▷ 26. Prove that three points with homogeneous coordinates (a_1, a_2, a_3), (b_1, b_2, b_3), and (c_1, c_2, c_3) are collinear if and only if

$$\begin{vmatrix} a_1 & a_2 & a_3 \\ b_1 & b_2 & b_3 \\ c_1 & c_2 & c_3 \end{vmatrix} = 0$$

▷ 27. Assign homogeneous coordinates to the points of the finite geometry of 4 lines and develop some theorems of that geometry analytically.

▷ 28. Assign homogeneous coordinates to the points of the finite geometry $PG(2, 2)$, Fano's geometry, and develop some theorems of that geometry analytically.

▷ 29. Investigate the relationship between Desarguesian and Pappian Planes, using Chapter 3 of Garner, *An Outline of Projective Geometry*.

▲ 7.7 Equations for Projective Transformations

In projective geometry, straight lines are transformed into straight lines, so the set of equations for projective transformations is a set of three simultaneous linear equations relating the homogeneous coordinates of a point to the homogeneous coordinates of its image.

Three points (x_1, x_2, x_3), (y_1, y_2, y_3), (z_1, z_2, z_3) are collinear if and only if there exist three numbers a, b, c, not all zero, such that $ax_i, +by_i + cz_i = 0$. This means that any point on a line is a linear combination of any other two distinct points on the line. For example, the point (8, 13, 23) is on the line with points (1, 2, 4) and (2, 3, 5), since

$$(2)(1) + (3)(2) + (-1)(8) = 0$$
$$(2)(2) + (3)(3) + (-1)(13) = 0$$
$$(2)(4) + (3)(5) + (-1)(23) = 0$$

In this example, $a = 2$, $b = 3$, and $c = -1$.

No three of the four basic points (1, 0, 0), (0, 1, 0), (0, 0, 1), (1, 1, 1) are collinear. These four points may be transformed by a projective transformation into four non-collinear points d, e, f, and $(d + e + f)$ whose coordinates are

$$(d_1, d_2, d_3)$$
$$(e_1, e_2, e_3)$$
$$(f_1, f_2, f_3)$$

and

$$(d_1 + e_1 + f_1, d_2 + e_2 + f_2, d_3 + e_3 + f_3)$$

This transformation can be accomplished by the following correspondence:

$$x'_1 = d_1x_1 + e_1x_2 + f_1x_3$$
$$x'_2 = d_2x_1 + e_2x_2 + f_2x_3$$
$$x'_3 = d_3x_1 + e_3x_2 + f_3x_3$$

This set of equations transforms any other point (k, l, m) into a point whose coordinates are given by the equations

$$x'_1 = d_1k + e_1l + f_1m$$
$$x'_2 = d_2k + e_2l + f_2m$$
$$x'_3 = d_3k + e_3l + f_3m$$

The new point may be indicated by $(dk + el + fm)$. This shows that it has coordinates referred to points d, e, f rather than the original vertices of the fundamental triangle. This is made clearer by the fact that (k, l, m) could have been indicated by the coefficients in the expression

$$([k \cdot 1] + [l \cdot 1] + [m \cdot 1])$$

The linear equations given are the simultaneous equations sought to represent the projective transformation. They provide a one-to-one correspondence of points in the projective plane. They preserve collinearity of points.

THEOREM 7.12 ➤ A set of equations representing a projective transformation is of the form

$$\begin{aligned}\theta x_1' &= d_1x_1 + e_1x_2 + f_1x_3\\ \theta x_2' &= d_2x_1 + e_2x_2 + f_2x_3\\ \theta x_3' &= d_3x_1 + e_3x_2 + f_3x_3\end{aligned}$$

The parameter $\propto$ is used because any multiple of the homogeneous coordinates results in another name for the same point. It is understood that the determinant of the coefficients is not zero.

The proof of one and the assumption of two additional theorems will help establish that the equations of Theorem 7.12 do provide for a projective transformation because they keep projective properties invariant.

THEOREM 7.13 ➤ A transformation in the plane is a projective transformation if it transforms one line projectively.

PROOF: In Figure 7.34, let m and m' be the given lines with the projective transformation and l and l' any other pair of corresponding lines. Let X and X' represent corresponding points on l and l', with A and B any other two distinct points not on any of the four lines. Line XAY is transformed into $X'BY'$, and since Y and Y' are projectively related, the following chain shows the projectivity connecting l and l':

$$X \overset{A}{\barwedge} Y \barwedge Y' \overset{B}{\barwedge} X'$$

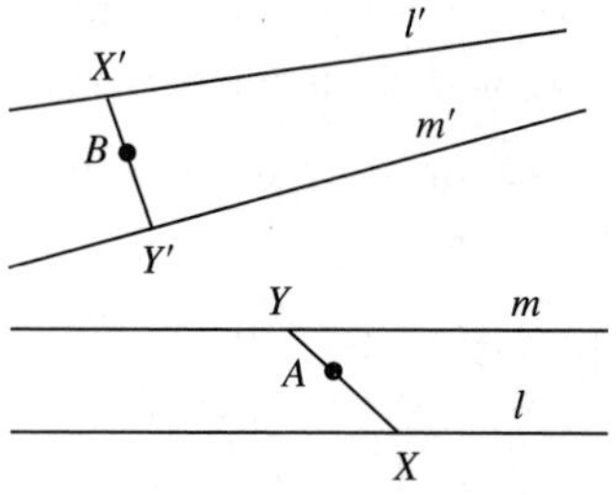

Figure 7.34

Homogeneous coordinates for three points on a line can be chosen so that they are (x_1, x_2), (y_1, y_2), and $(x_1 + y_1, x_2 + y_2)$. Then it can be proved that the harmonic conjugate of the third point with respect to the first two has the coordinates $(x_1 - y_1, x_2 - y_2)$. For example, the harmonic conjugate of $(4, 3)$ with respect to $(1, 1)$ and $(3, 2)$ is $(-2, -1)$.

The truth of this assertion can be explained, using the notation of Figure 7.35. Let $A = (x_1, x_2)$, $B = (y_1, y_2)$, $C = (x_1 + y_1, x_2 + y_2)$, and $F = (z_1, z_2)$. Then $E = (z_1 + x_1, z_2 + x_2)$. H is the intersection of $\overleftrightarrow{FC}$ and $\overleftrightarrow{EB}$, and has coordinates $(x_1 + y_1 + z_1, x_2 + y_2 + z_2)$. G, on both $\overleftrightarrow{AH}$ and $\overleftrightarrow{FB}$, has coordinates $(y_1 + z_1, y_2 + z_2)$. Finally, D is on $\overleftrightarrow{AB}$ and $\overleftrightarrow{EG}$. Its coordinates must satisfy the equation $x_i + ky_i = (x_i + z_i) + l(y_i + z_i)$ for some values of k and l. But l must equal negative one to eliminate z_i, which means $x_i + ky_i = x_i - y_i$, and k equals negative one. Thus, the coordinates of D are $(x_1 - y_1, x_2 - y_2)$.

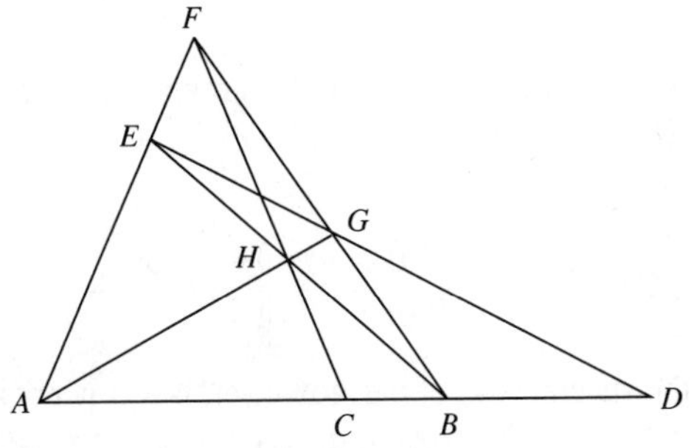

Figure 7.35

A generalization of the concept of harmonic set is the idea of the *cross ratio* of four points.

DEFINITION ➤ The cross ratio r of four collinear points is the number r when the coordinates of the four points are written in the form (x_1, x_2), (y_1, y_2), $(x_1 + y_1, x_2 + y_2)$, $(rx_1 + y_1, rx_2 + y_2)$.

For example, the cross ratio for $A(2, 3)$, $B(5, 6)$, $C(7, 9)$, $D(9, 12)$, with the pairing as in the definition, is 2, since $9 = (2 \cdot 2) + 5$ and $12 = (2 \cdot 3) + 6$.

Instead of using the projective definition given previously, you may calculate the cross ratio as a ratio of ratios. If $H(AB, CD)$ is a harmonic set, as in Figure 7.36a, then $(AC/CB)/(AD/DB) = -1$. This ratio of ratios is a specific case of a cross ratio. Thus, in Euclidean geometry, a harmonic set of points consists of the endpoints of a segment and two points that divide the segment internally and externally in the same ratio.

In Figure 7.36b, the four points are *not* a harmonic set. In this case, the cross ratio, denoted (AB, CD), is still defined to be $(AC/CB)/(AD/DB)$, but the value is some number other than -1. The notation for the cross ratio shows how the points are paired. The cross ratio may have any real number for its value.

Figure 7.36

Example ➤ Find the numerical value of (AC, BD) in Figure 7.37.

$$(AC, BD) = \frac{(AB/BC)}{(AD/DC)} = \frac{\frac{3}{2}}{-\frac{8}{3}} = \frac{3}{2} \cdot -\frac{3}{8} = -\frac{9}{16}$$

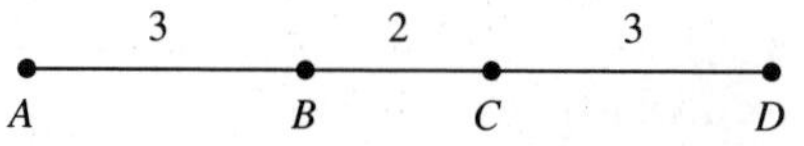

Figure 7.37

Example ➤ Using the same points as in the previous example, find the numerical value of (AB, CD).

$$(AB, CD) = \frac{(AC/CB)}{(AD/DB)} = \frac{-\frac{5}{2}}{-\frac{8}{5}} = -\frac{5}{2} \cdot -\frac{5}{8} = \frac{25}{16}$$

Cross ratio, and the special case of harmonic set, is a concept associated with four points on a projective line. If the cross ratio (AB, CD) is negative, the pairs A, B and C, D are said to separate each other, as in Figure 7.36a. *Separation* in projective geometry is a more general term that takes the place of "between." Intuitively, if one pair of points separates another, you cannot get from one point of the first pair to the second point of that pair by any path along the line without encountering one or the other of the points of the second set. The concept of "between" is no longer adequate since there are two paths from A to B along the line; thus, it would not mean anything to say C was between A and B if it was on the segment joining them.

Two additional theorems, stated without proof, will make a conclusion possible.

THEOREM 7.14 ➤ The cross ratio of four points is an invariant under projection.

THEOREM 7.15 ➤ A transformation that preserves the cross ratio of every four collinear points is a projective transformation.

Theorem 7.14 and Theorem 7.15 may be used for the line $x_1 = 0$. The set of simultaneous equations given in Theorem 7.12 transforms a point, $(0, x_2, x_3)$ on $x_1 = 0$, into $(e_1x_2 + f_1x_3, e_2x_2 + f_2x_3, e_3x_2 + f_3x_3)$. The four points $(0, 0, 1)$, $(0, 1, 1)$, $(0, 1, 0)$, and $(0, x_2, x_3)$ have a cross ratio of x_2/x_3, since

$$(0, x_2, x_3) = \left(0, \frac{x_2}{x_3}, 1\right)$$

It can be verified that these four points are projected into four points with this same cross ratio. Since the four sets of coordinates chosen may represent any four points on the line, the line is transformed projectively.

It has now been established that every projective transformation in the projective plane has equations of the form

$$x_1' = a_1x_1 + a_2x_2 + a_3x_3$$

$$x_2' = b_1x_1 + b_2x_2 + b_3x_3$$

$$x_3' = c_1x_1 + c_2x_2 + c_3x_3$$

with the stipulation that the determinant of the coefficients,

$$\begin{vmatrix} a_1 & a_2 & a_3 \\ b_1 & b_2 & b_3 \\ c_1 & c_2 & c_3 \end{vmatrix}$$

is not zero.

As a matrix equation, this set of equations can be interpreted as

$$(x_1' \quad x_2' \quad x_3') = (x_1 \quad x_2 \quad x_3)\begin{pmatrix} a_1 & b_1 & c_1 \\ a_2 & b_2 & c_2 \\ a_3 & b_3 & c_3 \end{pmatrix}$$

Recall that special cases of this general matrix have been encountered previously. For example, each similarity matrix has the form

$$\begin{pmatrix} a_1 & \pm(-a_2) & 0 \\ a_2 & (\pm a_1) & 0 \\ a_3 & a_4 & 1 \end{pmatrix}$$

with $a_1^2 + a_2^2 \neq 0$, and every motion in a plane has the same form with $a_1^2 + a_2^2 = 1$.

The application of transformation matrices with homogeneous coordinates in three dimensions is important in computer graphics, particularly in engineering applications. The 4×4 matrices may be partitioned into submatrices to show the effect of various elements:

$$\left(\begin{array}{c|c} A & C \\ \hline B & D \end{array}\right)$$

Submatrix A	3×3	Scaling, shearing, rotation, reflection
Submatrix B	1×3	Translation
Submatrix C	3×1	Perspective drawing
Submatrix D	1×1	Overall scaling

Examples involving elements in submatrix A are given first:

$$\begin{pmatrix} 2 & 0 & 0 & 0 \\ 0 & 3 & 0 & 0 \\ 0 & 0 & 4 & 0 \\ 0 & 0 & 0 & 1 \end{pmatrix} \quad \text{Local Scaling}$$

$$\begin{pmatrix} 1 & 0 & 3 & 0 \\ 0 & 1 & 0 & 0 \\ 2 & 0 & 1 & 0 \\ 0 & 0 & 0 & 1 \end{pmatrix} \quad \text{Shear}$$

$$\begin{pmatrix} 1 & 0 & 0 & 0 \\ 0 & 1 & 0 & 0 \\ 0 & 0 & -1 & 0 \\ 0 & 0 & 0 & 1 \end{pmatrix} \quad \text{Reflection about plane } z = 0$$

The next matrix shows how a value in submatrix D is used for overall scaling:

$$\begin{pmatrix} 1 & 0 & 0 & 0 \\ 0 & 1 & 0 & 0 \\ 0 & 0 & 1 & 0 \\ 0 & 0 & 0 & 3 \end{pmatrix} \quad \text{Overall scaling}$$

The following matrix equation shows how the elements in submatrix B are used to perform a translation in three dimensions:

$$(2 \quad 1 \quad -3 \quad 1)\begin{pmatrix} 1 & 0 & 0 & 0 \\ 0 & 1 & 0 & 0 \\ 0 & 0 & 1 & 0 \\ 2 & 3 & 4 & 1 \end{pmatrix} = (4 \quad 4 \quad 1 \quad 1)$$

Check to see that this transformation matrix transforms any point $(x, y, z, 1)$ into $(x + 2, y + 3, z + 4, 1)$.

The only concept not illustrated so far is the use of nonzero elements in submatrix C to give a perspective drawing. For a three-dimensional drawing of a box, such as the one shown in Figure 7.38a, parallel lines in the original object remain parallel in the drawing. On the other hand, in perspective drawings in art and drafting, lines that are parallel in the original converge in the drawing at points called vanishing points. If lines originally parallel to only one coordinate axis converge, the result is a one-point perspectivity. If lines originally parallel to either of two coordinate axes converge, the result is a two-point perspectivity, and if lines originally parallel to either of the three coordinate axes converge, the result is a three-point perspectivity. Examples of these three types of perspective drawings are shown in Figure 7.38b–d.

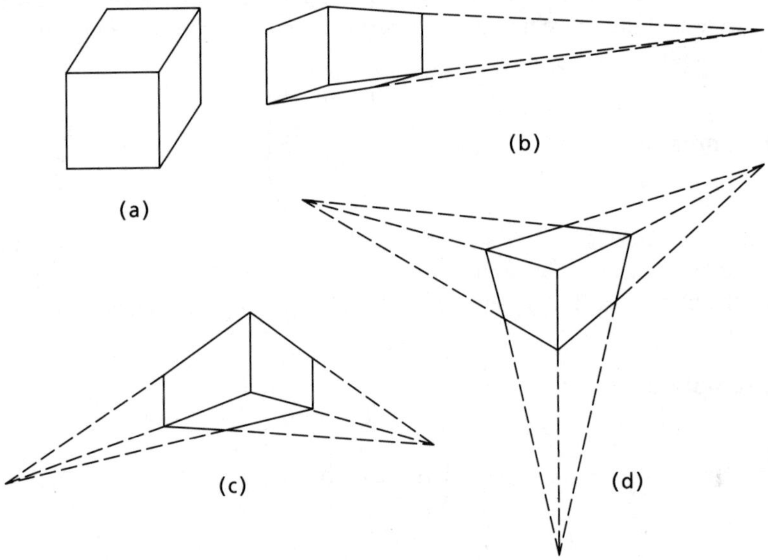

Figure 7.38

The transformation matrices to transform a box into a figure with pairs of edges meeting at one or more vanishing points has one or more nonzero elements in submatrix C. For example, the matrix

$$\begin{pmatrix} 1 & 0 & 0 & 0 \\ 0 & 1 & 0 & 0 \\ 0 & 0 & 1 & 3 \\ 0 & 0 & 0 & 1 \end{pmatrix}$$

causes lines parallel to the z-axis to meet at the vanishing point $(0, 0, \frac{1}{3}, 1)$. Examples of matrices for two- and three-point perspectives are

$$\begin{pmatrix} 1 & 0 & 0 & 0 \\ 0 & 1 & 0 & 4 \\ 0 & 0 & 1 & 3 \\ 0 & 0 & 0 & 1 \end{pmatrix} \quad \text{and} \quad \begin{pmatrix} 1 & 0 & 0 & 2 \\ 0 & 1 & 0 & 4 \\ 0 & 0 & 1 & 3 \\ 0 & 0 & 0 & 1 \end{pmatrix}$$

One additional practical example of matrices in three-dimensional projective geometry is the axonometric projections used in engineering. Axonometric projections are used to produce various two-dimensional views (called auxiliary views) of a three-dimensional object. Because they cause a change of dimension, axonometric projections technically are mappings rather than transformations, and the determinant of the matrix is zero, so there is no inverse.

The matrix for an axonometric projection may be considered the product of two other matrices. The first is a transformation used to rotate or translate, and the second is a projection of the entire figure onto a particular plane, called a viewing plane. Two examples follow.

Example ➤

$$\begin{pmatrix} 1 & 0 & 0 & 0 \\ 0 & 1 & 0 & 0 \\ 0 & 0 & 0 & 0 \\ 0 & 0 & 2 & 1 \end{pmatrix}$$

This matrix results in a projection onto the plane $z = 2$. You may verify that any point $(x, y, z, 1)$ is transformed into $(x, y, 2, 1)$, and hence, all image points are on the plane $z = 2$. You should also verify that the determinant of the matrix is zero.

Example ➤

$$\begin{pmatrix} 1 & 0 & 0 & 0 \\ 0 & 0 & 0 & 0 \\ 0 & -1 & 0 & 0 \\ 0 & 0 & 0 & 1 \end{pmatrix}$$

This matrix results in a rotation of 90 degrees about the x-axis, then a projection onto the plane $z = 0$.

▲ EXERCISES 7.7 ▲

1. Find numbers a, b, c to show that the line through $(1, 1, 1)$ and $(3, 2, 1)$ also passes through $(7, 4, 1)$.

For Exercises 2 and 3, find the harmonic conjugate of the third point with respect to the first two.

2. $(3, 1), (-2, 5), (1, 6)$
3. $(-1, -2), (3, 2), (2, 0)$

For Exercises 4–7, find the value of each cross ratio indicated, using Figure 7.39.

4. (AB, CD)
5. (AB, BD)
6. (EF, HG)
7. (EH, GF)

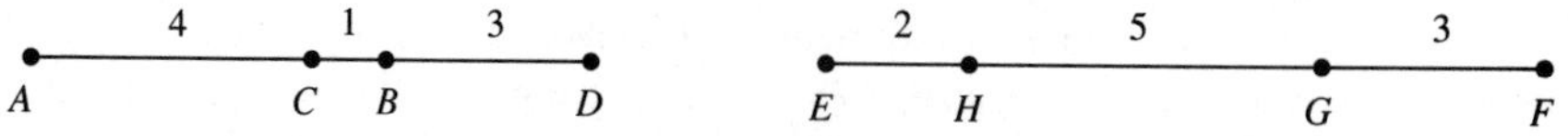

Figure 7.39

For Exercises 8–10, use the projective transformation

$$x_1' = x_1 + x_2 + x_3$$

$$x_2' = x_1 - x_2 + x_3$$

$$x_3' = x_1 + x_2 - x_3$$

8. Find the images of the points (0, 0, 1), (1, 0, 0), (1, 0, 1), and (3, 0, 5).

9. Find the cross ratio of the four original points in Exercise 8.

10. Verify that the cross ratio of the images is the same as the original cross ratio in Exercise 9.

For Exercises 11–14, what type of transformation, if any, is represented by each matrix?

11. $\begin{bmatrix} 2 & -3 & 0 \\ 3 & 2 & 0 \\ 4 & 1 & 1 \end{bmatrix}$ 12. $\begin{bmatrix} 1 & 4 & 7 \\ 2 & 5 & 8 \\ 3 & 6 & 9 \end{bmatrix}$

13. $\begin{bmatrix} 1 & 4 & 0 \\ 2 & 5 & 0 \\ 3 & 6 & 5 \end{bmatrix}$ 14. $\begin{bmatrix} 1 & 0 & 0 \\ 0 & 1 & 0 \\ 2 & 3 & 1 \end{bmatrix}$

For Exercises 15–24, identify each matrix for a transformation or mapping, then use matrix multiplication to find the image of the points (1, 0, 1, 1), (0, 1, 1, 1), (0, 0, 0, 1), and (1, 1, 0, 1).

15. $\begin{bmatrix} 3 & 0 & 0 & 0 \\ 0 & 2 & 0 & 0 \\ 0 & 0 & 4 & 0 \\ 0 & 0 & 0 & 1 \end{bmatrix}$ 16. $\begin{bmatrix} 1 & 0 & 0 & 0 \\ 0 & 1 & 0 & 0 \\ 0 & 0 & 1 & 0 \\ 3 & 4 & 2 & 1 \end{bmatrix}$

17. $\begin{bmatrix} 1 & 0 & 0 & 3 \\ 0 & 1 & 0 & 2 \\ 0 & 0 & 1 & 4 \\ 0 & 0 & 0 & 1 \end{bmatrix}$ 18. $\begin{bmatrix} 1 & 0 & 0 & 0 \\ 0 & 1 & 0 & 0 \\ 0 & 0 & 0 & 0 \\ 0 & 0 & 4 & 1 \end{bmatrix}$

19. $\begin{bmatrix} 1 & 0 & 0 & 0 \\ 0 & -1 & 0 & 0 \\ 0 & 0 & 1 & 0 \\ 0 & 0 & 0 & 1 \end{bmatrix}$ 20. $\begin{bmatrix} 1 & 2 & 3 & 0 \\ 0 & 1 & 0 & 0 \\ 4 & 0 & 1 & 0 \\ 0 & 0 & 0 & 1 \end{bmatrix}$

21. $\begin{bmatrix} 1 & 0 & 0 & 0 \\ 0 & 1 & 0 & 0 \\ 0 & 0 & 1 & 0 \\ 0 & 0 & 0 & 2 \end{bmatrix}$ 22. $\begin{bmatrix} 1 & 0 & 0 & 2 \\ 0 & 1 & 0 & 3 \\ 0 & 0 & 1 & 0 \\ 0 & 0 & 0 & 1 \end{bmatrix}$

23. $\begin{bmatrix} 1 & 0 & 0 & 0 \\ 0 & 0 & 0 & 0 \\ 0 & 0 & 1 & 0 \\ 0 & -2 & 0 & 1 \end{bmatrix}$ 24. $\begin{bmatrix} 1 & 0 & 0 & 0 \\ 0 & 1 & 0 & 0 \\ 0 & 0 & -1 & 0 \\ 0 & 0 & 0 & 1 \end{bmatrix}$

25. Prove that if three points (x_1, x_2, x_3), (y_1, y_2, y_3), (z_1, z_2, z_3) are collinear, there exist three numbers a, b, c, not all zero, such that $ax_i + by_i + cz_i, = 0$.

26. Prove that the transformation matrix causes

$$\begin{bmatrix} 1 & 0 & 0 & 0 \\ 0 & 1 & 0 & a \\ 0 & 0 & 1 & 0 \\ 0 & 0 & 0 & 1 \end{bmatrix}$$

lines parallel to the y-axis to meet at the vanishing point $(0, 1/a, 0, 1)$.

▷ **27.** Write a formal definition of separation.

▷ **28.** Show that the concept of the cross ratio as a ratio of ratios is consistent with the original definition given.

▷ **29.** Prove Theorems 7.14 and 7.15, or study the proofs found in a reference.

▲ 7.8 Special Projectivities

This section considers some special cases of projectivities in one and two dimensions.

A one-dimensional projectivity for points on a line renames the points on the same line.

DEFINITION ➤ A one-dimensional projectivity is called *elliptic*, *parabolic*, or *hyperbolic* if the number of invariant points is zero, one, or two, respectively. If there are three invariant points, the projectivity is the identity transformation.

The fundamental theorem of projective geometry can be used to show that:

1. A hyperbolic projectivity is uniquely determined when both invariant points and one other set of corresponding points are given.
2. A parabolic projectivity is uniquely determined when its invariant point and two other sets of corresponding points are given.

This second statement can be sharpened significantly, however, by proving the following theorem:

THEOREM 7.16 ➤ A parabolic projectivity is uniquely determined when its invariant point and one other set of corresponding points are given.

PROOF: In Figure 7.40, let a parabolic projectivity be given, with A the invariant point and B and B' any other pair of corresponding points. An arbitrary point S can be chosen so that $ABB' \overset{S}{\barwedge} ACC'$, where A, C, C' are on a line l through A not containing S. A second center of perspectivity S' can be chosen so that $ACC' \overset{S'}{\barwedge} AB'C''$. The image of any other point D on $\overleftrightarrow{AB}$ can be found by using the two perspectivities established. From what has been said so far, it has not been proved that the projectivity established is indeed parabolic rather than hyperbolic. That is, for some point D on line AB, the two perspectivities might result in D as the image, giving a second invariant point. For this to happen, D, S, and S' must be collinear, a situation that can be avoided simply by inserting the restriction that S and S' lie on the same line through A.

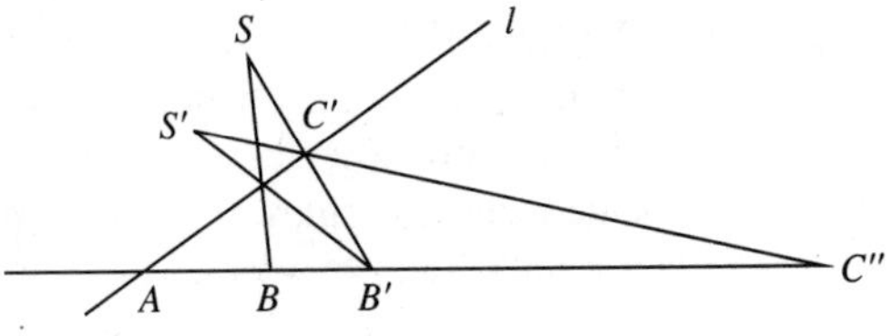

Figure 7.40

Some special projectivities are periodic in the sense that they will result in the identity transformation after a finite number of repeated applications.

DEFINITION ➤ A projectivity of *period n* is one that must be repeated n times before it first results in the identity transformation.

DEFINITION ➤ An *involution* is a projectivity of period two.

It can be seen intuitively that a one-dimensional involution simply interchanges pairs of points. The following theorem establishes a minimum condition for a projectivity to be an involution.

THEOREM 7.17 ➤ A one-dimensional projectivity that exchanges one pair of distinct points is an involution.

PROOF: In Figure 7.41, let ABX be any three given distinct collinear points, with $A'B'Y$ their images in a projectivity that interchanges A and B. These six points determine a unique projectivity that can be represented by these perspectivities: $ABXY \overset{F}{\barwedge} EDXG \overset{A}{\barwedge} FCYG \overset{D}{\barwedge} BAYX$, so $ABXY \barwedge BAYX$. This projectivity interchanges all arbitrary pairs of points XY and is an involution.

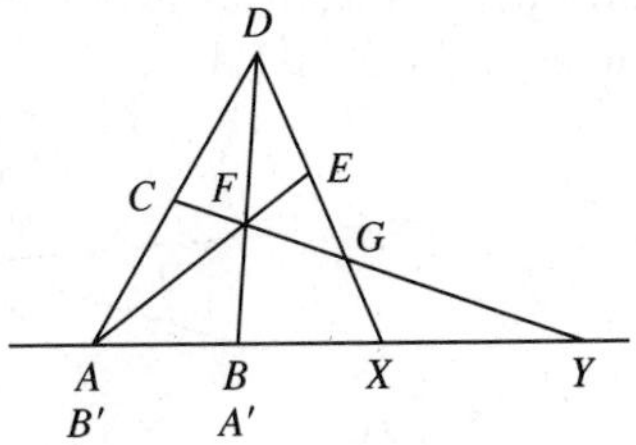

Figure 7.41

Some special one-dimensional transformations have been mentioned, and one-dimensional transformations are still involved in the discussion of the projective geometry of the plane. A two-dimensional projectivity transforms every one-dimensional set projectively. That is, a two-dimensional projective transformation may involve every point of the plane, but each line in the plane is transformed into another line so that a projectivity is established by the points on the two lines.

THEOREM 7.18 ➤ A two-dimensional projective transformation that leaves the four lines of a complete quadrilateral invariant is the identity transformation.

As illustrated in Figure 7.42 for quadrilateral a, b, c, d, the six vertices lie three on each line, and these vertices are invariant points. There is a projectivity between two corresponding sides, and every point on these sides must be invariant since three pairs are. Any other line in the plane meets the sides of this quadrilateral in invariant points, so there are three invariant points, and every point on it must be invariant, so the transformation is the identity.

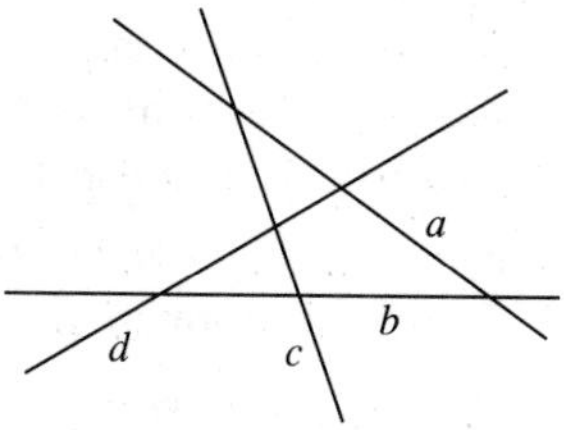

Figure 7.42

Special types of two-dimensional projectivities include one relating two given perspective triangles, called a *perspective collineation*. That is, the two triangles are images under a projectivity and are also perspective. The point and line of perspectivity are the *center* and *axis* of the transformation. The special case for which the center lies on the axis is called an *elation*, while all other transformations of this type are *homologies*. A special case is a homology in which the harmonic conjugate of the center with respect to pairs of corresponding points is on the axis. This transformation is called a *harmonic homology* and is illustrated by the example in Figure 7.43. Triangles ABC and $A'B'C'$ are perspective from point S and from line l. Furthermore, the harmonic conjugate of S with respect to pairs of corresponding vertices is on line l. For example, S' is the harmonic conjugate of S with respect to A and A'.

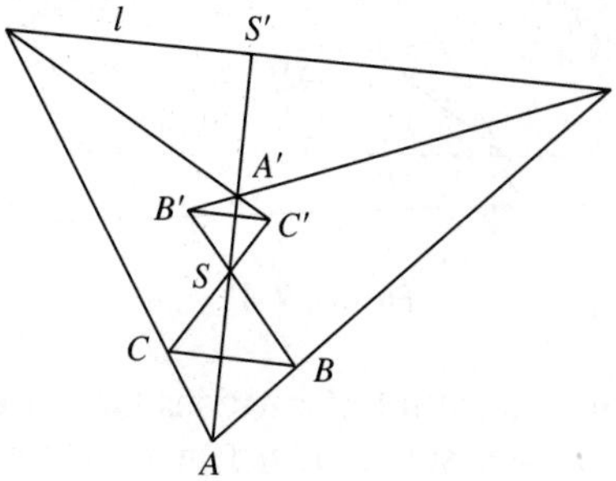

Figure 7.43

If the center and axis are given, the pairs of corresponding points in a harmonic homology can be determined, since they are harmonic conjugates with respect to the center and a collinear point on the axis. This type of transformation is of period two, since repeating it will result in the identity. In fact, it can be established that every two-dimensional transformation of period two is a harmonic homology.

▲ EXERCISES 7.8 ▲

1. Explain why a reflection is an example of an involution.

2. In general, is a glide reflection an involution?

For Exercises 3–5, give an example of a rotation that is of period

3. Two **4.** Three **5.** Four

6. Show how the image of any other point on the line can be constructed if the invariant point and one other set of corresponding points are given for a one-dimensional parabolic projectivity.

7. Show how the image of any other point on the line can be constructed if the invariant points and one other set of corresponding points are given for a one-dimensional hyperbolic projectivity.

8. State and prove a theorem about the information needed to determine a unique one-dimensional involution.

9. State the plane dual of Theorem 7.18 and prove it by writing the dual of the proof of Theorem 7.18.

10. Sketch an example of an elation.

11. Could an elation be a harmonic homology?

12. Sketch a homology that is not a harmonic homology.

13. In Figure 7.44, let S and l determine a harmonic homology. Construct the images of points A, B, C.

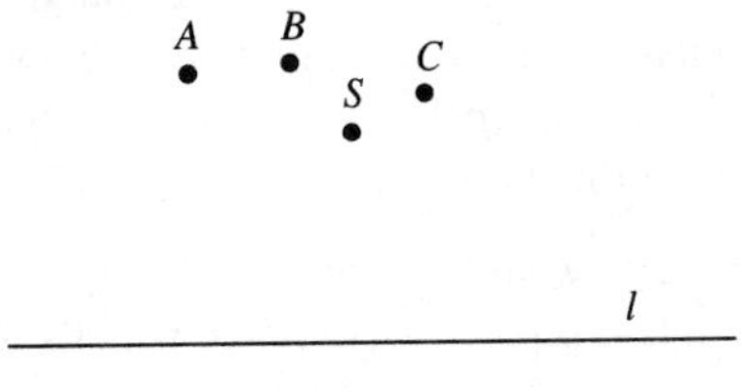

Figure 7.44

For Exercises 14–16, use matrix multiplication to find nonzero real values for a, if possible, so that the transformation is an involution.

14. $\begin{bmatrix} 1 & 0 & 0 & 0 \\ 0 & 1 & 0 & a \\ 0 & 0 & 1 & 0 \\ 0 & 0 & 0 & 1 \end{bmatrix}$ 15. $\begin{bmatrix} 1 & 0 & 0 & 0 \\ 0 & a & 0 & 0 \\ 0 & 0 & 1 & 0 \\ 0 & 0 & 0 & 1 \end{bmatrix}$ 16. $\begin{bmatrix} 1 & 0 & 0 & 0 \\ 0 & 1 & 0 & 0 \\ a & a & 1 & 0 \\ 0 & 0 & 0 & 1 \end{bmatrix}$

▷ **17.** Investigate, using powers of the matrix, the possibility of imposing conditions so that some other transformations besides rotations might have periods of three or greater. Can you discover any?

▲ 7.9 Conics

The various conics of Euclidean geometry can all be described as sections of cones of two nappes, as illustrated in Figure 7.45. From a different point of view, Figure 7.45 shows that a circle can be projected into an ellipse (Figure 7.45a), a parabola (Figure 7.45b), or a hyperbola (Figure 7.45c). The property of being a conic is an invariant under the group of projective transformations, but the property of being an ellipse, a parabola, or a hyperbola is not. The invariance of the conic explains why it is included in this chapter.

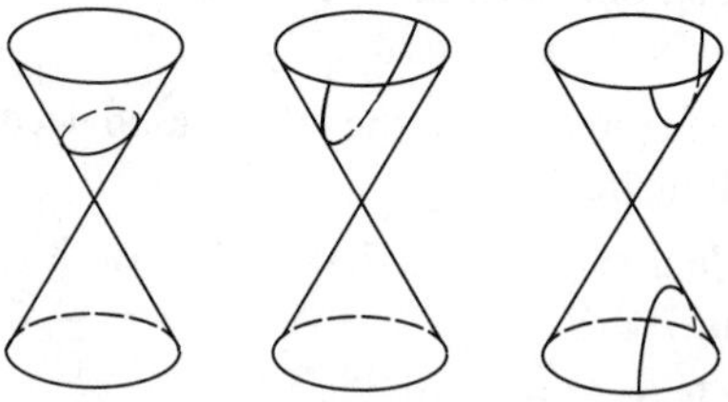

(a) (b) (c)

Figure 7.45

Just as there is more than one way to define conics in Euclidean analytic geometry, so there is more than one possibility in projective geometry. The first definitions chosen here are credited to Jacob Steiner, a Swiss mathematician who wrote (in 1832) on harmonic division and conic sections in projective geometry; the definitions are used because they emphasize clearly the relationship between conics and the projectivities already studied. In the definitions, *pencils of lines* are sets of concurrent lines, and *ranges of points* are sets of collinear points.

DEFINITION ➤ A *point conic* is the set of points that are intersections of corresponding lines in two projectively related pencils of lines in the same plane.

DEFINITION ➤ A *line conic* is the set of lines that join corresponding points in two projectively related ranges of points in the same plane.

It would be well to reflect on these remarkable definitions. Observe that several previous figures in this chapter have included points on a conic, although the text has not called attention to the fact. While a complete reconciliation of these definitions with intuitive ideas of what a conic is might be extremely difficult, it is not hard to see that the definitions do provide for the fact that no three points of a point conic are collinear. Figures 7.46a and b show a point conic and a line conic, with the connections between them and the projectivities that determine them emphasized. In Figure 7.46a, A and B are the centers of two projectively related pencils, with C, D, E, F points of intersection of corresponding lines so that the projectivity is indicated by

$$(\overleftrightarrow{AC}, \overleftrightarrow{AD}, \overleftrightarrow{AE}, \overleftrightarrow{AF} - \barwedge \overleftrightarrow{BC}, \overleftrightarrow{BD}, \overleftrightarrow{BE}, \overleftrightarrow{BF} -)$$

In Figure 7.46b, a and b are lines with two projectively related ranges of points A, A'; B, B'; C, C' are pairs of corresponding points so that the projectivity is indicated by $(A, B, C - \barwedge A', B', C'-)$.

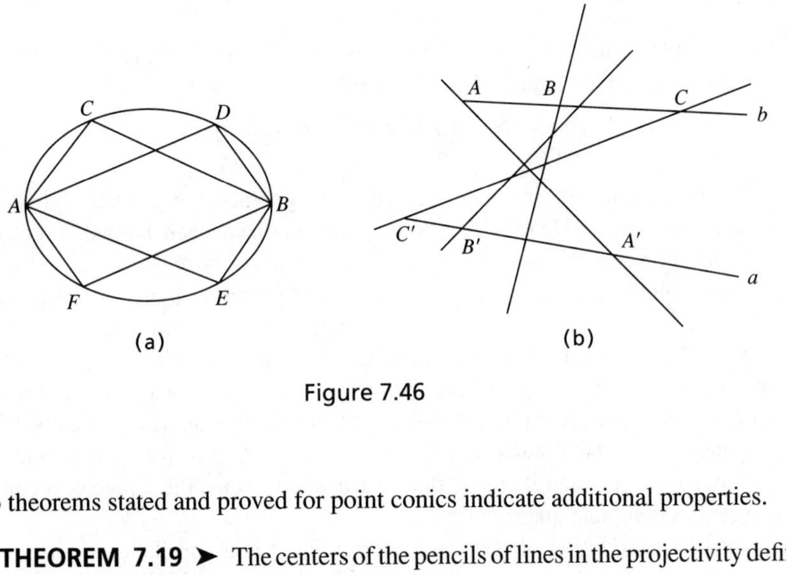

Figure 7.46

Two theorems stated and proved for point conics indicate additional properties.

THEOREM 7.19 ➤ The centers of the pencils of lines in the projectivity defining the point conic are also points of the conic.

Proof: In Figure 7.47, let A and B be the given centers. If $\overleftrightarrow{AB}$ is considered as one of the lines in the pencil with A as center, then its corresponding line m is one of the lines in the pencil with B as center. The two lines intersect at B, a point on the conic. Similarly, A lies on $\overleftrightarrow{AB}$, considered as a line through B, and on its corresponding line l through A.

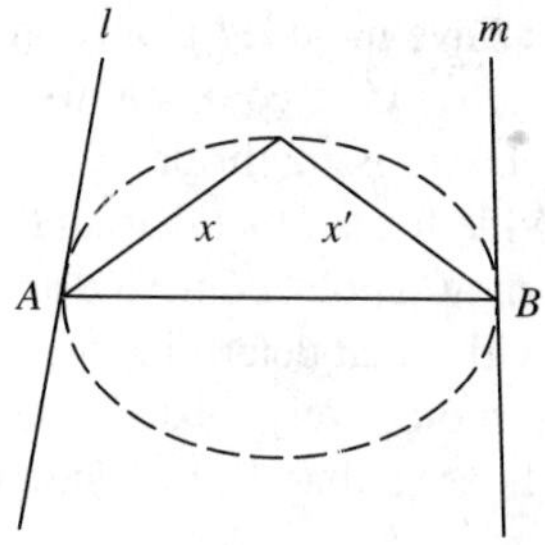

Figure 7.47

DEFINITION ➤ A *tangent* to a point conic is a line in the plane of the conic having exactly one point in common with the conic.

THEOREM 7.20 ➤ The lines corresponding to the common line of the two pencils of lines determining a point conic are the tangents at the centers of the two pencils of lines.

PROOF: In Figure 7.47, each line x through A has two points in common with the conic, A and the intersection of x and its corresponding line. But by Theorem 7.19, if x' is $\overleftrightarrow{AB}$ considered as a line through B, then x and x' intersect at A, so x must be the tangent l.

A second approach to defining conics in projective geometry requires the introduction of several new concepts. Up to this point, projectivities have been defined in such a way that they pair points with points and lines with lines. A *collineation* may be defined as a point-to-point or line-to-line transformation that preserves incidence. A projective collineation is a collineation that transforms every pencil or range projectively. It is possible to generalize the definition to include correspondence between points and lines. The original type of projectivity is a *collineation*, and the newer type now introduced is a *correlation*. The set of equations representing a correlation would have line coordinates for the primed letters but otherwise would be the same form as for a collineation. In a correlation, an element and its plane dual correspond. Thus, the image of a complete quadrangle is a complete quadrilateral.

The projective correlation of particular interest here is a correlation of period two, called a *polarity*. A polarity pairs a point called the *pole* and a line called its *polar*, such as A and a in Figure 7.48. Any point X on a is paired with a line x on A. Points such as A and X are *conjugate points*, since the polar of one passes through the other. Similarly, x and a are conjugate lines.

In Figure 7.48, conjugate points A and X are distinct. A polarity is called a *hyperbolic polarity* if a point can be self-conjugate. A self-conjugate point lies on its polar. Figure 7.49 shows various lines in a hyperbolic polarity, with the self-conjugate points indicated. These concepts can now be used to give an alternate definition for a conic.

The study of sets of conjugate points with respect to a conic can be applied to the transformation called *polar reciprocation*, which results in the correspondence of poles and polars with respect to a conic (see Exercise 15I).

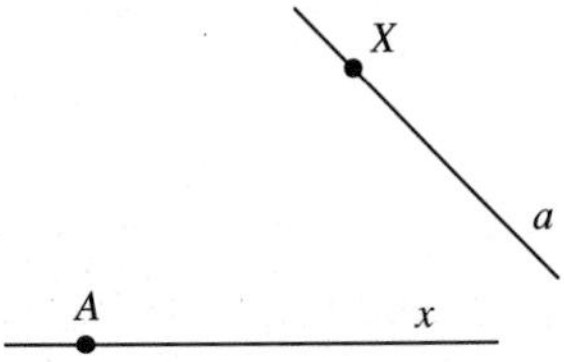

Figure 7.48

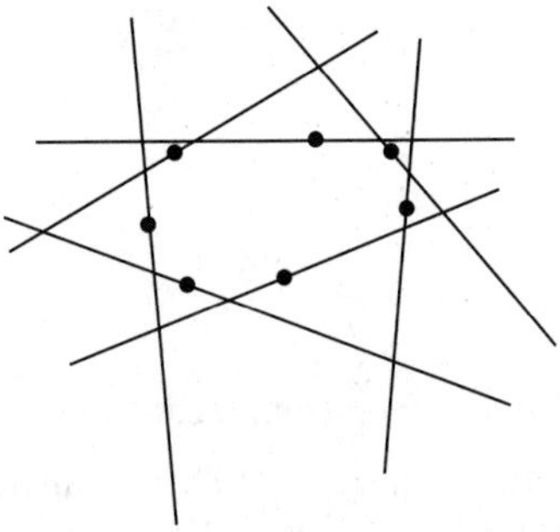

Figure 7.49

DEFINITION ➤ A point conic is the set of self-corresponding points in a hyperbolic polarity.

The section concludes with two major theorems, plane duals of each other, that give an additional property of any conic.

DEFINITION ➤ A *simple hexagon* is a set of six points in a plane, no three collinear, and the lines joining them in a particular order. For a hexagon with consecutive vertices A, B, C, D, E, F, opposite sides are $\overleftrightarrow{AB}$ and $\overleftrightarrow{DE}$, $\overleftrightarrow{BC}$ and $\overleftrightarrow{EF}$, and $\overleftrightarrow{CD}$ and $\overleftrightarrow{AF}$. This definition allows hexagons in which the opposite sides intersect within the conic.

THEOREM 7.21 ➤ *Pascal's theorem.* If a simple hexagon is inscribed in a point conic, the intersections of the three pairs of opposite sides are collinear.

Let $ACBEFD$ in Figure 7.50 represent any inscribed simple hexagon. The pairs of opposite sides are $\overleftrightarrow{AC}$ and $\overleftrightarrow{EF}$, $\overleftrightarrow{CB}$ and $\overleftrightarrow{FD}$, and $\overleftrightarrow{BE}$ and $\overleftrightarrow{DA}$. Let the points of intersection be P_1, P_2, P_3. It is required to show that these three points are collinear. The method of proving the three points collinear is to show that two of them are corresponding points in a perspectivity in which the third one is the center.

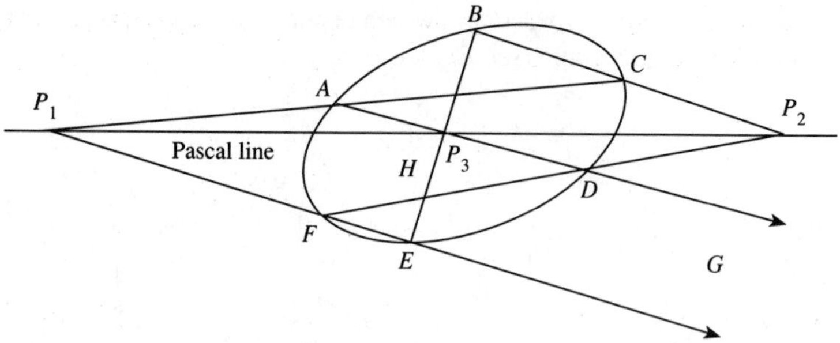

Figure 7.50

Two points A and B can be chosen as centers of pencils of lines in a projectivity determining the conic. There is also a projectivity established on $\overleftrightarrow{EF}$ and $\overleftrightarrow{FD}$ by considering the points of intersection of these two lines and pairs of corresponding lines in the previous projectivity. This means that if $\overleftrightarrow{AD}$ and $\overleftrightarrow{EF}$ meet at G, and $\overleftrightarrow{BE}$ and $\overleftrightarrow{FD}$ meet at H, then $P_1EFG \,\overline{\wedge}\, P_2HFD$. These two sets of points have their common element F self-corresponding, and they are related by a perspectivity (see Exercise 8, Exercises 7.9). The center of this perspectivity is P_3, which implies that P_1 and P_2 lie on a line through P_3, as was to be proved.

The six vertices of a simple hexagon determine 60 different hexagons, found by connecting the points in different orders. Each of these hexagons in turn has a different Pascal line. The 60 Pascal lines associated with six given points on a conic are known as *Pascal's mystic hexagram*.

Pascal's theorem was proved by Blaise Pascal (1623–1662), but the plane dual was not proved until much later, by C. J. Brianchon (1785–1864), after the development of the concept of duality.

THEOREM 7.22 ➤ *Brianchon's theorem.* If the six lines of a simple hexagon are lines of a line conic, then the three lines connecting pairs of opposite vertices of the hexagon are concurrent.

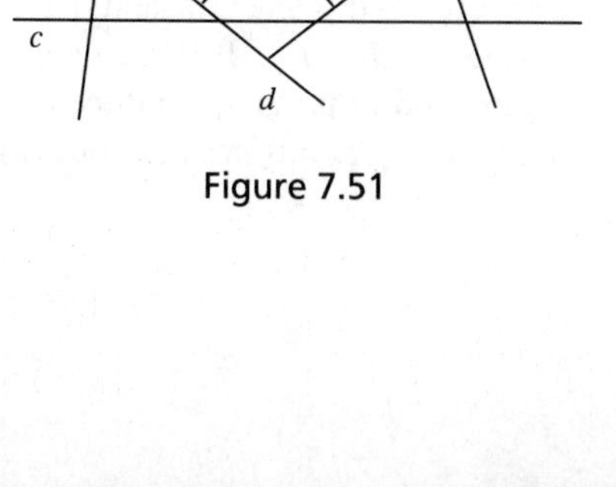

Figure 7.51

If the six consecutive lines are a, b, c, d, e, f, then opposite vertices are points of intersection of a, b and d, e; b, c and e, f; c, d and f, a. Figure 7.51 illustrates the theorem. The point of concurrency P is called a Brianchon point.

▲ EXERCISES 7.9 ▲

1. What would be the effect of allowing the pencils to be perspective in the definition of a point conic?
2. Point out figures in this chapter, before Figure 7.45, in which points on a conic could be located.
3. State the plane dual of Theorem 7.19.
4. State the plane dual of Theorem 7.20.
5. Write the equation of the line that is the image of the point (1, 1, 1) under the correlation with equations

$$X_1' = x_1 + x_2 + x_3$$
$$X_2' = x_1 - x_2 + x_3$$
$$X_3' = x_1 + x_2 - x_3$$

6. Prove that the line joining two self-conjugate points in a polarity cannot be a self-conjugate line (passing through its pole).
7. In Figure 7.48, let the intersection of a and x be point Y. A triangle such that each vertex is the pole of the opposite side is called a self-polar triangle. Is triangle AXY necessarily a self-polar triangle?
8. Prove that if two sets of points in a projectivity on two lines have their common point self-corresponding, they are related by a perspectivity.
9. Explain why six points, no three collinear, determine 60 hexagons.
10. Prove Brianchon's theorem directly.
11. Write the general equation for a conic in projective geometry by rewriting the equation $ax^2 + bxy + cy^2 + dx + ey + f = 0$ in homogeneous coordinates.
12. Find the points of intersection of the line $x_2 - 2x_3 = 0$ and the conic $x_1^2 - x_2x_3 = 0$.
13. Prove that any two conjugate points A and B on a secant to a conic are harmonic conjugates with respect to the points of intersection C and D.
14. Prove that if a quadrangle is inscribed in a conic, its diagonal points form a self-polar triangle.
15. ▷ Study (in Coxeter, *Introduction to Geometry*, 2d edition, or another reference) the concept and uses of polar reciprocation with respect to a general conic.

16. Read in a history of mathematics text about the early life of Pascal and his discovery of Pascal's theorem for conics.

17. Take six points on an ellipse and construct the 60 lines of Pascal's mystic hexagram.

18. In Figure 7.51, construct additional Brianchon points.

19. Study (using Coxeter, *Projective Geometry*, 2d edition, or another reference) the axiomatic development of the finite projective geometry $PG(2, 5)$ (see Exercises 1.6).

20. Study Section 3.7 of Hopkins and Hails, *An Introduction to Plane Projective Geometry*, to see how equations of conics can be derived from sections of cones of two nappes.

21. Study Section 8.1 of Hopkins and Hails, *An Introduction to Plane Projective Geometry*, to see how projective definitions of conics can be reconciled with the equations for the conics.

7.10 Construction Of Conics

Because conics have been defined in terms of projectivities, any number of points or lines on a conic can be constructed using only a straightedge, if enough information is given to determine a projectivity. This procedure is analyzed through examples.

Example ➤ Given five points on a point conic, construct other points.

Let A, B, C, D, E in Figure 7.52 be the five given points. Choose any two of these, say A and D, as centers of the pencils of lines for a projectivity. Pairs of corresponding lines meet at B, C, E. Lines in the pencil with center A are labeled $b, c, e, \ldots$, and lines in the pencil with center D are labeled $b', c', e', \ldots$.

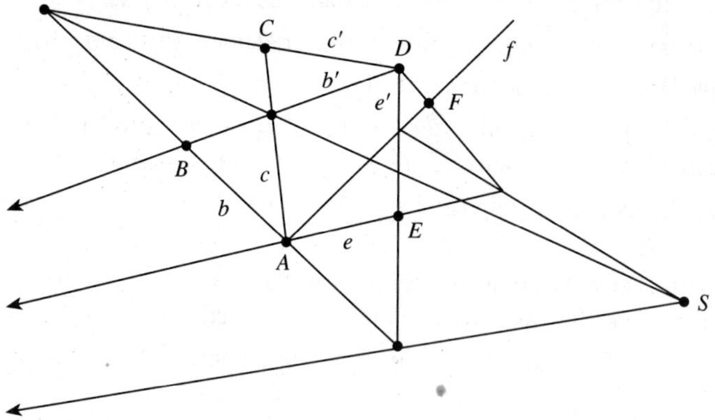

Figure 7.52

The lines connecting the intersections of the cross joins determine S, the center of homology.

Let f be another line through A. Since fe' and ef' must meet at S, f' can be constructed. The intersection of f and f' is point F, one of the required points.

Figure 7.53 shows points on a parabola and a hyperbola as intersections of corresponding lines of projectivites of two points on the conic.

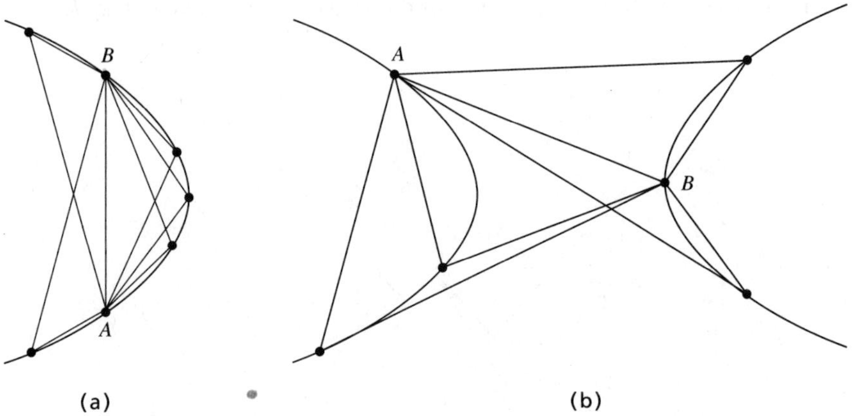

Figure 7.53

Example ➤ Given four points on a conic and the tangent at one of them, construct another point of the point conic.

If points A, B, C, D, and tangent a are given as in Figure 7.54, the projectivity can be determined by the three pairs of lines shown. The cross joins $a'b$ and $b'a$ (see Figure 7.54b) both lie on a, so that the center of projectivity is on the tangent. A second pair of cross joins bc' and cb' determine line e. The center of projectivity is the point of intersection of a and e.

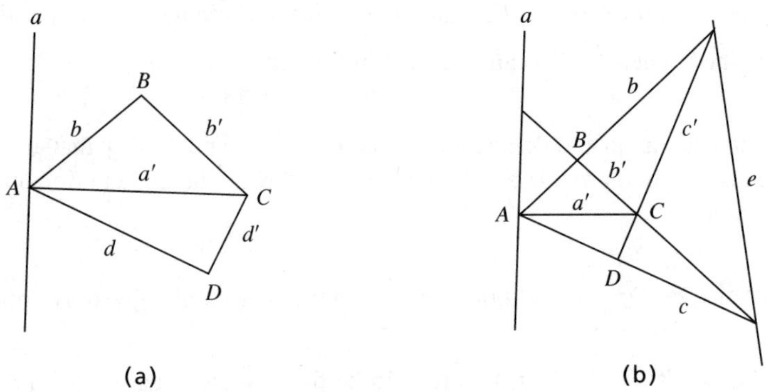

Figure 7.54

Example ➤ Construct additional lines of a line conic, given five lines, no three concurrent.

Let the five given lines be a, b, c, d, e in Figure 7.55. Two of these, say a and b, can be chosen as the lines to contain the corresponding points in the projectivity, while the other three determine three pairs of corresponding points. The intersections of pairs of cross joins determine the axis of homology l.

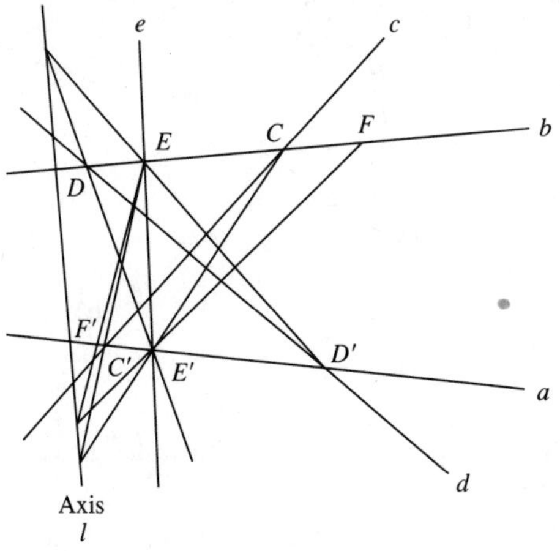

Figure 7.55

Choose any point F on b. $\overleftrightarrow{FC'}$ and $\overleftrightarrow{CF'}$ must meet on the axis so that F' can be determined. Then $\overleftrightarrow{FF'}$ is another line of the conic.

The theorems of Pascal and Brianchon, which include the possibility that all six given points are not distinct, provide an alternative way of constructing additional points or lines on a conic.

Example ➤ Find one additional point on the same conic, given five points.

Let A, B, C, D, E in Figure 7.56 be the five given points, with l any line through E not intersecting one of the given points. The sixth point F will be located on l, so that A, B, C, D, E, F will be points of an inscribed hexagon.

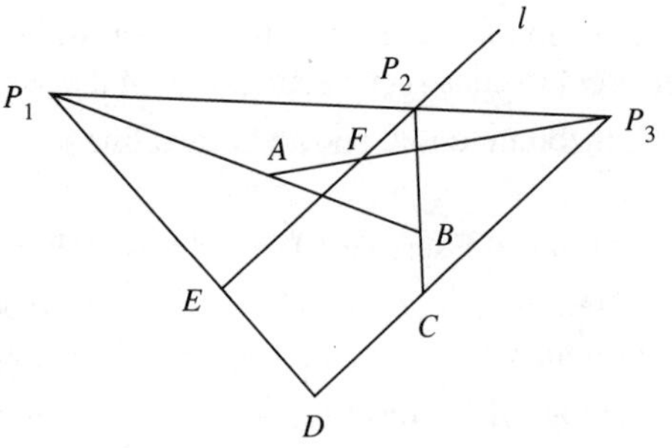

Figure 7.56

The given information is enough to determine two of the points on the Pascal line and, hence, the line itself. Let $\overleftrightarrow{AB}$ and $\overleftrightarrow{DE}$ meet at P_1, and let $\overleftrightarrow{BC}$ and l meet at P_2. Then $\overleftrightarrow{CD}$ and $\overleftrightarrow{AF}$ must meet at a point P_3 on $\overleftrightarrow{P_1P_2}$, so that F is the intersection of AP_3 and l.

This section concludes the presentation of projective geometry as a more general geometry that includes Euclidean geometry as a special case. In this chapter you have encountered homogeneous coordinates and their applications in computer graphics, constructions with only a straightedge, and projective transformations. You have studied some of the invariant properties in a geometry in which you did not already know what was invariant and what was not invariant. In the next chapter, a much more general geometry is introduced, and projective geometry will be just a special subgeometry.

▲ EXERCISES 7.10 ▲

1. Locate five new given points, as in Figure 7.52, and use the method described in the text to construct a sixth point on the conic.
2. Follow the instructions in Exercise 1, but choose five different new points.
3. Locate new given points and a given line, as in Figure 7.54 and use the method described in the text to construct a sixth point on the conic.
4. Follow the instructions in Exercise 3, but choose different new points and line.
5. Locate five new given lines in Figure 7.55 and use the method described in the text to construct a sixth line on the conic.
6. Follow the instructions of Exercise 5, but choose different new lines.
7. Construct another line of a conic, given four lines and the point of contact on one.

8. Follow the instructions of Exercise 7, but choose different lines and point.
9. Choose five points, no three collinear, that appear to be on a hyperbola, then use Pascal's theorem to construct a sixth point on the same conic.
10. Use Brianchon's theorem to find one additional line of a line conic if five lines are given.
11. Follow the instructions of Exercise 10, but choose different lines.
12. Use Pascal's theorem with a degenerate hexagon to construct one more point on a conic if four points and the tangent at one of these are known.
13. Follow the instructions of Exercise 12, but choose different points and tangent.
14. Given five points on a point conic, use any method with a straightedge alone to construct ten additional points on the conic.
15. Start with 5 points on an ellipse and construct 20 more points on the same conic. Explain a pattern that allows you to construct additional points by drawing a minimum number of additional lines.
16. Extend the construction problems to those involving given data consisting of some points and some lines, such as three points and two tangents.

▲ CHAPTER 7 REVIEW EXERCISES ▲

For Exercises 1–3, which properties are invariant under the transformation of projection?

1. Property of being a circle
2. Property of three points being collinear
3. Property of a point and line being incident
4. Are there properties invariant in projective geometry that are not invariant in Euclidean geometry? Explain your answer.
5. Can a triangle be projected into a quadrangle?
6. In a complete quadrangle, two diagonal points are harmonic conjugates with respect to what other pair of points?
7. A complete quadrangle has how many pairs of opposite sides?
8. Write the plane dual of this statement: A complete quadrangle has three pairs of opposite sides.
9. In order to uniquely determine all the points of a harmonic net, what minimum number of points of the net must be given?

10. Find the point of intersection of the lines [1, 3, 2] and [2, 7, −1].

11. Name the ideal point on the line $3x_1 + 4x_2 + 2x_3 = 0$.

12. Find the correct third coordinate so that the point (2, 1,__) lies on the line [−1, −2, 3]

13. Find the numerical value of the cross ratio (AB, CD) for the points on the line in Figure 7.57.

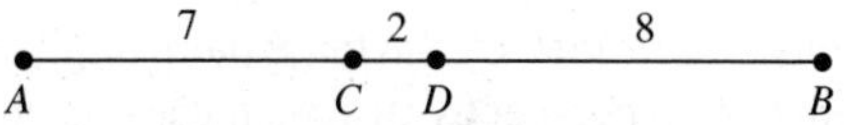

Figure 7.57

14. If a 3 x 3 matrix represents a plane motion, then exactly how many zeros must there be on the third column of the matrix?

15. Find the image of the point (1, 2, 3) under the projectivity with these equations given:

$$x_1' = x_1 + x_2 + 2x_3$$
$$x_2' = x_1 - x_2 + 3x_3$$
$$x_3' = 2x_1 + x_2 - x_3$$

16. Review the equation $2x - 5y + 7 = 0$ in homogeneous coordinates.

17. A rotation of 40 degrees about a point is an example of a projectivity with what period?

18. Can the following matrix be used to represent a projective transformation?

$$\begin{pmatrix} 1 & 3 & 2 \\ -1 & 3 & 2 \\ 2 & 3 & -1 \end{pmatrix}$$

For Exercises 19-21, classify the matrix for the transformation or mapping and find the image of the points (2, 1, 3, 0) and (1, 2, -1, 1).

19. $\begin{bmatrix} 2 & 0 & 0 & 0 \\ 0 & 3 & 0 & 0 \\ 0 & 0 & 1 & 0 \\ 0 & 0 & 0 & 1 \end{bmatrix}$

20. $\begin{bmatrix} 1 & 0 & 0 & 0 \\ 0 & 1 & 0 & 0 \\ 0 & 0 & 1 & 0 \\ -4 & 5 & -1 & 1 \end{bmatrix}$

21. $\begin{bmatrix} 1 & 0 & 0 & 2 \\ 0 & 1 & 0 & 0 \\ 0 & 0 & 1 & 1 \\ 0 & 0 & 0 & 1 \end{bmatrix}$

22. Write the homogeneous equation for the point (2, 3, 4).

23. By construction with a straightedge, locate the harmonic conjugate of C with respect to A and B, using Figure 7.58.

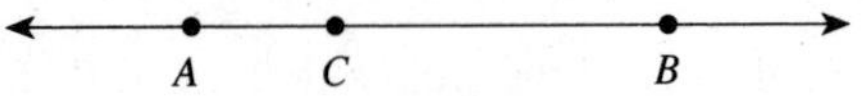

Figure 7.58

24. For the prjectivity determined by the three pairs of given points in Figure 7.59, establish a chain of two perspectivities, then use this chain to find the image of point X.

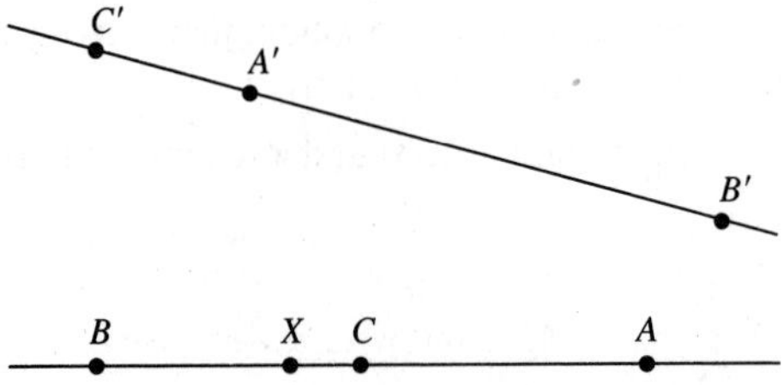

Figure 7.59

25. For the conic determined by the five given points in Figure 7.60, construct a sixth point of the conic that also lies on the given line.

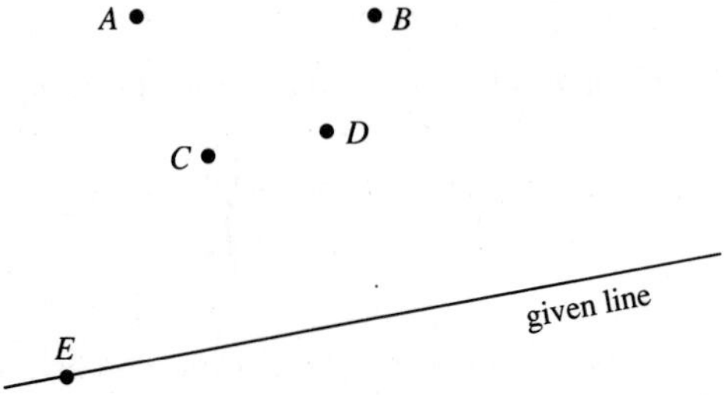

Figure 7.60

Additional chapter review exercises may be selected from any exercises not already assigned from previous exercise sets in this chapter.

8 Introduction to Topological Transformations

▲ 8.1 Topological Transformations

This chapter provides an introduction to a group of transformations far more general than those of projective geometry, and including projective transformations as special cases. One of the reasons for studying topological transformations is to investigate some of the invariant properties that exist, even though many of the familiar invariant properties from other groups of transformations are lost. Although the chapter contains some material studied in beginning courses in topology and graph theory, it is not meant to be a systematic introduction to either area of mathematics—the historic connections with geometry determine the content.

Visualize sets of points on a thin sheet of rubber, as in Figure 8.1. The sheet may be stretched and twisted, but not torn or placed so that two distinct points actually coincide. This description of points on a sheet of rubber provides an intuitive (but not totally correct) idea of the geometry called topology. It can be seen that the image of a circle could be an ellipse, a triangle, or a polygon, for example. Straight lines are not necessarily changed into straight lines. Most of the common properties of Euclidean geometry are no longer preserved under the set of topological transformations. Can you discover any properties that are invariant?

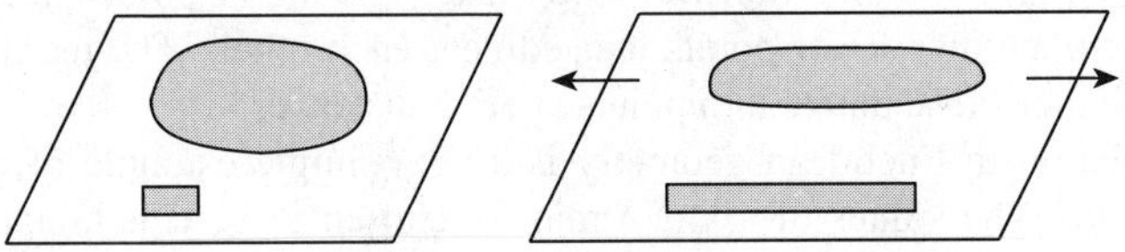

Figure 8.1

Much of the basic vocabulary of topology has already been introduced, especially in Chapter 3 on convexity. The reader who has not studied Chapter 3 should refer back, when necessary, to terms explained in that chapter. Historically, topology and convexity are closely connected and share many of the same fundamental concepts, such as neighborhood of a point and interior, exterior, and boundary points.

The term *topology* was introduced in 1847 by J. B. Listing (1808–1882), a pupil of Gauss to replace the earlier name "analysis situs." Other outstanding names in the history of topology include A. F. Möbius (1790–1868), Felix Hausdorff (1868–1942), S. Lefschetz (1884–1972), and Bernhad Riemann (1826–1866). The study of topology continues to grow and develop. Courses in topology are common at the graduate and undergraduate levels. Intuitive concepts from topology are frequently used in enrichment activities in both secondary and elementary schools.

Topology began as a branch of geometry, but now is itself a fundamental division of mathematics. Intuitively, topology may be considered as the mathematical study of continuity. The definition and discussion given here will be limited to some geometric aspects of topology.

The definition of a *continuous* transformation is reviewed briefly, since its definition makes use of the topological concept of neighborhood of a point. A transformation of points in the plane region S_1 onto points in the plane region S_2 is continuous if for every point of S_2 and each positive number ε there is a positive number δ such that the image of any point of S_1 that is in the neighborhood of point A with a radius of δ is in the neighborhood of the image of A with a radius of ε. See Figure 8.2.

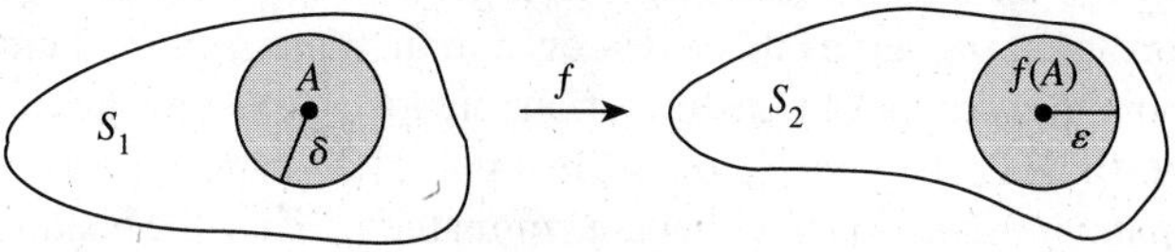

Figure 8.2

DEFINITION ➤ A transformation f is *bicontinuous* if and only if f and f^{-1} are both continuous. A bicontinuous transformation is also called a *homeomorphism*.

DEFINITION ➤ Topology is the study of those properties of a set of points invariant under the group of bicontinuous transformations of a space onto itself.

Intuitively, a continuous transformation takes one set of points sufficiently near each other into another set of points located near each other. The tearing apart of a surface could take close points into points at great distances.

A translation in Euclidean geometry is a very simple example of a continuous transformation. The values of ε and δ may be chosen to be equal, since the points in any circular region have images in another congruent circular region. Another example of a continuous transformation is $x \to x^3 + 2x - 1$, whereas an example of a transformation that is not continuous is $x \to \tan x$ for x the set of real numbers. The equation $x' = (1 - x)/x$ is an example of a transformation that is continuous in a range — for example, $-2 < x < -0.5$ — but its inverse $x = 1/(x' + 1)$ is not continuous in that same range.

Topology may be considered as a generalization of both Euclidean geometry and projective geometry, since the group of plane motions and the group of projective transformations are both proper subgroups of the group of topological transformations.

THEOREM 8.1 ➤ The set of topological transformations of a space onto itself is a group of transformations.

Questions about the proof of Theorem 8.1 can be found in Exercises 8.1.

It was suggested earlier in the section that you think about which properties are preserved under the group of topological transformations. This section concludes with the exploration of one of these properties that is fundamental.

A basic concept in topology is that of a *connected set* of points.

DEFINITION ➤ A set is connected if and only if any two points of the set can be joined by some curve lying wholly in the set.

Recall that in Chapter 3 a curve was defined as the graph of a set of equations of the form $x = f(t)$, $y = g(t)$, for f and g continuous functions and the domain of t an interval of real numbers.

All convex sets are connected, since any two of their points can be joined by segments in the set. The requirement for a set to be connected is a much looser requirement than for a set of points to be convex in the following sense: If all the curves in the definition of a connected set must be segments, then the set is convex.

Example ➤ Some common examples of connected sets are two intersecting circles, a square, a line, and a pyramid. Examples of sets that are not connected include a hyperbola, two parallel lines, and two nonintersecting circles. Figure 8.3a shows examples of connected sets that are not convex, whereas Figure 8.3b shows a set that is not connected.

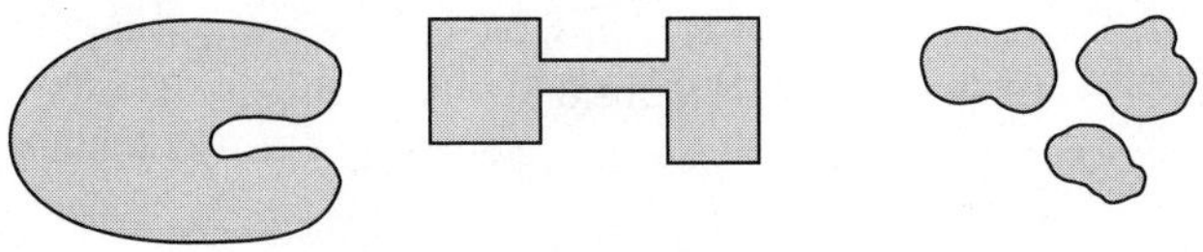

(a) (b)

Figure 8.3

The property of being a connected set is a topological invariant, but not all connected sets of points are topologically equivalent. In other words, it is not always possible to find a topological transformation such that any two given connected sets of points are images of each other. There are different types of connectivity, however, and a set of points and its image for any topological transformation have the same type of connectivity. The analysis of types of connectivity in this section is confined to two-dimensional sets of points with interior points.

DEFINITION ➤ A set is *simply connected* if any closed curve in the set can be continuously deformed to a single point in the set. A connected set that is not simply connected is *multiply connected.*

An example of a simply connected set is shown in Figure 8.4a; a multiply connected set is shown in Figure 8.4b.

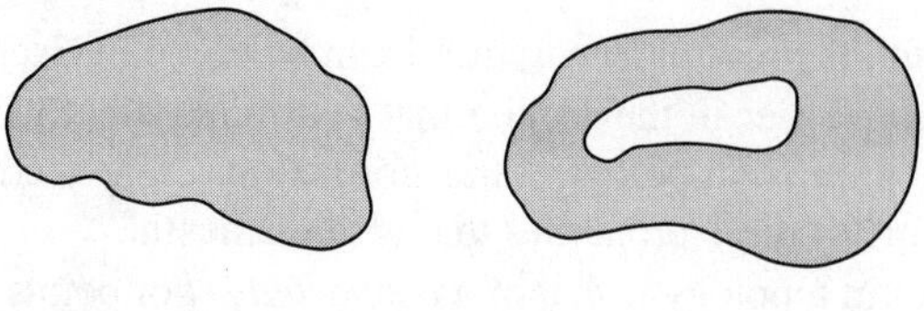

(a) (b)

Figure 8.4

What is meant by being continuously deformed to a single point can be made clearer by considering a closed curve, like a rubber band, lying entirely within the region in Figure 8.4a. The band could be shrunk without any point going outside the region. On the other hand, suppose the rubber band was in the region in Figure 8.4b, but wrapped around the hole. There would be no way to shrink it to a point without going through points in the hole. A multiply connected set, such as the one in Figure 8.4b, can be converted into a simply connected set by making one cut, as shown in Figure 8.5. Verify this intuitively by considering the rubber band situation again. However, a connected set with two holes, as shown in Figure 8.6a, requires two cuts (such as those in Figure 8.6b) to convert it into a simply connected set.

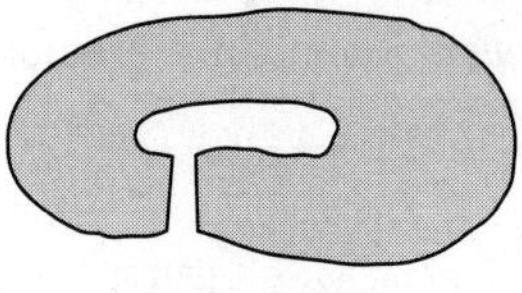

Figure 8.5

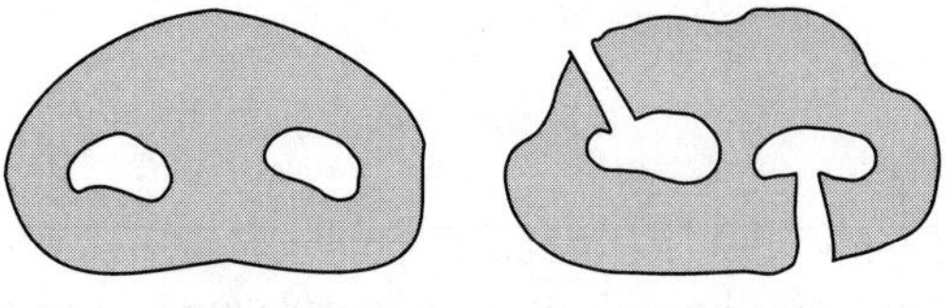

(a) (b)

Figure 8.6

DEFINITION ➤ In general, if $n - 1$ nonintersecting cuts are needed to convert a set into a simply connected set, the domain is *n-tuply connected.*

For example, the degree of connectivity is three if two cuts are necessary. The degree of connectivity of a set of points is an invariant under the group of topological transformations. For example, Figure 8.7 shows two plane sets of points that are topologically equivalent, both with degree of connectivity of four. These two sets of points are images of each other for some topological transformation.

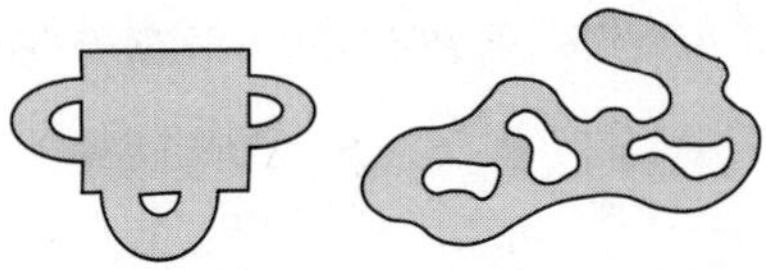

Figure 8.7

▲ EXERCISES 8.1 ▲

1. Give one additional example of a continuous transformation and one additional example of a transformation that is not continuous.

Exercises 2 and 3 concern the proof of Theorem 8.1.

2. Explain why the inverse of a bicontinuous transformation is always a bicontinuous transformation.
3. Outline a proof that the product of two bicontinuous transformations is a bicontinuous transformation.
4. Name some invariants under the group of plane motions that are not topological invariants.
5. Write a definition of convex set as a special kind of connected set.

For Exercises 6–15, state which of the sets of points are always connected sets.

6. Line
7. Circle
8. Polygon
9. Angle
10. Hyperbola
11. Reuleaux triangle
12. Two concentric circles
13. Half-plane
14. Open set
15. Bounded set

For Exercises 16–21, state which of the sets are always connected sets.

16. Sphere
17. Hyperboloid of two sheets
18. Ellipsoid
19. Tetrahedron
20. $\{x, y, z | x > y\}$
21. $\{x, y, z | x, y, \text{and } z \text{ are rational numbers}\}$

For Exercises 22–24, give a practical example of a flat object from the physical world resembling a set of points with connectivity of degree:

22. One **23.** Two **24.** Five

25. Draw two other sets of points topologically equivalent to those of Figure 8.7.

For Exercises 26–29, state which of the properties seem to be topological invariants.

26. Intersection of curves **27.** Cross ratio

28. Midpoint **29.** Convexity

▷ **30.** Cut a portion of rubber from the top part of an ordinary rubber glove and prepare models to demonstrate properties that are or are not invariant in topology.

▷ **31.** Give additional examples of sets of points that are connected and that are not connected.

▷ **32.** Investigate and give specific examples to show whether or not the union and the intersection of connected sets are connected sets.

▷ **33.** Read more about the history of topology.

▲ 8.2 Simple Closed Curves

The concept of simple closed curve was used in Chapter 3 on convexity. The definitions of Section 3.1 should be studied if necessary. The reason for extending the study of simple closed curves in this chapter is that the property of being a simple closed curve is a topological invariant. The image of a simple closed curve under a topological transformation is a simple closed curve.

Although the idea seems relatively simple, actually proving that one curve is the image of another under topological transformations is difficult. A special case is given to illustrate that you must prove continuity for both the transformation and its inverse.

THEOREM 8.2 ➤ A simple closed curve that is the boundary of a two-dimensional convex body is the bicontinuous image of a circle.

Proof: In Figure 8.8, K is the convex body. K contains an interior point A, and there is a circle with center A lying wholly within K. Each ray with endpoint A intersects the circle and the boundary of the convex body in exactly one point each, so there is a transformation of the boundary onto the circle. This transformation must be shown to be continuous. Let B and C be corresponding points, one on the boundary of K and the other on the circle. For ε any positive number, a number δ can be found small enough so that the segment joining any point of $N(B, \delta)$ to A will intersect the circle in a point of $N(C, \varepsilon)$. Then the transformation is continuous.

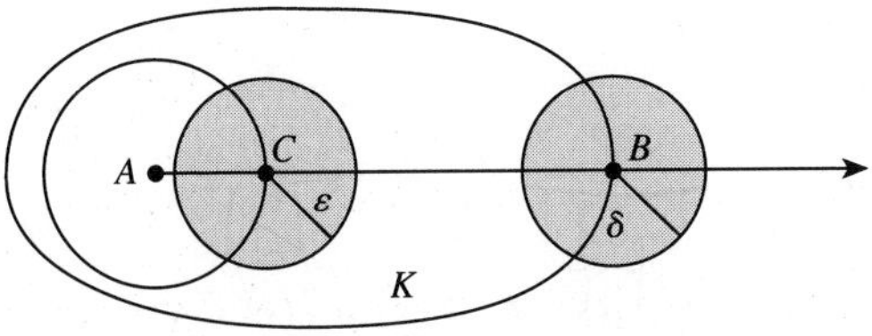

Figure 8.8

For the transformation to be bicontinuous, the inverse transformation must also be continuous. In Figure 8.9, ε is any positive number representing the radius of the neighborhood of point B, on the boundary of the convex body.

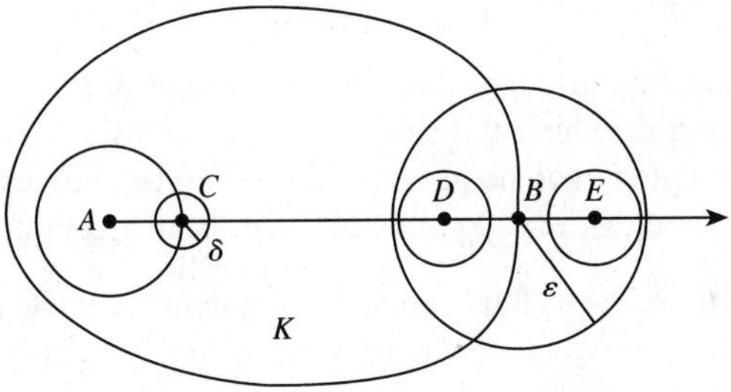

Figure 8.9

On ray $\overrightarrow{AB}$, it is possible to find a point D with neighborhood $N(D, \alpha)$ in the interior of both K and $N(B, \varepsilon)$ and an exterior point E with neighborhood $N(E, \beta)$ in the exterior of K but in the interior of $N(B, \varepsilon)$, for some positive numbers α and β.

Now, δ can be chosen small enough for $N(C, \delta)$ that any ray with endpoint A passing through a point of $N(C, \delta)$ will also pass through both $N(D, \alpha)$ and $N(E, \beta)$. Since one of the neighborhoods is interior to K and one is exterior, a boundary point of K lies on the ray and also lies in $N(B, \varepsilon)$. It follows that the inverse transformation meets the definition of a continuous function.

Some proofs of the fundamental theorem of algebra depend on an application of topology using ideas very similar to those in the proof of Theorem 8.2.

Consider the intuitive notion of tracing a simple closed curve by moving a pencil over a piece of paper, subject only to the restriction that you cannot cross a previous path and that you must return to the starting point. The result might be a complicated drawing such as the one in Figure 8.10. It is hard to distinguish between interior and exterior points, let alone to visualize a simple interior. A significant theorem about simple closed curves proved in topology, but assumed in earlier chapters and in elementary geometry, is called the Jordan curve theorem, named for Camille Jordan (1838–1922).

Figure 8.10

THEOREM 8.3 ➤ *Jordan curve theorem.* Any simple closed curve in the plane partitions the plane into three disjoint connected sets such that the set that is the curve is the boundary of both the other sets.

The proof of the Jordan curve theorem is outlined in this section only for the following special case in which the simple closed curves are polygons. Although the outline seems complex, it contains some interesting techniques and illustrates that proofs in geometric topology may be quite different from algebraic proofs.

THEOREM 8.4 ➤ Any simple closed polygon in the plane partitions the plane into three disjoint connected sets such that the set that is the polygon is the boundary of the other two sets.

Let $\overrightarrow{OX}$ be some fixed ray in the plane, not parallel to one of the sides of the polygon. Figure 8.11 shows a typical example. Any point P of the plane may be regarded as the endpoint of a ray parallel to $\overrightarrow{OX}$. In Figure 8.11, three such rays are shown, with endpoints P, P', and P''. The number of points of intersection of the rays and the polygon are one, three, and two, respectively. If the number of intersections is odd, the endpoint P is said to have an *odd parity*; if the number of intersections is even, the endpoint has an *even parity*. Thus, P and P' have an odd parity, while P'' has an even parity.

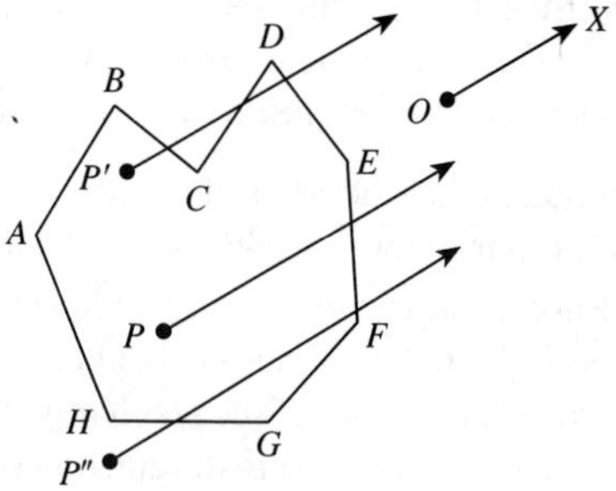

Figure 8.11

To find the parity of each point in the plane, it is necessary to consider the possibility of a ray passing through one or more vertices of the polygon. Two possible relationships of adjacent sides and the vertex of intersection are illustrated in Figure 8.12. In Figure 8.12a the vertex is not counted as an intersection. For Figure 8.12b the vertex is counted. The vertex is not counted when the two adjacent sides of the polygon are on the same side of the ray, but it is counted when they are on opposite sides of the ray.

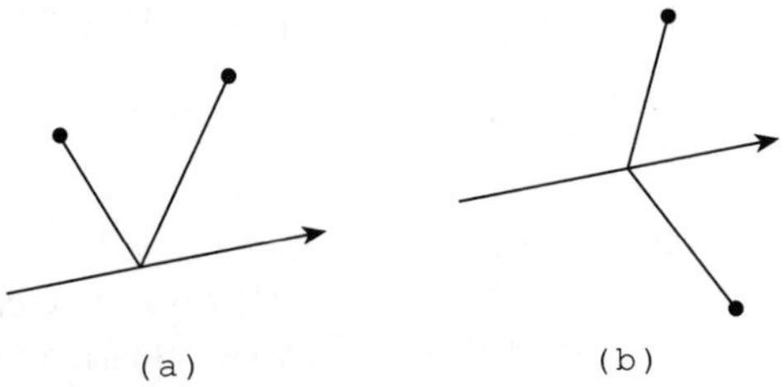

Figure 8.12

Let *set* S be the set of all points in the plane not on the polygon such that the parity is odd, and let S' be the set of all points in the plane not on the polygon such that the parity is even. Let $\overline{PP'}$ be any segment not intersecting the polygon. Then every point on $\overline{PP'}$ has the same parity, and the segment lies entirely in S or S'.

Now it can be shown that any two points P, P' of the same set S or S' can be joined by a polygonal curve that does not intersect the polygon. If $\overline{PP'}$ does not intersect the polygon, the statement is obvious. Otherwise, let X be the point of intersection closest to P, and let X' be the point of intersection closest to P', as shown in Figure 8.13. Let P'' be a point of $\overline{PP'}$ near P' and with the same parity as P'.

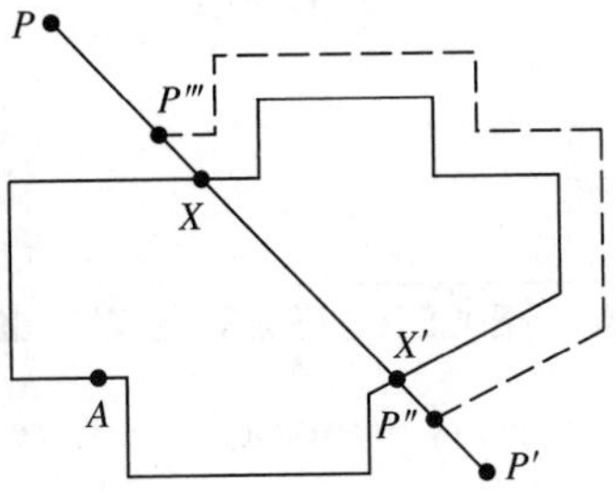

Figure 8.13

It is possible to trace a polygonal path with sides close to the sides of the original polygon until a point P''' on $\overline{PP'}$ near P is reached. P''' is between P and X, and the polygonal path from P'' to P''' is entirely composed of points with the same parity.

The original polygon is the boundary of both set S and set S'. If A is any point of the polygon, then there are points B and C on either side of A, but arbitrarily close, such that their parity is different. Since every neighborhood of A contains points of both S and S', the polygon is the boundary of both S and S'. Points in the exterior of the polygon have an even parity, and points in the interior have an odd parity.

The following two theorems, whose proofs are left as Exercises 14 and 15 of Exercises 8.2, give additional properties of simple closed curves and their intersections with one-dimensional convex sets in the same plane.

THEOREM 8.5 ➤ For one point A in the interior and one point B in the exterior of a simple closed curve S, $\overline{AB}$ and S have at least one point in common.

THEOREM 8.6 ➤ Every ray with an endpoint in the interior of a simple closed curve intersects the curve.

The Jordan curve theorem is involved in various puzzle-type problems in topology. One of these is illustrated in Figure 8.14a. The problem is to draw curves connecting the three pairs of points with the same numbers in such a way that the curves do not intersect each other or any of the other lines given in the figure. The problem is impossible to solve, as shown by Figure 8.14b. The given lines and the paths drawn from 1–1 and 3–3 result in a shaded region representing a simple closed curve and its interior. One of the points labeled 2 is an interior point and the other is an exterior point, so any curve connecting them must intersect the boundary.

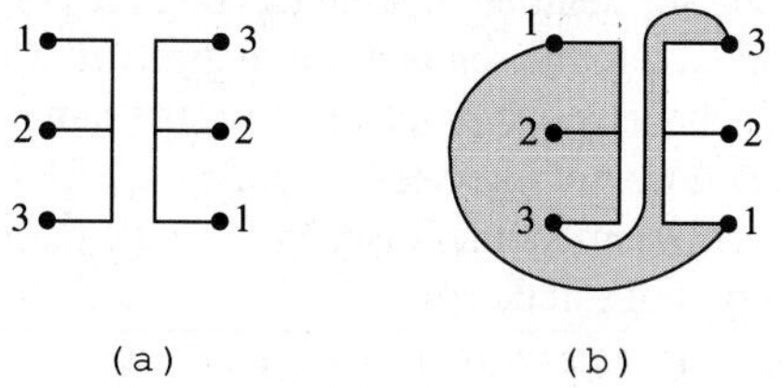

Figure 8.14

▲ EXERCISES 8.2 ▲

For Exercises 1–4, indicate which of the sets of points are topologically equivalent to any simple closed curve.

1. Triangle
2. Hyperbola
3. Two concentric circles
4. Boundary of Reuleaux triangle
5. List additional sets of points that are topologically equivalent to a simple closed curve.

For Exercises 6–9, indicate which of the plane regions are topologically equivalent to any simple closed curve and its interior.

6. Triangular region

7. Circular region
8. Plane region with degree of connectivity one
9. Plane region with degree of connectivity two
10. List additional sets of points that are topologically equivalent to a simple closed curve and its interior.
11. In Figure 8.11, explain what happens when vertices C, E of the polygon are encountered by rays drawn parallel to $\overrightarrow{OX}$ with points on $\overline{P'P}$ as endpoints.
12. Do the points in the interior of a simple closed curve have odd or even parity?
13. Is the exterior of a simple closed polygon simply connected or multiply connected?
14. Prove Theorem 8.5.
15. Prove Theorem 8.6.
16. In Figure 8.15, can curves be drawn connecting the three pairs of points with the same numbers in such a way that the curves do not intersect each other or any of the other lines given in the figure? Explain why or why not.

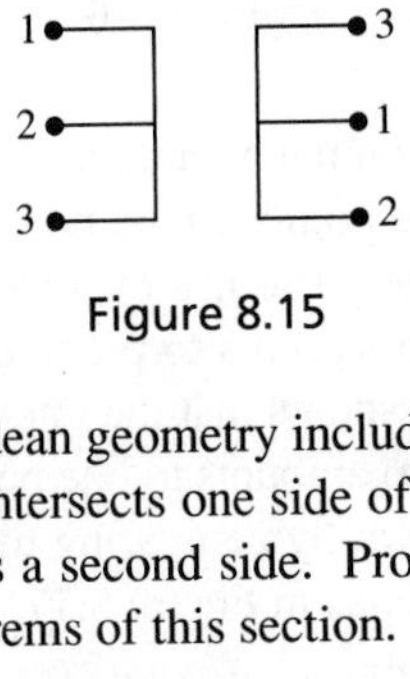

Figure 8.15

17. Pasch's axiom for Euclidean geometry includes the statement that a line in the plane of a triangle that intersects one side of the triangle at a point other than the vertex also intersects a second side. Prove the statement as a theorem in topology, using the theorems of this section.
18. Find a proof of the fundamental theorem of algebra that uses ideas very similar to those in the proof of Theorem 8.2 and explain their use.
19. Make up a topological puzzle similar to the one in Figure 8.14, and show when it is impossible to solve.
20. Find in a reference and outline a general proof of the Jordan curve theorem.

▲ 8.3 Invariant Points and Networks

Invariant points are less common under the transformations of topology than for the motions of a plane. One of the simplest examples of a theorem about invariant points under topological transformations is the Brouwer fixed point theorem for a circular region. This theorem is named for L. E. J. Brouwer, a famous twentieth-century Dutch mathematician (1882–1966).

THEOREM 8.7 ➤ *Brouwer fixed point theorem.* If the points of a circular region undergo a continuous transformation so that each image is a member of the set, then there is at least one fixed point.

A simple example of Theorem 8.7 is a circle rotated about its center, with the center the fixed point. The theorem can be proved indirectly, and an outline of this proof follows. Assume there is no fixed point. The transformation may be visualized by associating a vector with each point, as in Figure 8.16a. The initial point of each vector is the original point, and the terminal point is its image.

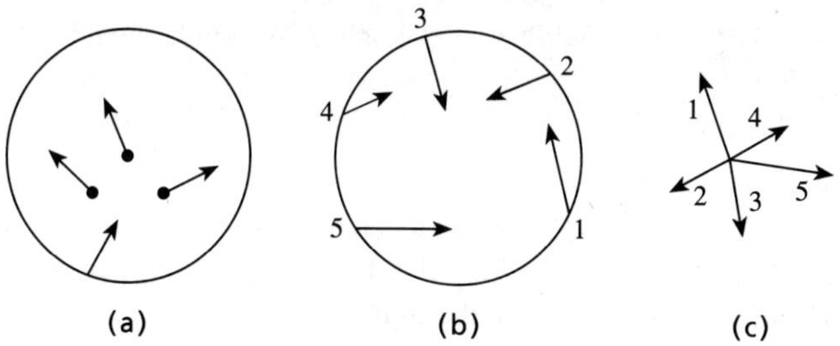

Figure 8.16

All of the vectors for points on the boundary of the region point into the circle, as in Figure 8.16b. Suppose that the points on the boundary are considered in a counterclockwise order around the circle. The use of a vector diagram, as in Figure 8.16c, shows that the sequence of vectors seems to progress around a circle and come back to the original position. It can be shown that the total algebraic change in the direction of the vectors (not the vector sum) amounts to one positive revolution (the *index* of the vectors is one). This can be shown by comparing this set of vectors with the tangent vectors for all points on the circle, as in Figure 8.17.

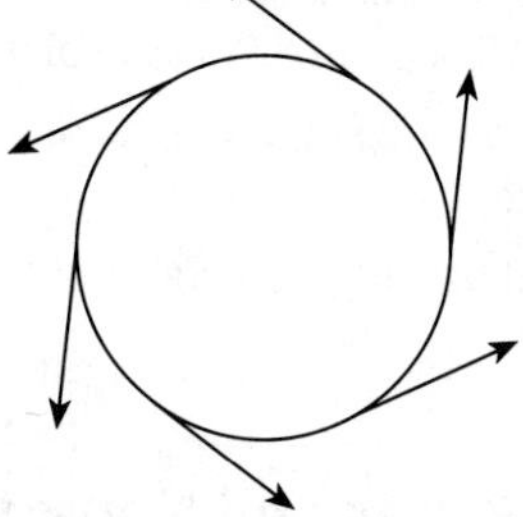

Figure 8.17

The tangent vectors make one complete revolution, and if the vectors for the points on the circumference turn through some different angle, the difference is a multiple of 2π. This means the vectors of the transformation must turn completely around the tangents at least once. But since the vectors turn continuously, at some time the transformation vector and the tangent vector must point in the same direction. This is impossible, since all the transformation vectors point inside the circle.

For any circle concentric to the given circle and contained within it, the index of the transformation vectors must also be one. This is true because passing continuously from one circumference to smaller ones results in a continuous change in the transformation vector index. This means that the changes cannot take place, since the vector index may assume only integral values. No matter how small the concentric circle is, the index of the vectors of the transformation will always be one. But this is impossible; the vectors on a sufficiently small circle will all point in approximately the same direction as the vector at the center of the circle, since the transformation is continuous. If the index of the vectors remains an integer and becomes smaller than one, it will be zero. Thus, the assumption of a vector for every point has led to a contradiction, so there is at least one fixed point.

A recent application of invariant points and topology is in the field of oceanography. In an attempt to explain the theory of continental drift, a mathematical model is used to illustrate what happens when two rigid plates on a sphere spread out from a ridge crossed by fracture zones. The plates must revolve about an invariant point called the *pole of spreading*. The rotation also takes place around an *axis of spreading*, which passes through the pole of spreading and the center of the earth. For more information, read *Planet Earth* by Jonathan Weiner.

The second example in this section of a topological invariant is related to the idea of a *network*. The word *network* is used to describe a connected set of vertices, segments, and curves such as those shown in Figure 8.18. These may be thought of as a set of vertices with paths connecting them. Networks such as these furnish examples of what are known as *tracing puzzles*. Can the networks be traced without lifting the pencil from the paper and without repeating a portion of the path? Try to trace the networks in Figure 8.18a–c, and to discover how to decide in advance whether they can be traced or not, before reading the next paragraph.

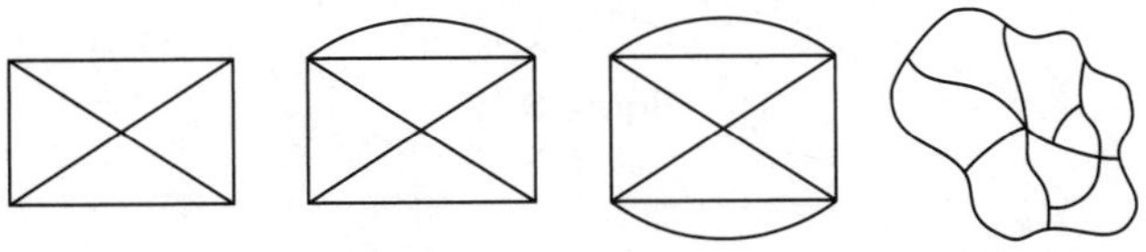

(a) (b) (c) (d)

Figure 8.18

The network in Figure 8.18a cannot be traced, whereas those in Figures 8.18b and 8.18c can be traced. Try to decide what is the essential difference between networks that can be traced and those that cannot.

The number of paths leading to each vertex in a network is crucial in trying to decide whether or not the network can be traced. If a vertex has an even number of paths to it (an *even vertex*), then these may be used in pairs in going to and leaving the vertex. If there are an odd number of paths to a point (an *odd vertex*), they cannot be used in pairs, and it is necessary to begin or end at that point. If the network has two or fewer vertices with an odd number of paths, it can be traced. The network will be a curve in this case. If there are no vertices with an odd number of paths, then the network can be traced by beginning and ending at the same point. The network will be a closed curve in this case. Now check Figure 8.18d. This network has more than two vertices with an odd number of paths, so it cannot be traced. The information about tracing puzzles is summarized in Table 8.1.

Table 8.1

Number of Vertices with Odd Number of Paths	**Network Can Be Traced**	**Name of Network**
more than two	no	not a curve
two or fewer	yes	curve
zero	yes	closed curve

The essential characteristics of a network in terms of the number of even and odd vertices is a topological invariant. In other words, for a particular network, the number of paths leading to a vertex is not changed by a topological transformation.

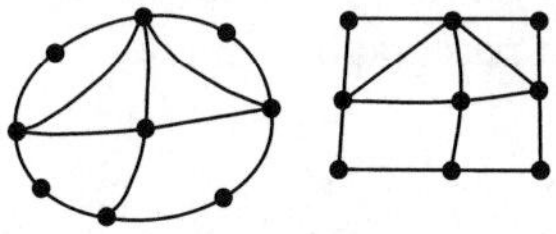

Figure 8.19

Example ➤ The two networks in Figure 8.19 are topologically equivalent.

Much of the work on networks from this section, as well as material from the remaining sections of the chapter, is also incorporated in a typical course in *graph theory*. In this context, a graph can be depicted by a diagram in which points represent nodes and arrows represent arcs. Paths are finite sequences of arcs, and a cycle is a path whose endpoints coincide. Graph theory, like topology, has become an established branch of mathematics in its own right, with applications in an amazing number of modern fields of study.

▲ EXERCISES 8.3 ▲

1. Could the continuous transformation in Theorem 8.7 be a glide reflection? Why?
2. Outline briefly the major steps in the proof of Theorem 8.7.

For Exercises 3–10, tell whether or not the networks in Figure 8.20 can be traced.

For Exercises 11–18, tell whether the networks in Exercises 3–10 are curves.

For Exercises 19–26, tell whether the networks in Exercises 3–10 are closed curves.

For Exercises 27–34, tell whether the networks in Exercises 3–10 are simple closed curves.

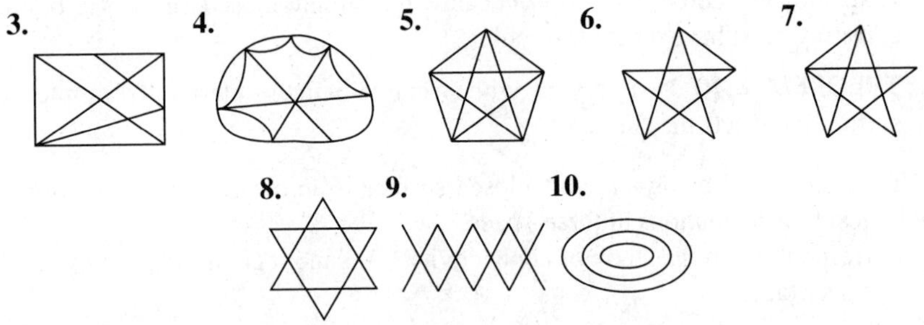

Figure 8.20

▷ **35.** Is it possible to have a network with exactly one odd vertex? Explain your answer.

▷ **36.** Rewrite the material on networks in this section, using the terms and definitions from graph theory.

▲ 8.4 Introduction to the Topology of Surfaces

This section introduces some basic theorems for simple closed surfaces, analyzes a topological invariant of surfaces, and gives an example of a recently solved problem in topology.

The following theorem is the Jordan curve theorem stated for three-space.

THEOREM 8.8 ➤ Any simple closed surface in three-space partitions the space into three disjoint connected sets such that the set that is the surface is the boundary of both of the other sets.

The following theorem and corollary can also be proved for simple closed surfaces; they are analogous to the theorem proved for the simple closed curve.

THEOREM 8.9 ➤ For one point A in the interior and one point B in the exterior of a simple closed surface S, $\overline{AB} \cap S$ is not empty (see Figure 8.21).

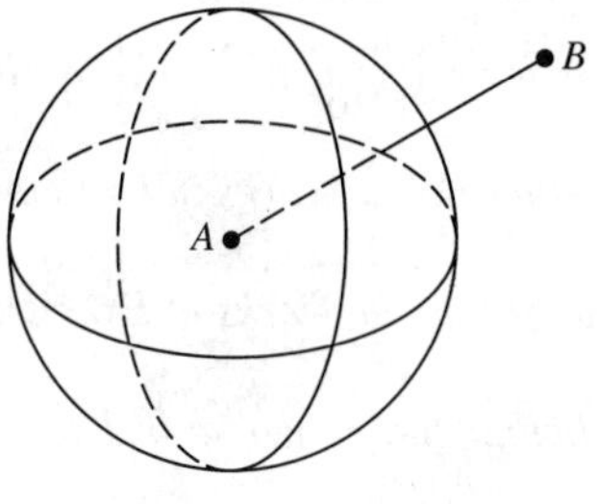

Figure 8.21

The following corollary is true because the simple closed surface is bounded; hence, the ray must have an exterior point.

THEOREM 8.10 ➤ Every ray with an endpoint in the interior of a simple closed surface intersects the surface.

The property of being a simple closed surface is an invariant under the group of topological transformations in three-space.

Corresponding to the study of connectivity for plane regions is the analysis of the *genus* of a surface.

DEFINITION ➤ The genus of a surface is a whole number giving the largest number of non-intersecting simple closed curves that can be drawn on the surface without separating it into two unconnected parts.

For example, the genus of a sphere is zero, since any simple closed curve separates it into two unconnected parts, as shown in Figure 8.22. Similarly, many common surfaces such as an ellipsoid and a convex polyhedron are also of genus zero.

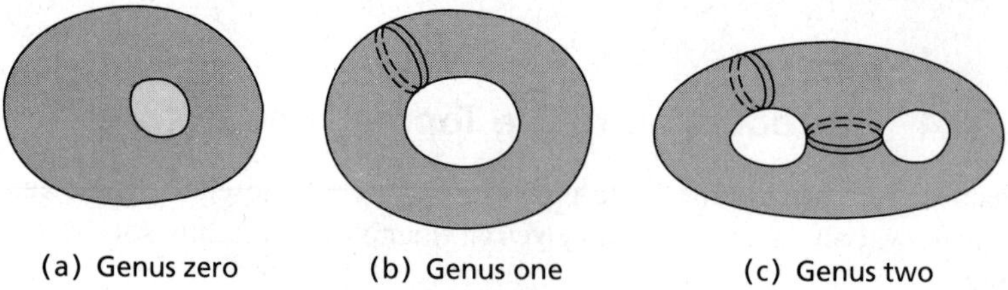

(a) Genus zero (b) Genus one (c) Genus two

Figure 8.22

Other surfaces exist besides those with a genus of zero. Figure 8.22b shows a *torus* with genus one, and Figure 8.22c shows a surface with genus two. For the figure of genus one, a second cut would separate the surface into two unconnected parts. Similarly, for the figure of genus two, a third cut would separate the surface into unconnected parts.

The genus of a surface is another example of an invariant under any topological transformation. Also, any surface of a particular genus may be changed into any other surface of the same genus by a topological transformation. The effect of the two previous statements is that two surfaces with the same genus are topologically equivalent. Figure 8.23 shows several topologically equivalent surfaces of genus one.

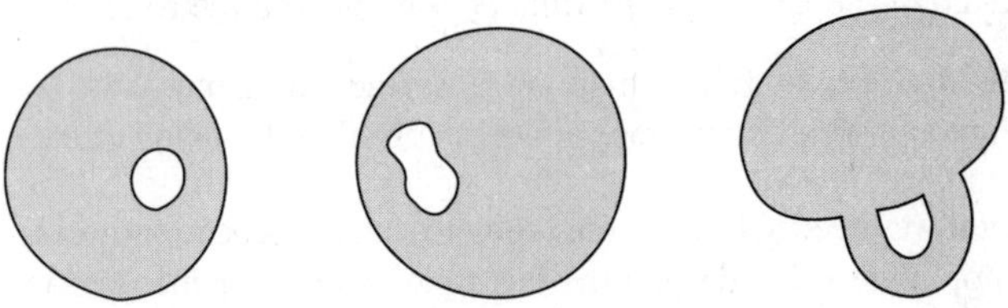

Figure 8.23

One application of the topology of surfaces is in what is called *catastrophe theory*. (See Woodcock and Davis, *Catastrophe Theory*.)

An interesting topological problem closely associated with networks, with simple closed curves, and with invariance concerns the coloring of maps.

For two-dimensional maps such as those shown in Figure 8.24, or for similar ones on a sphere, only four colors are necessary if each two countries with a common boundary must have a different color. It is assumed that two countries may have the same color if the boundaries intersect in a single point. This is the case for the two countries numbered 1 in Figure 8.25.

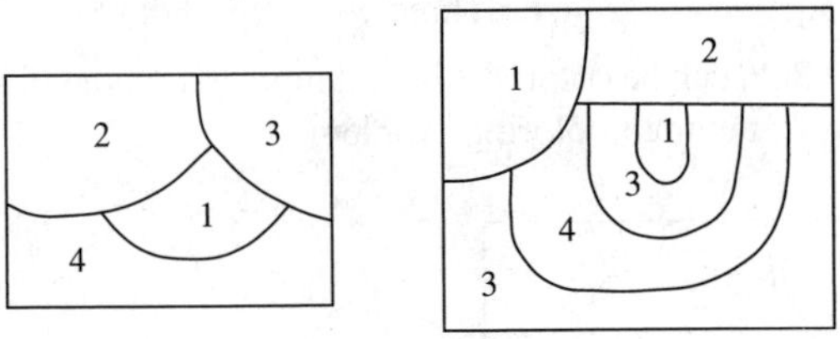

Figure 8.24

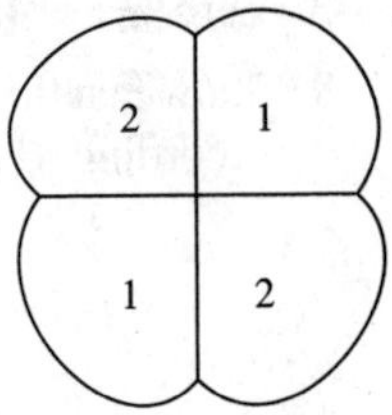

Figure 8.25

The general statement that any map can be colored with four colors has been extremely difficult to prove, although Möbius proposed proving the statement as early as 1840. About 50 years later, the following theorem was proved.

THEOREM 8.11 ➤ Every map on a sphere can be colored according to the rules for the map-coloring problem by using at most five colors.

The countries in Theorem 8.11 are assumed to be simply connected regions. The rules for the map-coloring problem are that two countries with a common boundary must have a different color but that they may have the same color if the boundaries intersect in a single point.

As was pointed out at the beginning of the discussion, the map-coloring problem is related to other topological concepts. The boundaries of the countries can be any simple closed curve, so the exact shape is unimportant. The map-coloring problem changes if the genus of the surface changes. Theorem 8.11 applies to any surface of genus zero, but not to a torus, for example.

The map-coloring problem was finally solved in July 1976, by Kenneth Appel and Wolfgang Haken of the University of Illinois, who proved the following theorem.

THEOREM 8.12 ➤ Every map on a surface of genus zero can be colored according to the rules of the map-coloring problem by using at most four colors.

A computer was needed to solve the problem, but the proof continues to be accepted, generally, if not universally, despite the fact that some mathematicians still hope for a proof of a different sort.

▲ EXERCISES 8.4 ▲

1. Prove Theorem 8.9.
2. What is the genus of a simple closed surface?

For Exercises 3–7, name or describe other specific representations of surfaces of the same genus besides those mentioned in this section.

3. Zero 4. One 5. Two 6. Three 7. Four

8. Show that the map in Figure 8.26 can be colored with not more than four colors, according to the restrictions of the map-coloring problem.

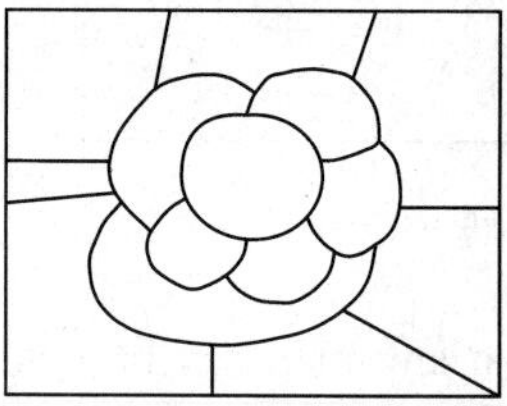

Figure 8.26

9. What is the minimum number of colors required to color each map in Figure 8.24 according to the rules of the map-coloring problem?
10. Draw a map with 12 countries that requires only three colors, yet stays within the map-coloring requirements.

11. Draw a map with 12 countries that requires only two colors to fulfill the map-coloring requirements.
12. Explain why it is important that the countries on a map in the map-coloring problem be simply connected regions.
13. Explain why it is important in the map-coloring problem that countries with a single point in common can have the same color.
14. Prove Theorem 8.8, or study the proof in a reference.
15. Outline the proof of Theorem 8.11, using a reference if necessary.
16. Study the basic ideas of catastrophe theory and explain the significance of the theory.
17. Study and outline Appel and Haken's proof of the four-color map problem.
18. Investigate the map-coloring problem for a torus. How many colors will suffice?

▲ 8.5 Euler's Formula and Special Surfaces

This section continues the study of the topology of surfaces by introducing Euler's formula for surfaces that do not have the same genus and by suggesting special surfaces with strange properties unlike those of ordinary Euclidean geometry.

The development of the subject of topology has come during the past one hundred years, but the subject had its beginning even before that. One of the earliest ideas that is actually toplogical in nature was first discovered by Descartes, then rediscovered by Euler in 1752. It has since become known as *Euler's formula.* Euler's formula relates the number of faces, vertices, and edges of a simple polyhedron. Recall that a *polyhedron* is a closed surface consisting of a number of faces, each of which is a polygonal region. If the surface has no holes in it and can be deformed into a sphere by a continuous transformation, it is a *simple* polyhedron. A simple polyhedron has a genus of zero.

The Greeks showed particular interest in five polyhedra—those that were *regular*, with congruent faces and angles. The five regular polyhedra, which were treated in Book XIII of Euclid's *Elements*, are shown in Figure 8.27. The names and some of the numbers of *vertices*, *faces*, and *edges* are listed in Table 8.2. The completion of the table is left as Exercise 1, in Exercises 8.5.

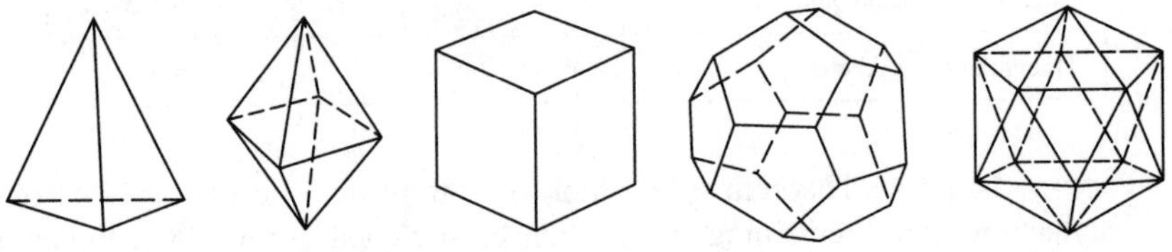

(a) (b) (c) (d) (e)

Figure 8.27

What is the relationship between the number of vertices, faces, and edges for these polyhedra? Take various other examples of polyhedra, if necessary, to arrive at Euler's formula on your own before continuing.

Table 8.2

Figure Number	Name of Polyhedron	V	F	E
8.27a	tetrahedron	4	4	6
8.27b	octahedron	6	8	12
8.27c	cube	8	?	?
8.27d	dodecahedron	?	12	?
8.27e	icosahedron	?	?	?

THEOREM 8.13 ➤ For a simple closed polyhedron, $V + F = E + 2$, where V is the number of vertices, E the number of edges, and F the number of faces.

Example ➤ Figure 8.28 shows two additional examples of polyhedra that can be used to verify Euler's formula. The results are summarized in Table 8.3.

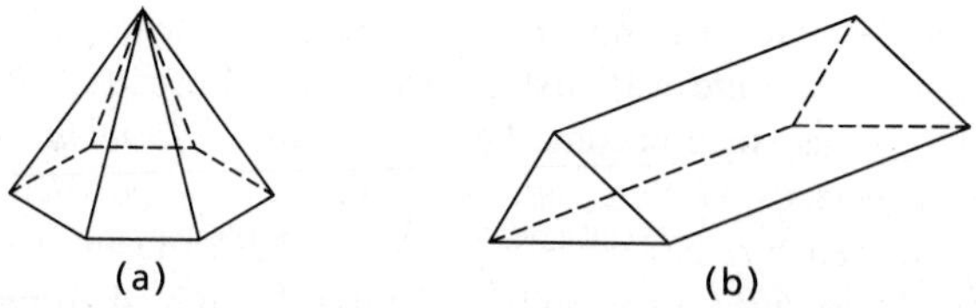

Figure 8.28

Table 8.3

	V	F	E	$V + F = E + 2$
Pyramid with hexagonal base	7	7	12	$7 + 7 = 12 + 2$
Triangular right prism	6	5	9	$6 + 5 = 9 + 2$

PROOF: To prove Theorem 8.13, think of cutting out one of the faces of the polyhedron so the remaining surface can be stretched out flat by a topological transformation. The network of vertices and edges in the plane will have the same number of vertices and edges as in the original polyhedron, but there is one less face, since one has been removed. Figure 8.29 shows an example, using a cube.

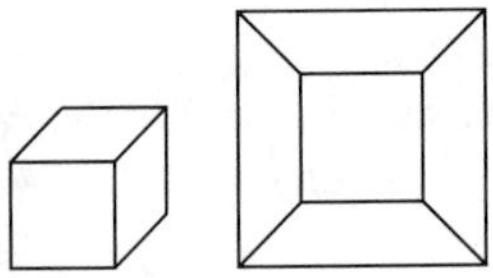

Figure 8.29

Now triangulate the plane network by drawing diagonals connecting vertices until a triangular decomposition has been achieved. In this triangulation, the value of $V - E + F$ is not changed, since drawing each diagonal adds one edge and one face.

Once the triangulation has been completed, the triangles may be removed one at a time until a single one remains. (This process is illustrated for the cube in Figure 8.30.) Some of the triangles will have edges on the boundary of the network. First, remove any edge of a boundary triangle not an edge of another triangle.

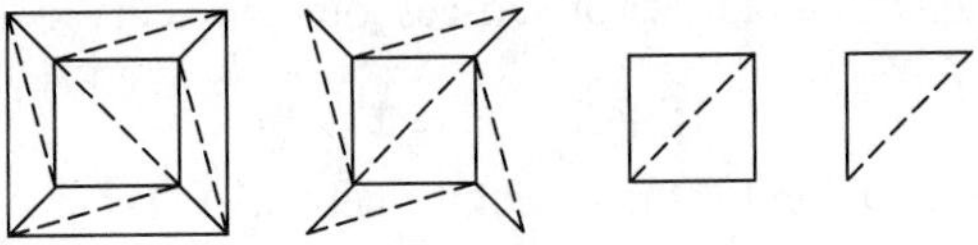

Figure 8.30

1. If the triangle has only one edge on the boundary, removing that one edge reduces both E and F by one, so $V - E + F$ is not changed.
2. If the triangle has two edges on the boundary, removing them reduces V by one, E by two, and F by one, so $V - E + F$ is unchanged.

By a continuation of this process, the boundary can be changed until all that is left of the triangulation network is a single triangle with three edges, three vertices, and one face. For this triangle, $V - E + F = 1$. But since $V - E + F$ is an invariant in this process of removing triangles, the formula $V - E + F = 1$ holds for the original plane network. The formula $V - E + F = 2$ applies to the original polyhedron, which had one more face.

Euler's formula $V - E + F = 2$ is an example of a topological property of a figure, since it is unchanged under a topological transformation. Of course, the polyhedron can be transformed into any simple closed surface so that the system of vertices, faces, and edges becomes a network of points, regions, and paths on the surface.

Euler's formula does not hold unless the genus of the surface is zero. Furthermore, an agreement on how to count the faces is necessary. For example, the set of points in Figure 8.31 has genus one. To count the faces requires an agreement that face $AGFD$ is counted but $IJKL$ is then subtracted so that the total number of faces is eight. Check to see that Euler's formula does not hold for this surface. An alternate way of counting faces is suggested in Exercise 21I.

It is interesting to realize that the concept of the Euler formula can be generalized for any closed surface, regardless of genus.

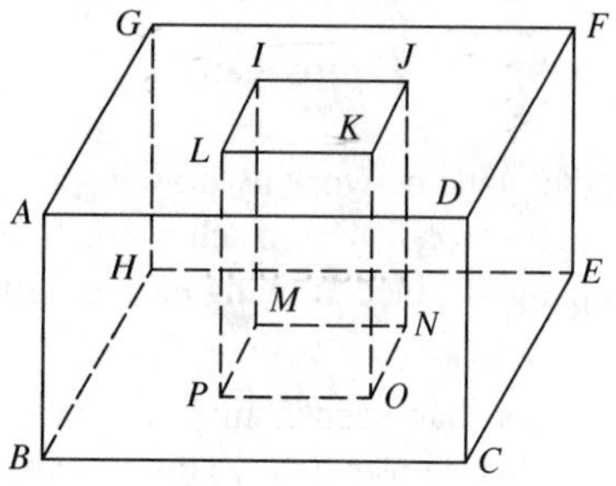

Figure 8.31

THEOREM 8.14 ➤ If a closed surface of genus n is partitioned into simply connected regions by a number of vertices joined by curved arcs, then

$$V - E + F = 2 - 2n$$

(F represents regions and E the arcs of the network). The number $2 - 2n$ is called the *Euler characteristic*. Note that this reduces to the ordinary Euler formula for a surface of genus zero.

The final topological topic to be introduced is that of a very special kind of surface unlike those studied up to this time. Ordinary surfaces have two sides, but Augustus Möbius (or Moebius), who had also written on analytic projective geometry, wrote a paper about 1865 on polyhedral surfaces that led to the realization that it is possible to have surfaces with only one side. For a surface of one side, a path from one point to another does not need to intersect the edge.

In Figure 8.32, which shows a surface of two sides, you could not draw a line from A to B, for example, without crossing over an edge.

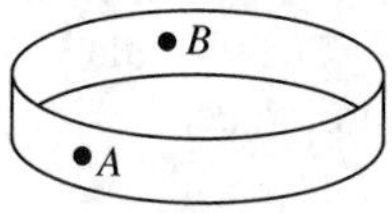

Figure 8.32

The simplest one-sided surface is called a *Möbius strip*. It is pictured in Figure 8.33. A representation of a Möbius strip can be formed from a rectangular strip of paper. Give the paper a half-twist and tape the ends together. A Möbius strip has only one side and one edge. You can trace a curve from any one point on the surface to any other points without crossing an edge. It is fascinating to experiment with a Möbius strip to see that it behaves unlike a two-sided surface. For example, if a Möbius strip is cut down a line through the middle, parallel to the edge, it remains in one piece.

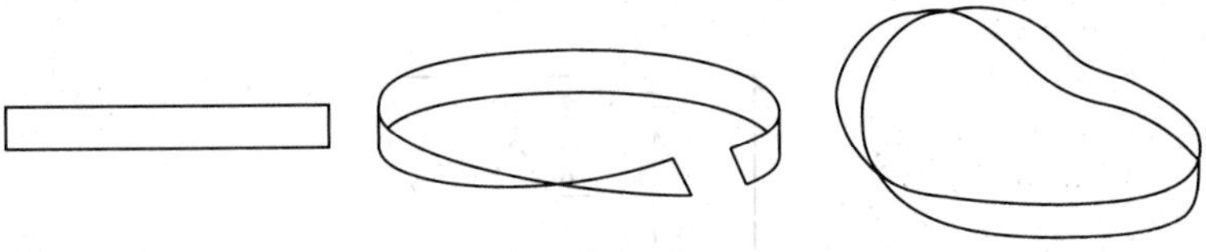

Figure 8.33

A second example of a one-sided surface is a Klein bottle, shown in Figure 8.34. This intersecting surface has no inside or outside. An unusual application of one-sided surfaces for women's dresses appeared in Jean J. Pedersen's article, "Dressing up Mathematics," in *The Mathematics Teacher*, February 1968. A possibly more practical application of the Möbius strip is shown in Figure 8.35. The shaping of a belt in the form of a Möbius strip connecting two wheels allows the belt to wear out at the same rate everywhere, not just on one side as in the usual arrangement.

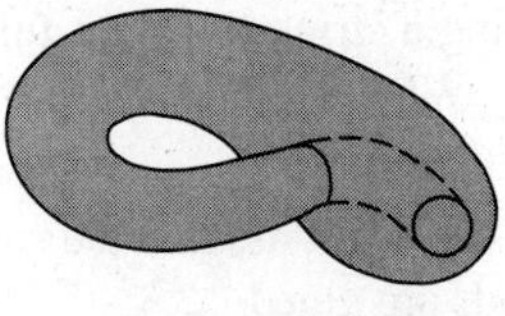

Figure 8.34

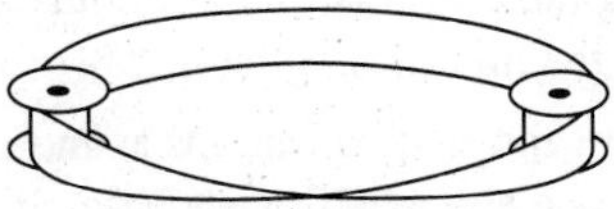

Figure 8.35

The property of being a two-sided or a one-sided surface is a topological invariant, which explains why the topic is included in this section. For example, a Möbius strip cannot be changed to a two-sided surface by a topological transformation. A related application of topology that should be mentioned briefly here is the classification of various kinds of *knots*. Two knots are topologically equivalent if one can be transformed into the other with no intermediate self-intersections.

▲ EXERCISES 8.5 ▲

1. Complete Table 8.2, showing the number of vertices, faces, and edges for the five regular polyhedra.
2. Verify Euler's formula for each of the five regular polyhedra.
3. Verify Euler's formula for an:

   a. Hexagonal right prism   b. Octagonal right prism

4. Make drawings similar to Figures 8.29 and 8.30 for a tetrahedron.

5. Find the Euler characteristic for the set of points in Figure 8.31.

For Exercises 6–9, what is the numerical value of the Euler characteristic for a surface of the genus indicated?

6. Zero? **7.** One?

8. Two? **9.** Three?

For Exercises 10–13, what is the genus corresponding to the given value of the Euler characteristic? If the answer is not a whole number, state that finding an answer is impossible.

10. −8 **11.** −12 **12.** 3 **13.** 4

14. Draw a figure similar to Figure 8.31, but with a genus of two. Find the number of vertices, faces, and edges; then determine the Euler characteristic.

15. By constructing a model, verify that if a Möbius strip is cut down a line through the middle, it remains in one piece.

16. What is the result when a Möbius strip is cut lengthwise, beginning one third of the way from the edge, rather than down the middle?

17. Suppose a surface is constructed by giving a strip of paper a full twist before gluing the ends. Is this a Möbius strip? What happens when the band is cut down the middle lengthwise?

18. Answer the same questions as in Exercise 17, but assume the strip of paper is given one and a half twists before the ends are glued.

19. A Möbius strip is a set of points in n-space, for n equal to what number?

20. Carry out the activities of Exercises 15–18 for surfaces formed by more than one and a half twists. Search for patterns and possible generalizations that help predict what will happen when the surface is cut in various ways.

21. Try to discover an alternate method of counting faces for a surface of genus other than zero by adding new edges to the existing figure.

22. In the proof of Theorem 8.13, use an octahedron as an example and show diagrams for the steps in the proof.

23. Study, then outline, a proof of Theorem 8.14.

24. Explain what it means to say a Möbius strip is an example of a non-orientable surface.

25. Study and report on topological classification of knots. You may use the article, "The Theory of Knots," by Lee Neuwirth, in the June 1979, issue of *Scientific American*.

26. Investigate the group of rotations for each of the five regular polyhedra, using a book on modern abstract algebra or modern applied algebra.

▲ CHAPTER 8 REVIEW EXERCISES ▲

1. Would it be possible for a transformation to exist that was bicontinuous, yet whose inverse was not continuous?

For Exercises 2–5, indicate which of the following sets of points are always connected sets.

2. Set of 25 distinct points on surface of sphere

3. Set of points with degree of connectivity of 25

4. Set of 25 distinct lines with one point in common

5. Polygon with 25 sides

6. If a plane had a degree of connectivity of three, then how many cuts would be required to change it to a simply connected region?

7. If two plane regions both have the same degree of connectivity, then is it possible to transform one into the other by a topological transformation?

8. For one point in the interior and one point in the exterior of a simple closed curve, the segment with the two points as endpoints has what minimum number of points in common with the curve?

For Exercises 9 and 10, for a surface of genus zero, use Euler's formula.

9. Find the number of vertices if the number of faces is 7 and the number of edges is 25.

10. Find the number of faces if the number of vertices is 16 and the number of edges is 34.

11. A surface with a genus of two has how many holes through it?

12. If the Euler characteristic for a surface is -8, what is the genus of the surface?

13. Could a one-sided surface be the image of a two-sided surface under some topological transformation?

14. Suppose that a Möbius strip is formed from a rectangular piece of paper 2 cm wide and 12 cm long, without any loss due to overlapping. Find the total surface area of the resulting Möbius strip.

For Exercises 15–19, check the number of odd vertices; then tell which describes the drawings in Figure 8.36 most precisely—not a curve, a curve, or a closed curve.

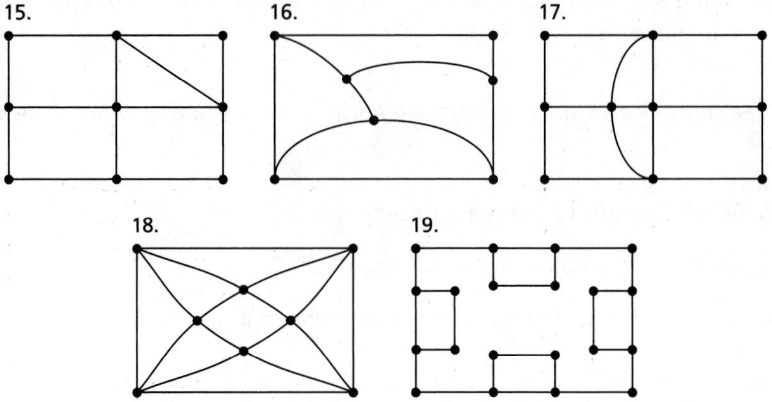

Figure 8.36

Additional chapter review exercises may be selected from any exercises not already assigned from previous exercise sets in this chapter.

9 Non–Euclidean Geometries

▲ 9.1 Introduction to Hyperbolic Geometry

Chapters 2–5 extended the study of Euclidean geometry by introducing many modern ideas. Chapters 6–8 incorporated modern transformations that resulted in very general geometries or a geometry that differed from Euclidean. The geometries featured in this last chapter were anticipated in Chapter 1 during the discussion of various sets of axioms and their properties.

Non-Euclidean geometries are specific examples of modern geometries with different sets of axioms from those of Euclidean geometry. The term *non-Euclidean geometry* is used in a very restricted sense. Non-Euclidean geometry differs from the geometry of Euclid because it substitutes an alternative for his fifth postulate on parallels. Studying non-Euclidean geometry is much like studying philosophy, because the profound revelations arrived at through pursuing the consistent axiomatic systems may disturb long-held views of the nature of mathematics and of the universe itself.

The form of the fifth postulate as stated by Euclid was considerably more complex than the form of the other postulates and axioms: "If a transversal falls on two lines in such a way that the interior angles on one side of the transversal are less than two right angles, then the lines meet on that side on which the angles are less than two right angles." According to Euclid, $\overleftrightarrow{AD}$ and $\overleftrightarrow{BC}$ meet to the right in Figure 9.1 if the sum of the measures of $\angle DAB$ and $\angle ABC$ is less than π radians.

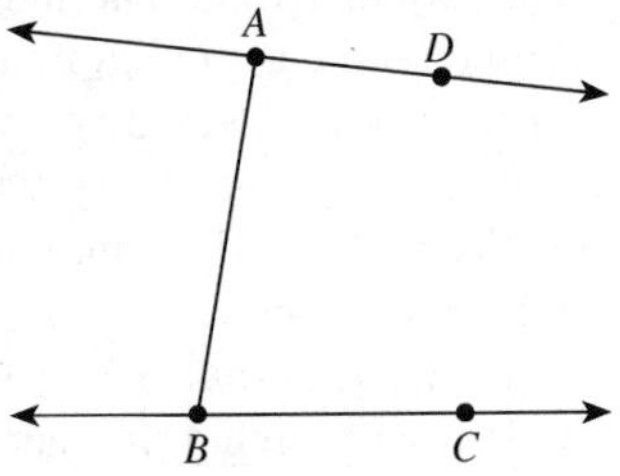

Figure 9.1

The word *parallel* does not appear in Euclid's fifth postulate. The wording of Euclid's fifth postulate that is most commonly used is called *Playfair's axiom* (named after John Playfair, 1748–1819, a Scottish physicist and mathematician): Through a point not on a given line, exactly one line can be drawn in the plane parallel to the given line. The word *parallel* as used here means not intersecting or having no Euclidean point in common. Playfair's axiom and the original fifth postulate are logically equivalent. This means that either one can be used, along with the other assumptions of Euclidean geometry, to prove the second.

Non-Euclidean geometry provided Einstein with a suitable model for his work on relativity. While it also occurs in differential geometry and elsewhere, it is worthwhile for other reasons as well. The real understanding of the concept of postulate sometimes comes only when a person begins with postulates that do not seem self-evident or practical. Also, non-Euclidean geometry, unlike projective geometry or topology, is significant in that it is something other than a generalization of Euclidean geometry. Non-Euclidean geometry does not include ordinary geometry as a special case.

From the time Euclid stated his postulates, about 300 B.C., controversies arose concerning the fifth postulate. Many substitutes (including Playfair's axiom) were suggested for it, but more significantly, mathematicians felt it was not a postulate at all. For almost 2,000 years, many people attempted to show that the statement was actually a theorem that could be proved from the rest of the postulates. The controversy continued through the eighteenth century. The work of Saccheri and Lambert, to be studied briefly in Section 9.3, continued the attempt to prove the fifth postulate. These two people used the indirect method of proposing an alternate parallel postulate and attempting to reach a contradiction. The fifth postulate was never proved, nor was a contradiction reached.

Shortly after 1800, mathematicians such as Carl Friedrich Gauss (1777–1855) began to realize that the fifth postulate could never be proved from the others, because it was indeed an independent postulate in the set of Euclidean postulates, not a theorem. Attempts to prove the fifth postulate by denying it had already produced strange theorems that had to be accepted as valid if some other substitute postulate was to be used.

As in the discovery of calculus, more than one person shares the credit for the discovery of non-Euclidean geometry. Though Gauss was aware of the significance of the subject, he did not publish any material. The first account of non-Euclidean geometry to be published was based on the assumption that, through a point not on a given line, more than one line can be drawn parallel to a given line in the plane. This type of geometry, called *hyperbolic* non-Euclidean geometry, was discovered independently by a Russian, Nikolai Lobachewsky (1793–1856), and, at about the same time, by a Hungarian, Johann Bolyai (1802–1860). The results were published about 1830. A second type of non-Euclidean geometry, *elliptic geometry*, is introduced briefly in Section 9.6. Felix Klein is credited with giving the names *hyperbolic*, *elliptic*, and *parabolic*, in 1871, to the two non-Euclidean geometries and to Euclidean geometry, respectively.

It would be possible to first approach the study of hyperbolic geometry through the use of models—perhaps a more intuitive way than the axiomatic approach that follows. In this chapter, a model for elliptic geometry is discussed in Section 9.6 and two models for hyperbolic geometry are used in Section 9.7 in the discussion of consistency.

The development of hyperbolic geometry in this chapter is based on all the assumptions and undefined terms of modern Euclidean geometry except for the following substitution for the parallel postulate, identified as the characteristic postulate of hyperbolic geometry.

Characteristic Postulate. Through a given point C, not on a given line $\overleftrightarrow{AB}$, passes more than one line in the plane not intersecting the given line.

Review Section 1.7 if necessary. Theorems from absolute geometry are included in non-Euclidean geometry as well, since they do not depend on any parallel postulate.

The relationship described in the characteristic postulate is pictured in Figure 9.2, if it is assumed that $\overleftrightarrow{CD}$ and $\overleftrightarrow{CE}$ are two distinct lines through C and that neither intersects $\overleftrightarrow{AB}$.

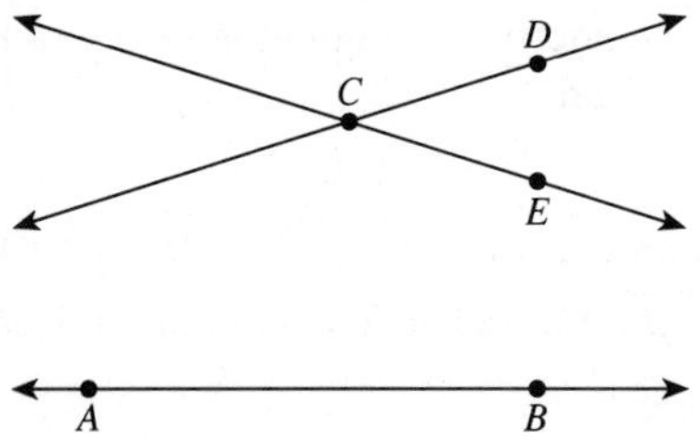

Figure 9.2

The idea of changing the parallel postulate may seem extremely strange, but there are many finite geometries where the parallel postulate of Euclid does not hold. For example, several of the finite geometries of Chapter 1 have an axiom stating that each two lines intersect at a point. The finite geometry of Desargues has a peculiarity concerning parallels; in that geometry, each point can have three lines through it parallel to one particular line, the polar of the point.

There is no finite geometry that has all the axioms of hyperbolic geometry, but some do have the characteristic postulate. A particular example is the geometry of thirteen points and twenty-six lines, represented by Table 9.1, in which each set of three points lies on a line.

Returning to hyperbolic geometry and its characteristic postulate, you should pause to consider some of the consequences of this change in postulates and see if you can predict some theorems before reading the development of theorems and proofs for the new plane geometry.

Table 9.1

A, B, C	B, D, F	C, D, G	D, H, I	E, F, K	F, G, M	G, I, K	H, J, L
A, D, E	B, E, I	C, E, J	D, J, K	E, G, L	F, I, J		
A, F, H	B, G, H	C, F, L	D, L, M	E, H, M			
A, G, J	B, J, M	C, H, K					
A, I, L	B, K, L	C, I, M					
A, K, M							

THEOREM 9.1 ➤ Through a given point C, not on a given line $\overleftrightarrow{AB}$, pass an infinite number of lines not intersecting the given line.

Appendix 6 lists the first 28 theorems of Euclid. None of them uses the parallel postulate, so all of them may be used in this chapter.

PROOF: There are an infinite number of lines passing through C and the interior of angle DCE in Figure 9.2. For Figure 9.3, let $\overline{CF}$ be the perpendicular from C, and assume that $\overleftrightarrow{CG}$, any one of these interior lines, does intersect $\overleftrightarrow{AB}$. This means that $\overleftrightarrow{CE}$ must also intersect $\overleftrightarrow{FG}$, by the axiom of Pasch, which continues to hold in hyperbolic geometry since only the axiom of parallels from Euclidean geometry has been replaced.

Recall that the axiom of Pasch states that a line entering a triangle at a vertex intersects the opposite side. In Figure 9.3, $\overleftrightarrow{CE}$ enters triangle CFG. The fact that $\overleftrightarrow{CE}$ intersects $\overleftrightarrow{AB}$ contradicts the assumption that it is parallel to $\overleftrightarrow{AB}$, so $\overleftrightarrow{CG}$ cannot intersect $\overleftrightarrow{AB}$.

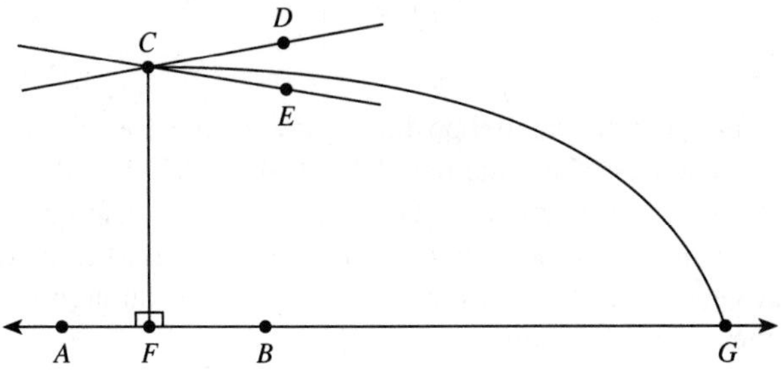

Figure 9.3

In Figure 9.4, the set of all lines in a plane passing through C is partitioned into two subsets: those that do intersect $\overleftrightarrow{AB}$ and those that do not. Because of the assumption of the one-to-one correspondence between sets of real numbers and sets of lines through a point, which was retained from Euclidean geometry, it is known that this partitioning is brought about by two different lines, shown as $\overleftrightarrow{CD}$ and $\overleftrightarrow{CE}$. These two lines must be either the last lines in either direction that do intersect $\overleftrightarrow{AB}$ or the first lines in either direction that do not intersect $\overleftrightarrow{AB}$. The assumption that there is a last intersecting line, say $\overleftrightarrow{CX}$, for example, is immediately contradicted by the statement that other points on $\overleftrightarrow{AB}$ to the right of X also determine intersecting lines. Then $\overleftrightarrow{CD}$ and $\overleftrightarrow{CE}$ are the first lines that do not intersect $\overleftrightarrow{AB}$.

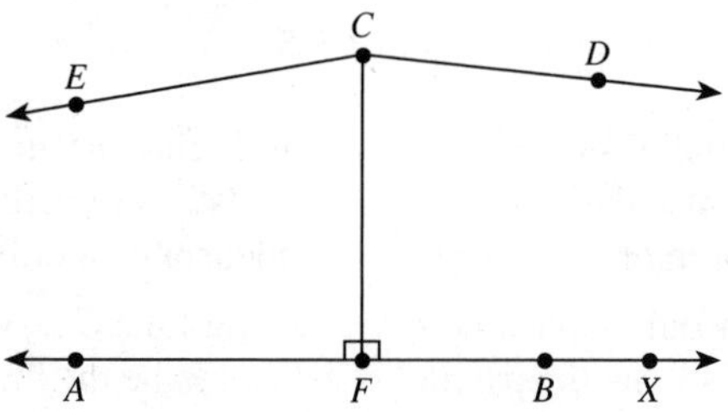

Figure 9.4

DEFINITION ➤ In hyperbolic geometry, the first lines in either direction through a point that do not intersect a given line are *parallel lines.*

DEFINITION ➤ All other lines through a point not intersecting the given line, other than the two parallel lines, are *nonintersecting lines.*

According to these technical definitions, there are exactly two parallels through C to $\overleftrightarrow{AB}$. $\overleftrightarrow{CD}$ is called the *right-hand parallel* and $\overleftrightarrow{CE}$ is the *left-hand parallel.* The angles FCD and FCE are *angles of parallelism* for the distance FC.

THEOREM 9.2 ➤ The two angles of parallelism for the same distance are congruent and acute.

PROOF: The proof is indirect, and consists of eliminating possibilities of not being congruent and not being acute. Assume that $\angle FCE$ and $\angle FCD$ in Figure 9.5 are angles of parallelism for CF, but are not congruent. Assume next that $\angle FCD$, for example, is greater. Then there is an angle, $\angle FCG$, congruent to $\angle FCE$ and such that $\overleftrightarrow{CG}$ is in the interior of $\angle FCD$. If $FH = FG$, then $\triangle FCG \cong \triangle FCH$, so $\angle FCH \cong \angle FCE$. This is a contradiction because $\overleftrightarrow{CE}$ has no point in common with $\overleftrightarrow{AB}$. Hence, $\angle FCD \cong \angle FCE$.

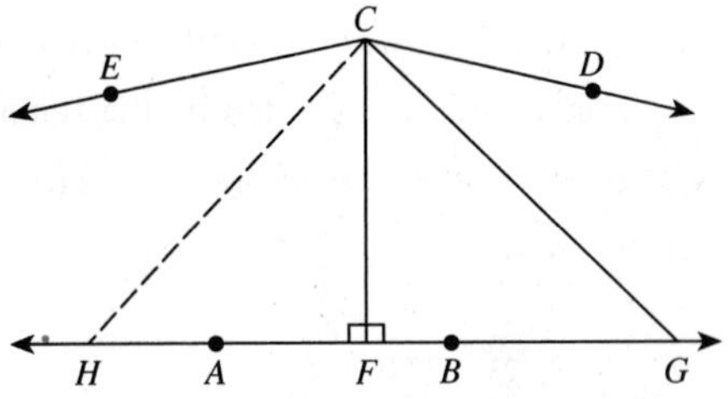

Figure 9.5

The angles of parallelism cannot be right angles, as in Figure 9.6a, because assuming this fact and also the characteristic postulate of hyperbolic geometry leads to a contradiction. Furthermore, the angle of parallelism cannot be obtuse, as in Figure 9.6b, because then it would have a non-intersecting line $\overleftrightarrow{CA}$ within the angle. This would contradict the fact that the parallel is defined to be the first noncutting line.

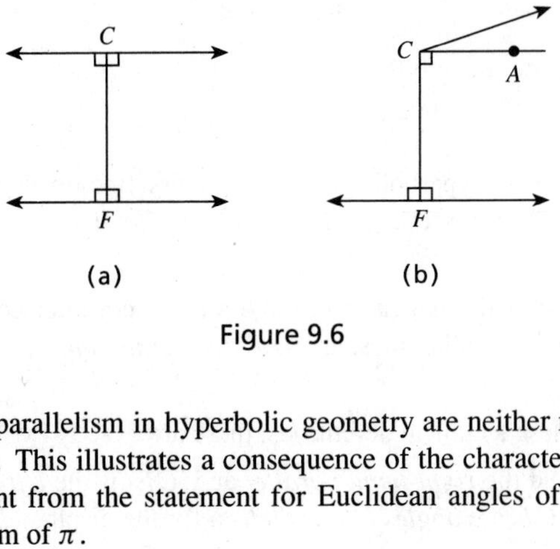

Figure 9.6

The angles of parallelism in hyperbolic geometry are neither right nor obtuse, so they must be acute. This illustrates a consequence of the characteristic postulate that is radically different from the statement for Euclidean angles of parallelism whose measures have a sum of π.

▲ EXERCISES 9.1 ▲

1. Prove Playfair's axiom, assuming Euclid's fifth postulate in the original form.
2. Prove the original statement of Euclid's fifth postulate, assuming Playfair's axiom.

It can be proved that each of the statements in Exercises 3–8 is equivalent to Euclid's fifth postulate. Reword each sentence so that it becomes a valid statement in non-Euclidean geometry.

3. If a straight line intersects one of two parallel lines, it will always intersect the other.
4. Straight lines parallel to the same straight line are always parallel to one another.
5. There exists one triangle for which the sum of the measures of the angles is π radians.
6. There exists a pair of similar but noncongruent triangles.
7. There exists a pair of straight lines the same distance apart at every point.
8. It is always possible to pass a circle through three noncollinear points.
9. For which of these finite geometries of Chapter 1 does Euclid's fifth postulate always hold?

   a. Geometry of Pappus **b.** Fano's geometry
   c. Four-line geometry **d.** Geometry of Desargues

10. For the thirteen-point finite geometry of this section, name all the lines through point A that do not have a point in common with line BDF.
11. Without using an axiom of parallelism, prove that if a transversal of two lines makes the alternate angles congruent, then the two given lines do not intersect.
12. In hyperbolic geometry, through a given point not on a given line, exactly how many lines can be drawn in that plane that are parallel to the given line?

For Exercises 13–15, prove that each statement is equivalent to Euclid's fifth postulate.

13. If a straight line intersects one of two parallel lines, it will always intersect the other.
14. Two straight lines parallel to the same straight line are always parallel to one another.
15. There exists one triangle for which the sum of the measures of the angles is π radians.
16. ▷ Read in a history of mathematics text about additional attempts to prove the fifth postulate.
17. ▷ Create another finite geometry that has the characteristic postulate of hyperbolic geometry, and prepare a table similar to Table 9.1.

▲ 9.2 Ideal Points and Omega Triangles

In hyperbolic geometry, two parallel lines do not have an ordinary point in common, but they are said to meet at an ideal point.

DEFINITION ➤ In hyperbolic geometry two parallel lines are said to intersect in an *ideal point*.

In Figure 9.7, the right- and left-hand parallels shown through point A to line $\overleftrightarrow{BC}$ meet that line in ideal points Ω (omega) and Ω'. It can be proved (see Exercise 1, Exercises 9.2) that Ω and Ω' are distinct points. An ordinary line in projective geometry has exactly one ideal point, but a line in hyperbolic geometry has two distinct ideal points.

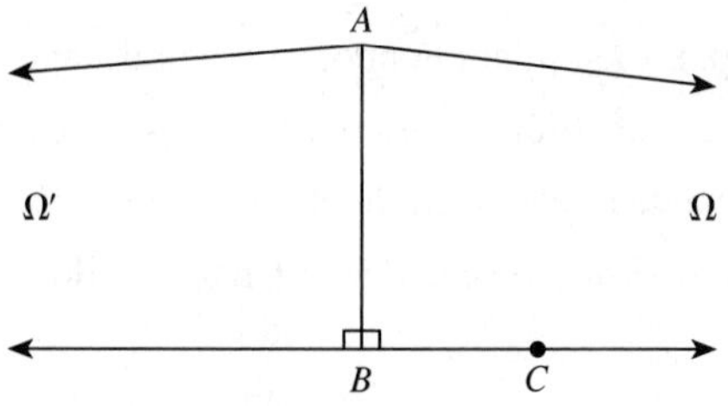

Figure 9.7

The development of the properties of parallel lines in hyperbolic geometry is continued by considering the *omega triangle*, a three-sided figure as in Figure 9.8, with one ideal vertex. Though not a triangle in the ordinary sense, an omega triangle does have some of the same properties as a triangle with three ordinary vertices.

THEOREM 9.3 ➤ The axiom of Pasch holds for an omega triangle, whether the line enters at a vertex or at a point not a vertex.

PROOF: In Figure 9.9, let C be any interior point of the omega triangle $AB\Omega$. Then $\overleftrightarrow{BC}$ and $\overleftrightarrow{AC}$ intersect the opposite side because $\overleftrightarrow{B\Omega}$ is the first noncutting line through B for $\overleftrightarrow{AD}$, and $\overleftrightarrow{AD}$ is the first noncutting line through A for $\overleftrightarrow{BE}$. (See Exercise 14I.) If a line $\overleftrightarrow{C\Omega}$ enters the omega triangle through the ideal vertex, it intersects $\overleftrightarrow{AB}$, because of Pasch's axiom applied to $\triangle ABD$. The second part of the proof is left as Exercise 6 of Exercises 9.2.

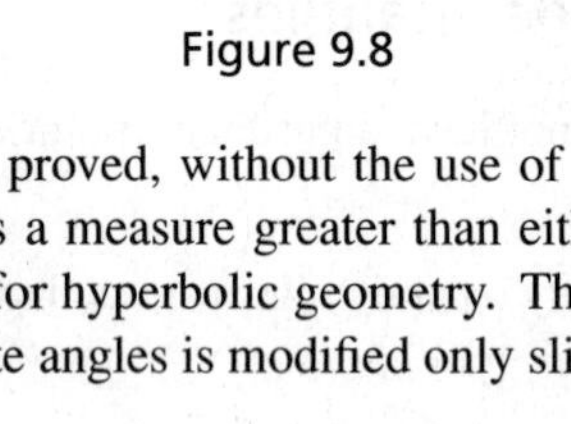

Figure 9.8

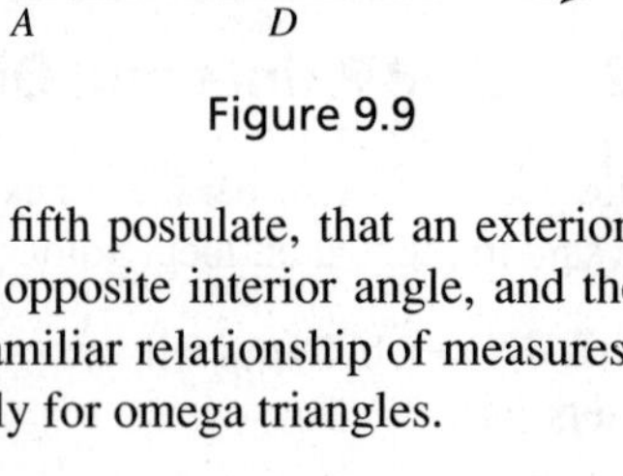

Figure 9.9

Euclid proved, without the use of the fifth postulate, that an exterior angle of a triangle has a measure greater than either opposite interior angle, and the statement also holds for hyperbolic geometry. The familiar relationship of measures of exterior and opposite angles is modified only slightly for omega triangles.

THEOREM 9.4 ➤ For any omega triangle $AB\Omega$, the measures of the exterior angles formed by extending $\overline{AB}$ are greater than the measures of their opposite interior angles.

Proof: This theorem may be proved indirectly by eliminating the other two possibilities. Suppose, in Figure 9.10, that $m(\angle CA\Omega) < m(\angle AB\Omega)$. Then a point D on $\overleftrightarrow{A\Omega}$ can be found such that $\angle CAD \cong \angle ABD$. But this is impossible, since $\triangle ABD$ is an ordinary triangle and the exterior angle cannot be congruent to an opposite interior angle. Suppose next, as in Figure 9.11, that $\angle CA\Omega \cong \angle AB\Omega$. Let D be the midpoint of $\overline{AB}$, let $\overline{DE}$ be perpendicular to $\overleftrightarrow{B\Omega}$, and let $FA = BE$. Then $\triangle FAD \cong \triangle EBD$, FDE is a straight line, and $\angle DFA$ is a right angle. But the angle of parallelism for the distance EF cannot be a right angle because of Theorem 9.2, which means that this is a contradiction and the assumption of congruence of angles must be rejected.

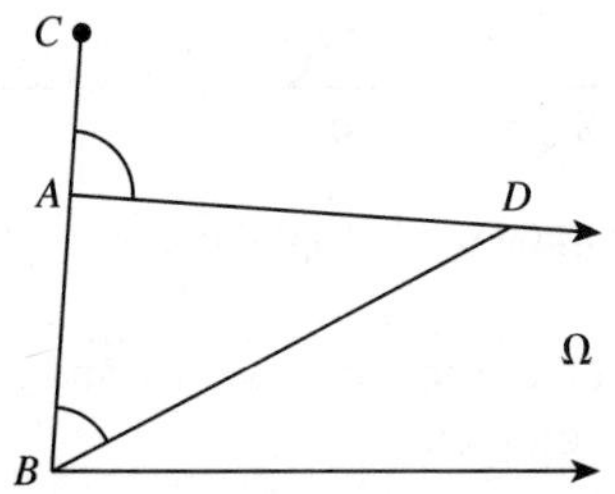

Figure 9.10

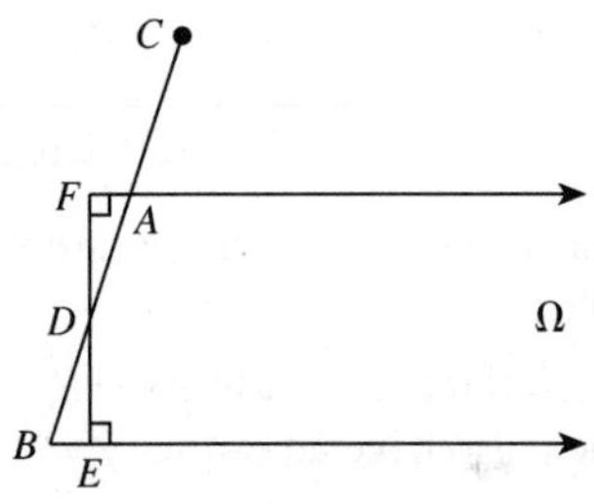

Figure 9.11

Congruence of omega triangles is somewhat simpler than that of ordinary triangles, since less information is required. One set of conditions for congruence is given in the following theorem.

THEOREM 9.5 ➤ Omega triangles $AB\Omega$ and $A'B'\Omega'$ are congruent if the sides of finite length are congruent and if a pair of corresponding angles at A and A' or B and B' are congruent.

Proof: The theorem is proved here by assuming that the remaining pair of angles at an ordinary vertex are not congruent, then arriving—on the basis of this assumption—at a contradiction. In Figure 9.12, assume that $\overline{AB} \cong \overline{A'B'}$ and that $\angle BA\Omega \cong \angle B'A'\Omega'$. Assume that one of the angles, say $\angle AB\Omega$, is greater than $\angle A'B'\Omega'$. Some point C can be located on $\overleftrightarrow{A\Omega}$ such that $\angle ABC \cong A'B'\Omega'$. If C' is located on $\overleftrightarrow{A'\Omega'}$ so that $\overline{A'C'} \cong \overline{AC}$, then $\triangle ABC \cong \triangle A'B'C'$. But this means that $\angle A'B'C' \cong \angle A'B'\Omega'$, which is a contradiction.

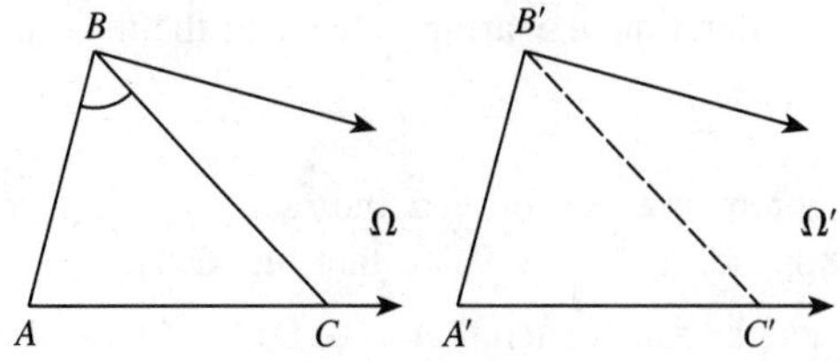

Figure 9.12

A second set of conditions for the congruence of omega triangles is stated here, but the proof is left as Exercise 8 of Exercises 9.2.

THEOREM 9.6 ➤ Omega triangles $AB\Omega$ and $A'B'\Omega'$ are congruent if the pair of angles at A and A' are congruent and the pair of angles at B and B' are congruent.

▲ EXERCISES 9.2 ▲

1. Explain why a line in hyperbolic geometry must contain two distinct ideal points.

2. Sketch three omega triangles, all with the same ideal vertex and each two of which also have an ordinary vertex in common.

3. Sketch a three-sided figure with two ideal vertices.

4. Sketch a three-sided figure with three ideal vertices.

5. Name all the omega triangles shown in Figure 9.9.

6. Prove that a line intersecting a side of an omega triangle at a point other than a vertex intersects a second side.

7. Prove that the sum of the measures of the two angles at ordinary vertices of an omega triangle is less than π. (*Hint:* Use Theorem 9.4.)

8. Prove Theorem 9.6. (*Hint:* Use Theorem 9.5.)

9. Prove that the angle of parallelism is constant for a given distance.

10. As the distance increases, does the angle of parallelism increase or decrease? Prove that your answer is correct.

11. Prove that if the two angles at ordinary vertices of an omega triangle are congruent, then the line from the ideal vertex to the midpoint of the opposite side is perpendicular to that side.

12. Prove the converse of the statement in Exercise 11.

▷ **13.** List as many ways as possible that (a) an omega triangle is like an ordinary triangle, and (b) an omega triangle is unlike an ordinary triangle.

▷ **14.** The *symmetric property* of parallelism (in Euclidean geometry) states that if a first line is parallel to a second, then the second is parallel to the first. Prove this property, or, if necessary, find and study the proof in a reference text.

▷ **15.** The *transitive property* of parallelism (in Euclidean geometry) states that if two lines are parallel to a third, they are parallel to each other. Prove this property, or, if necessary, find and study the proof in a reference text.

▷ **16.** Explain, using his *Elements*, how Euclid proved, without use of the fifth postulate, that an exterior angle of a triangle has a measure greater than either opposite interior angle.

▲ 9.3 Quadrilaterals and Triangles

The concept of omega triangles, developed in Section 9.2, leads very logically to the formulation of seemingly strange theorems for ordinary triangles and quadrilaterals in hyperbolic geometry.

Among the attempts to prove Euclid's fifth postulate, the most productive were those using the indirect method. By adopting a contradictory postulate and reaching valid conclusions based upon it, mathematicians were actually developing non-Euclidean geometry, even though they remained unaware of the significance of their work.

Girolamo Saccheri (1667–1733), an Italian Jesuit priest, published "Euclid Freed of Every Flaw" in 1733. In his attempt to prove the fifth postulate, Saccheri made use of a set of points now called a *Saccheri quadrilateral*. A Saccheri quadrilateral has two right angles and two congruent sides, as shown in Figure 9.13. $\overline{AB}$ is called the *base* and $\overline{CD}$ is called the *summit* of the quadrilateral. The two congruent segments are the *sides*. The next theorem shows that some properties of the corresponding figures in Euclidean geometry continue to hold in hyperbolic geometry.

THEOREM 9.7 ➤ The segment joining the midpoints of the base and summit of a Saccheri quadrilateral is perpendicular to both.

PROOF: In Figure 9.14, $\Delta DAE \cong \Delta CBE$, so $\Delta DFE \cong \Delta CFE$, and the proof follows.

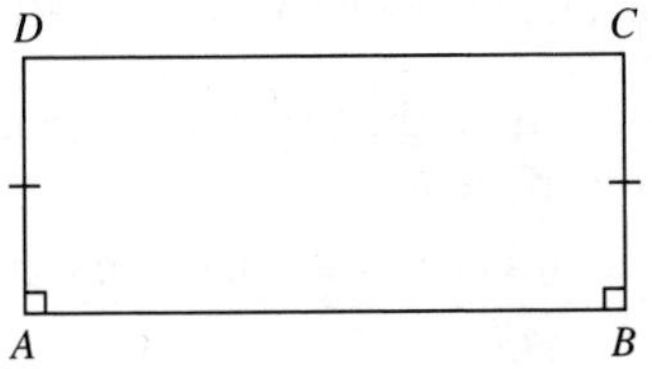

Figure 9.13

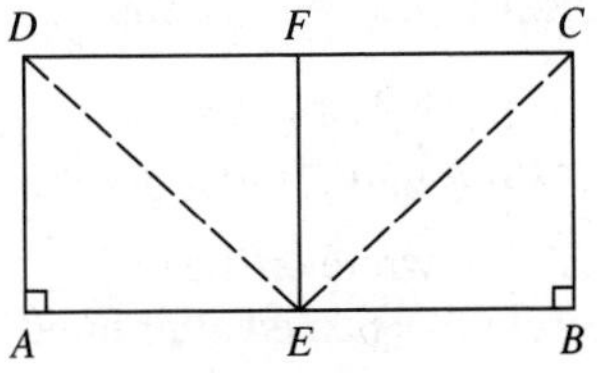

Figure 9.14

Euclid's work on congruence of triangles, which does not depend on the fifth postulate, continues to hold in non-Euclidean geometry.

Some of the properties of the Saccheri quadrilateral are unlike any for sets of points in Euclidean geometry.

THEOREM 9.8 ➤ The summit angles of a Saccheri quadrilateral are congruent and acute.

PROOF: The congruence of the summit angles is a consequence of the congruence of the pairs of triangles in Figure 9.14. The significant fact that the summit angles are acute is a consequence of established properties of the omega triangle. In Figure 9.15, $m(\angle EC\Omega) > m(\angle CD\Omega)$, since $\angle EC\Omega$ is an exterior angle for omega triangle $CD\Omega$. Since $\angle AD\Omega \cong BC\Omega$, $m(\angle BCE) > m(\angle ADC)$. But $\angle ADC \cong \angle BCD$, and therefore, $m(\angle BCE) > m(\angle BCD)$, so $\angle BCD$ is acute.

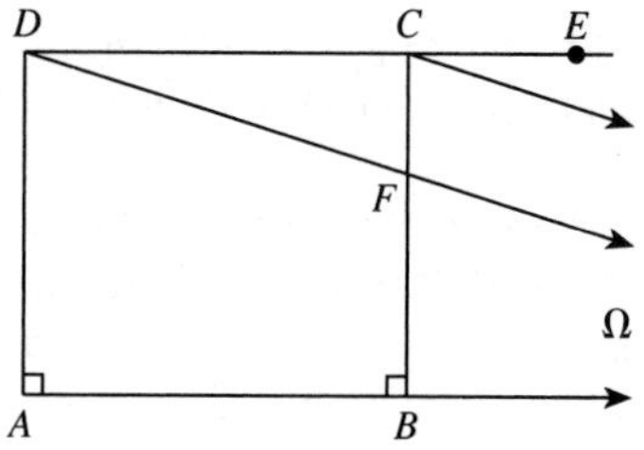

Figure 9.15

J. H. Lambert (1728–1777), like Saccheri, attempted to prove the fifth postulate by an indirect argument. He began with a quadrilateral with three right angles, now called a Lambert quadrilateral, shown in Figure 9.16.

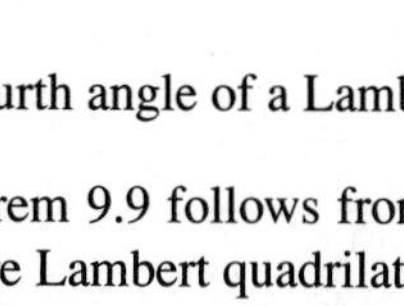

Figure 9.16

THEOREM 9.9 ➤ The fourth angle of a Lambert quadrilateral is acute.

PROOF: The proof of Theorem 9.9 follows from the observation that, in Figure 9.14, $EFCB$ and $EFDA$ are Lambert quadrilaterals.

Lambert also wrote on the theory of projections in making maps, and one common type of projection is named after him.

Theorem 9.9 is needed to prove the following, even more significant, theorem—one that serves to distinguish clearly between hyperbolic and Euclidean geometry.

THEOREM 9.10 ➤ The sum of the measures of the angles of a right triangle is less than π.

PROOF: In Figure 9.17, let $\triangle ABC$ be any right triangle, with D the midpoint of the hypotenuse. $\overline{DE}$ is perpendicular to $\overline{BC}$. Line $\overleftrightarrow{AF}$ is constructed so that $\angle FAD \cong \angle EBD$ and $\overline{AF} \cong \overline{BE}$. Then $\triangle AFD \cong \triangle BED$. This means that $\angle AFE$ is a right angle and $\angle ADF \cong \angle EDB$, so E, D, F is a straight line. The consequence is that $ACEF$ is a Lambert quadrilateral with the acute angle at A. The two angles at A, $\angle CAB$ and $\angle BAF$, have the sum of their measures equal to the sum of the measures of $\angle CAB$ and $\angle CBA$, so the sum of the measures of the three angles of $\triangle ABC$ is less than π.

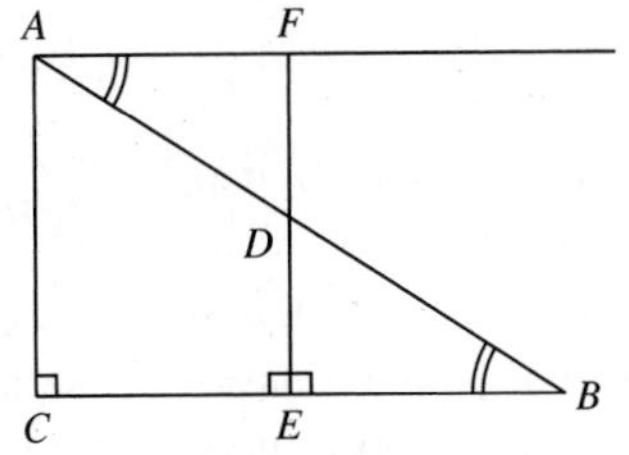

Figure 9.17

The following two theorems are corollaries of Theorem 9.10. The proofs are left as Exercises 13 and 14 of Exercises 9.3.

THEOREM 9.11 ➤ The sum of the measures of the angles of any triangle is less than π.

DEFINITION ➤ The difference between π and the angle sum of a triangle is called the *defect*.

For example, if the angle sum were $19\pi/20$, the defect would be $\pi/20$.

THEOREM 9.12 ➤ The sum of the measures of the angles of any convex quadrilateral is less than 2π.

You have found, in general, that the theory of congruence of triangles in hyperbolic geometry is much like the theory of congruence in Euclidean geometry. One significant difference results from an application of Theorem 9.12.

THEOREM 9.13 ➤ Two triangles are congruent if the three pairs of corresponding angles are congruent.

PROOF: The proof is by contradiction. Assume, as in Figure 9.18a, that $\triangle ABC$ and $\triangle ADE$ have three pairs of corresponding angles congruent. As a result, quadrilateral $BCED$ has the sum of the measures of its angles equal to 2π, but this is a contradiction of Theorem 9.12. Thus, the assumption of the existence of the similar but noncongruent triangles must be rejected for hyperbolic geometry. If $\overline{BC}$ and $\overline{DE}$ should intersect, as in Figure 9.18b, then a contradiction is also reached, because an exterior angle of a triangle is congruent to an opposite interior angle.

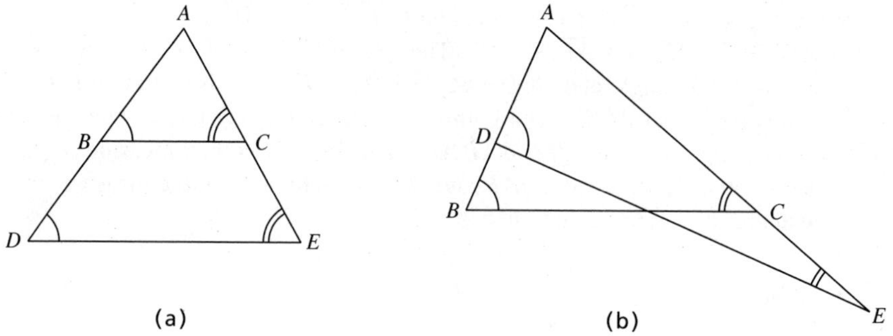

Figure 9.18

One consequence of Theorem 9.13 is that, in hyperbolic geometry, the shape and size of triangles are not independent. All triangles of the same shape are necessarily of the same size. Similar but noncongruent figures do not exist in hyperbolic geometry. The existence of a pair of such triangles is equivalent to Euclid's fifth postulate.

▲ EXERCISES 9.3 ▲

1. What is the maximum number of angles of a Saccheri quadrilateral that could be congruent to each other?
2. In hyperbolic geometry, why can there be no squares or rectangles?
3. Show that, for a figure such as Figure 9.19, if $AD > BC$, then $m(\angle BCD) > m(\angle ADC)$.

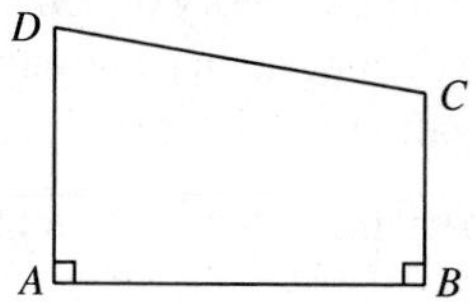

Figure 9.19

4. For a Lambert quadrilateral, which is longer, a side adjacent to the acute angle or the side opposite? Prove that your answer is correct.
5. What does the answer to Exercise 4 tell you about the distance between non-intersecting lines?

6. Which is longer, the base or the summit of a Saccheri quadrilateral?
7. Prove that your answer to Exercise 6 is correct.
8. Is the segment joining the midpoints of the sides of a Saccheri quadrilateral perpendicular to the sides? Prove that your answer is correct.
9. Suppose a triangle has a defect of $0.426°$. If two angles have measurements of $37.514°$ and $71.689°$, find the size of the third angle.
10. Let a triangular region be partitioned into two triangular regions by a segment through a vertex. Compare the defect of the original triangle with the defect of the two smaller triangles formed.
11. Let a triangular region be partitioned into any number of triangular regions by segments through the same vertex. State and prove a theorem about the defects of the original triangle and the defects of the subtriangles.
12. Point out in the proof of Theorem 9.10 exactly where the statements first differ from correct statements in Euclidean geometry.
13. Prove Theorem 9.11.
14. Prove Theorem 9.12.
15. State and prove a theorem giving minimum conditions that must be known if two Saccheri quadrilaterals are to be congruent.
16. State and prove a theorem giving a different set of minimum conditions from your answer to Exercise 15.
17. ▷ Summarize ways in which a Lambert quadrilateral is like an ordinary rectangle and ways in which it differs from an ordinary rectangle.
18. ▷ Prove, or study a proof, that if a line is a nonintersecting line through a given point to a given line, it is, at each of its points, a nonintersecting line to the given line.
19. ▷ Prove, or study a proof, that if a line is a parallel line through a given point to a given line, it is at each of its points a parallel line to the given line.

▲ 9.4 Pairs of Lines and Area of Triangular Regions

One more type of point besides ordinary and ideal points must be created for a complete discussion of the set of points in hyperbolic geometry. Two nonintersecting lines are said to meet at a *gamma point* (Γ). Another name for this point is an *ultra-ideal point*. Recall that nonintersecting lines do not include the parallel lines through a point to a line. For example, if $\overleftrightarrow{AB}$ and $\overleftrightarrow{AC}$ are the parallels through A to $\overleftrightarrow{DE}$, then $\overleftrightarrow{AF}$, $\overleftrightarrow{AG}$ and $\overleftrightarrow{AH}$ are three of the infinite number of nonintersecting lines with respect to $\overleftrightarrow{DE}$ passing through A. Each of these lines has a gamma point in common with $\overleftrightarrow{DE}$ (Figure 9.20).

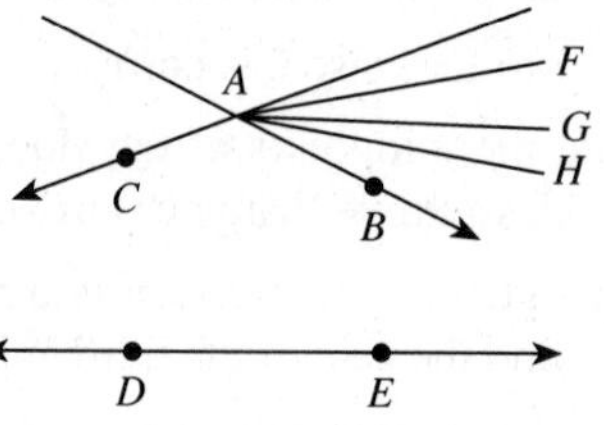

Figure 9.20

Besides having a gamma point in common, nonintersecting lines have, perhaps surprisingly, one of the properties of parallel lines in Euclidean geometry: they have a common perpendicular.

THEOREM 9.14 ➤ Two nonintersecting lines have a common perpendicular.

In the analysis figure shown in Figure 9.21, it can be seen that $\overline{AB}$ and $\overline{CD}$, on two nonintersecting lines, can be thought of as the base and summit of Saccheri quadrilateral $ABCD$; hence $\overline{EF}$, connecting the midpoints of the base and summit, is the common perpendicular. The problem is reduced to finding two congruent segments, such as $\overline{AD}$ and $\overline{BC}$, both perpendicular to one of the given nonintersecting lines, since the common perpendicular can then be found. An outline of the proof follows.

Assume, as in Figure 9.22, that $\overline{AD}$ and $\overline{BC}$ are perpendicular to the same line, but not congruent. Let $\overline{ED} \cong \overline{BC}$, and $\angle F'ED \cong \angle GBC$. Let F be the point of intersection of $\overleftrightarrow{EF'}$ and $\overleftrightarrow{AG}$. If $\overline{BG} \cong \overline{EF}$ and $\overline{CH} \cong \overline{DI}$, then quadrilateral $BGHC$ can be shown to be congruent to $EFID$, and $\overline{GH}$ is perpendicular at H; thus, $\overline{FI}$ and $\overline{GH}$ are the required sides for the Saccheri quadrilateral.

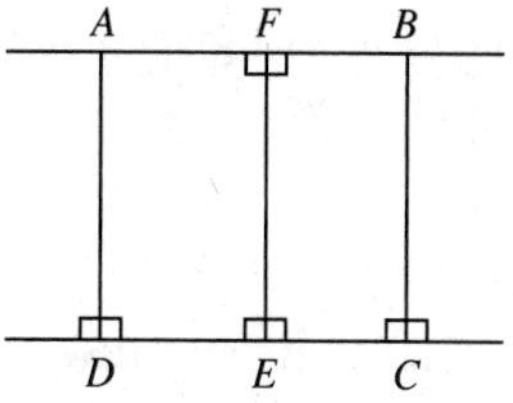

Figure 9.21

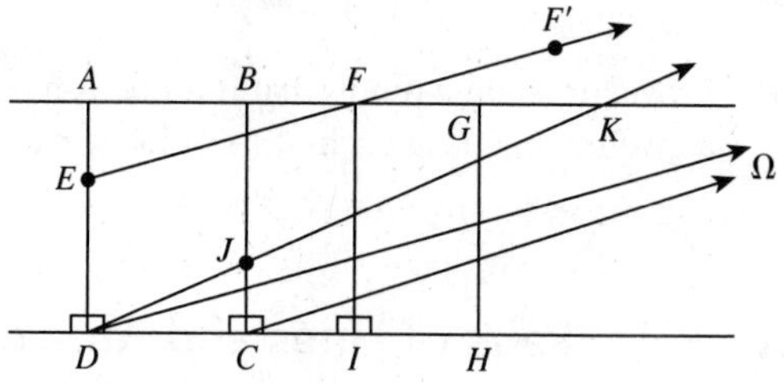

Figure 9.22

The fact that point F exists on line AB needs to be established. In Figure 9.22, assume that $D\Omega$ and $C\Omega$ are parallels to $\overleftrightarrow{AB}$. These lines lie within $\angle ADC$ and $\angle BCI$, respectively. For omega triangle $DC\Omega$, exterior angle $IC\Omega$ is greater than $\angle CD\Omega$, so that if $\angle CDJ \cong \angle IC\Omega$, then DJ must intersect AB at a point K. Now it can be shown by comparing figures $GBC\Omega$ and $F'EDJ$ that DJ is in fact parallel to EF'; hence, EF' must intersect $\overleftrightarrow{AB}$ in a point F.

Because of Theorem 9.14, all of the lines perpendicular to a given line can be said to have the same ultra-ideal point in common. It is left as an exercise to show that two lines in hyperbolic geometry cannot have two distinct common perpendiculars.

With the inclusion of the previous information about the gamma point and the common perpendicular for nonintersecting lines, Table 9.2 can be given summarizing the relationship between pairs of lines intersecting at each kind of point in hyperbolic geometry.

Table 9.2

Type of Point Common to Two Lines	Variation in Distance
Ordinary point	Lines diverge from their point of intersection
Ideal point	Lines converge in the direction of parallelism and diverge in the opposite direction
Ultra-ideal point	Lines diverge from their common perpendicular

You may find it interesting to observe that in none of the cases are the two lines always equidistant. You may speculate about the nature of a set of points with each member the same distance from a given line (see Section 9.5). Notice also in this connection that defining parallel lines as two lines everywhere equidistant is equivalent to assuming the fifth postulate.

The fact that there is no square in hyperbolic geometry means that a method of measuring the area of a plane region must be devised that does not depend on square units. Congruent triangles exist in hyperbolic geometry, and the theory of area can be based on this concept in a way very similar to the modern theory of area in Euclidean geometry.

DEFINITION ➤ Two polygons are called *equivalent* if they can be partitioned into the same finite number of pairs of congruent triangles.

For example, polygons $ABCD$ and $CDEF$ in Figure 9.23 are equivalent but not congruent. Two polygons both equivalent to another polygon are also equivalent to each other. Recall that in hyperbolic geometry, the difference between π and the sum of the measures of a triangle is defined as the *defect* of the triangle. The connection between the defect and equivalence of triangles is made clear in the following theorem.

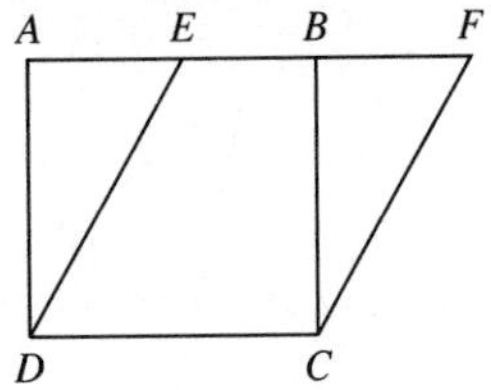

Figure 9.23

THEOREM 9.15 ➤ Two triangles are equivalent if and only if they have the same defect.

Proof: By definition, if two triangles are equivalent, they can be partitioned into a finite number of pairs of congruent triangles. The defect of each of the original triangles is equal to the sum of the defects of the triangles in the partitioning; hence, the original defects are equal. See Exercise 11, Exercises 9.3, for the special case in which all the segments in the partition have a vertex in common.

This establishes the conditional that if two triangles are equivalent, then they have the same defect. The converse of this conditional, if two triangles have the same defect then they are equivalent, is considerably more difficult to prove. Suppose that two triangles have the same defect. If they also have a pair of corresponding sides congruent, they can be shown to be equivalent to congruent Saccheri quadrilaterals and, hence, equivalent to each other.

In Figure 9.24, let triangles ABC and JHI have the same defect and congruent sides $\overline{BC}$ and $\overline{IJ}$. Triangle ABC is equivalent to Saccheri quadrilateral $BCGF$, where D and E are the midpoints of $\overline{AC}$ and $\overline{AB}$ (see Exercise 11, Exercises 9.4). Triangle HIJ is equivalent to Saccheri quadrilateral $IJNM$, where K and L are the midpoints of $\overline{HJ}$ and $\overline{HI}$. But the two Saccheri quadrilaterals are congruent because they have congruent summits and congruent summit angles. (See Exercise 12, Exercises 9.4). The proof that the summit angles of two Saccheri quadrilaterals are congruent is precisely where the hypothesis that the triangles have the same defects is used.

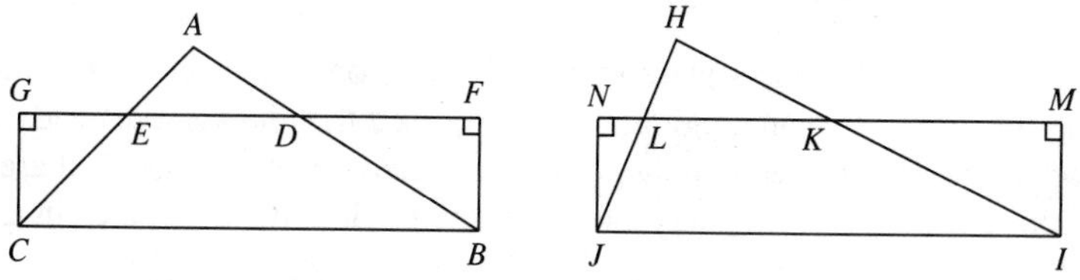

Figure 9.24

We have taken care of the special case of two triangles with the same defect having a congruent side, but now need to consider two triangles having the same defect but no pair of congruent sides.

Let triangles ABC and DEF be any two such triangles, as in Figure 9.25, with $DF > AC$.

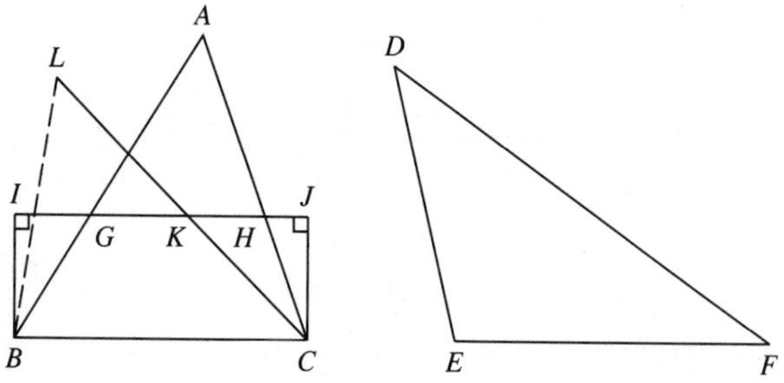

Figure 9.25

If G and H are the midpoints of $\overline{AB}$ and $\overline{AC}$, then $\triangle ABC$ is equivalent to Saccheri quadrilateral $BIJC$. Point K on $\overline{IJ}$ is a point located so that $KC = \frac{1}{2}DF$, and $\overline{LC} \cong \overline{DF}$. Triangles ABC and LBC can be shown to have the same defect and to be equivalent, since they have a pair of congruent sides. Triangles LBC and DEF also have the same defect and a pair of congruent sides; hence, they are equivalent. Since $\triangle DEF$ and $\triangle ABC$ are both equivalent to the same triangle, they are equivalent to each other.

In the proof of Theorem 9.15, the significance of the defect of a triangle was that it made possible the equivalence of triangles to Saccheri quadrilaterals that could be proved congruent. Since the equivalence of triangles depends on the defect, it is possible to define the measure of area of a triangle in hyperbolic geometry as a function of the defect.

DEFINITION ➤ $A = kd$, for d the defect and k a positive constant the same for all triangles in hyperbolic geometry. Thus equivalent polygons also have the same area.

The value of k depends on the particular triangle chosen to have a unit area. It should be evident that the measure of area increases as the defect increases. In other words, the larger the triangle, the smaller the sum of the measures of the angles.

Oddly enough, in hyperbolic geometry, triangles do not become larger and larger without limit. The triangle with three ideal vertices, shown in Figure 9.26, is the *triangle of maximum area*, and its angle sum could be considered zero.

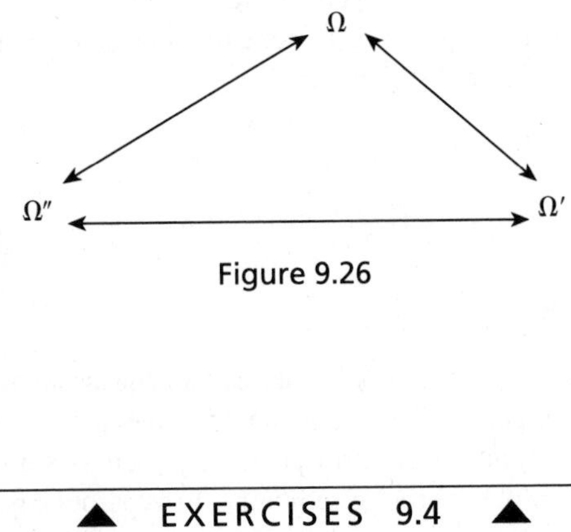

Figure 9.26

▲ EXERCISES 9.4 ▲

1. Explain why two distinct lines cannot have more than one common perpendicular.
2. Where are two nonintersecting lines closest together?
3. How many gamma points are on each line in hyperbolic geometry?
4. Do two lines in hyperbolic geometry always diverge from their point of intersection?

5. Prove that two parallel lines converge continuously in the direction of parallelism.
6. Prove that two polygonal regions equivalent to the same polygonal region are equivalent to each other.
7. Prove that a quadrilateral exists with four congruent angles.
8. For the quadrilateral in Exercise 7, are the four sides congruent? Prove that your answer is correct.
9. Prove that the segment joining the midpoints of two sides of a triangle has a length less than half the length of the third side.
10. In Figure 9.24, prove that $BCGF$ and $IJNM$ are Saccheri quadrilaterals.
11. Prove that $\triangle ABC$, Figure 9.24, is equivalent to Saccheri quadrilateral $BCGF$.
12. Why, in Figure 9.24, do the two Saccheri quadrilaterals have congruent summit angles?
13. Make a sketch to show how a hexagon and a triangle could be equivalent.
14. Show how, starting with any triangle, a triangle of maximum area can be drawn containing the given triangle in its interior.
15. Show, in Figure 9.22, that point F must exist (that is, that EF' and AG must intersect).
16. The proof in Exercise 8 depends on the fact that the perpendicular from A to BC lies inside triangle ABC. How do you know this is true?
17. Use this section, plus additional references on non-Euclidean geometry, to explain how, since square units do not exist, one can actually find the area of a plane region in hyperbolic geometry.

9.5 Curves

In hyperbolic geometry, a circle may be defined in the usual way—a set of points equidistant from a given point. The definition of circle and those properties relating to circles that do not depend on the fifth postulate, such as some having to do with perpendicularity, are parts of hyperbolic geometry. In hyperbolic geometry, the measure of an inscribed angle is no longer equal to half the measure of the intercepted arc, because that conclusion would depend on the fifth postulate.

Defining ideal and ultra-ideal points so that each two lines in hyperbolic geometry have a point in common leads to applications that are generalizations involving circles.

For any two points A and B on an ordinary circle, as in Figure 9.27a, $\angle OAB \cong \angle OBA$, where O is the center of the circle, because $\triangle ABO$ is isosceles. This leads to a generalized definition of *corresponding points*.

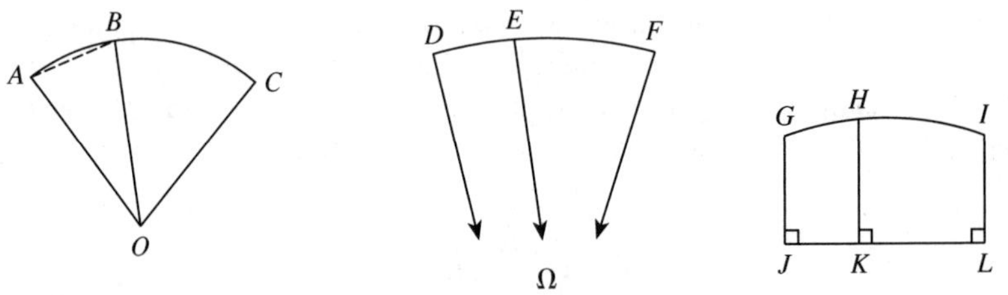

(a) (b) (c)

Figure 9.27

DEFINITION ➤ Two points, one on each of two lines, are called *corresponding points* if the two lines form congruent angles on the same side with the segment whose endpoints are the two given points.

Thus, A and B in Figure 9.27a are corresponding points because of the congruence of angles OAB and OBA. Points C and B, or C and A, similarly can be shown to be corresponding points. In fact, any two points on a circle are corresponding points. An ordinary circle may be defined as the set of all points corresponding to a given point (not the center) on a pencil of rays with an ordinary point as center. A pencil of rays is all rays in a plane with a common endpoint. Because of the fact that the tangent to a circle is perpendicular (orthogonal) to the radius at the point of contact, a circle can be considered the *orthogonal trajectory* of a pencil of rays with an ordinary vertex.

Two new curves, considered generalized circles in hyperbolic geometry, may now be defined.

DEFINITION ➤ A *limiting curve* (see Figure 9.27b) is the set of all points corresponding to a given point on a pencil of rays with an ideal point as vertex.

A limiting curve may be considered the orthogonal trajectory of a pencil of rays with an ideal vertex.

DEFINITION ➤ An *equidistant curve* (see Figure 9.27c) is the set of all points corresponding to a given point on a pencil of rays with a common perpendicular.

An equidistant curve may be considered the orthogonal trajectory of a pencil of rays with a common perpendicular.

A limiting curve in hyperbolic geometry has many of the properties of an ordinary circle. For example, a line perpendicular to a chord at its midpoint is a radius. Theorem 9.16 states a second common property.

THEOREM 9.16 ➤ Three distinct points on a limiting curve uniquely determine it.

In Figure 9.28, the perpendicular bisectors of $\overline{AB}$ and $\overline{BC}$, for A, B, C on a limiting curve, determine the unique point Ω, the center of the limiting curve. It should be emphasized that a part of the given condition was that the three points were on some limiting curve. This is not necessarily the case for any three points chosen at random. For example, in Figure 9.28, three points A, B, C lie on an ordinary circle if the perpendicular bisectors at D and E meet at a real point.

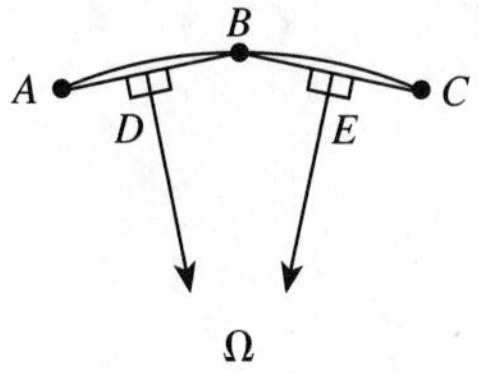

Figure 9.28

The following theorem states a property of limiting curves that is unlike any property of circles.

THEOREM 9.17 ➤ Any two different limiting curves are congruent.

Proof: Let $ABCD$ and $A'B'C'D'$ in Figure 9.29 be any two limiting curves. Suppose that points A, B, C, D are given and that A' is any point on a second limiting curve. If $\angle B'A'\Omega'$ is constructed congruent to $\angle BA\Omega$, and if $\overline{A'B'}$ is congruent to $\overline{AB}$, then point B' is located so that omega triangle $A'B'\Omega'$ is congruent to omega triangle $AB\Omega$. The result is that $\angle AB\Omega \cong \angle A'B'\Omega'$, and B' and A' are corresponding points with B' on the second limiting curve. In the same way, it can be shown that points C', D', ... on the limiting curve can be found so that $\overline{B'C'} \cong \overline{BC}$, $\overline{C'D'} \cong \overline{CD}$, and so on.

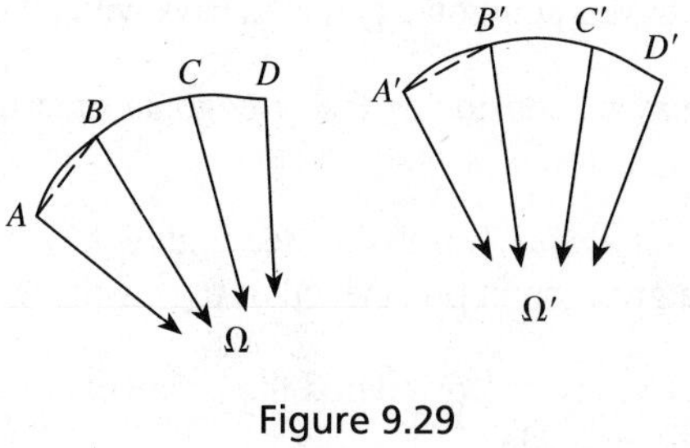

Figure 9.29

A corollary of Theorem 9.17 is the statement that congruent chords intercept congruent arcs and congruent arcs are intercepted by congruent chords for limiting curves.

The second new type of curve to be studied in hyperbolic geometry is the equidistant curve. Let A, B, C be any three points on an equidistant curve, as shown in Figure

9.30. The common perpendicular, $\overleftrightarrow{DF}$, is called the *baseline*. Quadrilateral $ABED$ is a Saccheri quadrilateral, since A and B are corresponding points and $\angle BAD \cong \angle ABE$. The name "equidistant curve" is justified because every point on the curve is the same perpendicular distance from the baseline.

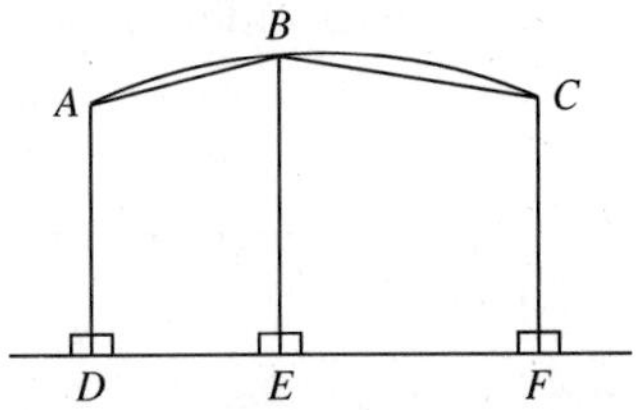

Figure 9.30

All the points in the plane the same distance from the baseline actually lie on an equidistant curve of two branches, as shown in Figure 9.31. In Exercises 9.5, properties of the figure $ABCD$, which are surprisingly like those of a parallelogram, are investigated. The equidistant curve has many other properties, a few of which are stated here. Though it is not as fundamental as the limiting curve in the analytic treatment of non-Euclidean geometry, it is simpler to study.

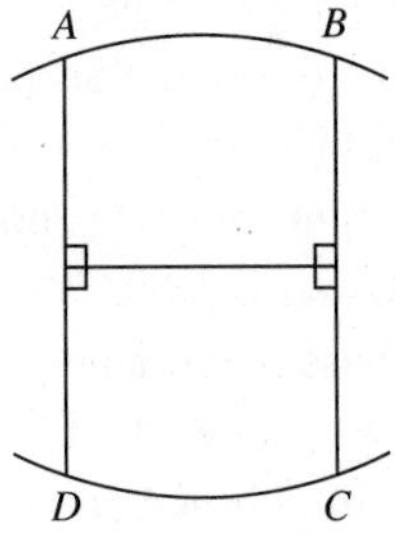

Figure 9.31

Three points on an equidistant curve determine the curve uniquely, since the baseline can be determined. Not all equidistant curves are congruent—only those whose points are the same distance from the baseline. For congruent equidistant curves, congruent chords intercept congruent arcs, and congruent arcs are intercepted by congruent chords. Given one line, an intersecting line diverges from the point of intersection; in addition, parallel lines and nonintersecting lines are at a nonconstant distance. It is only the equidistant curve in hyperbolic geometry that has the equidistant property characteristic of parallel lines in Euclidean geometry.

The statement that a circle can always be found that passes through three noncollinear points is equivalent to the fifth postulate of Euclid. In hyperbolic geometry, on the other hand, three noncollinear points may lie on a circle, a limiting curve, or on one branch of an equidistant curve.

▲ EXERCISES 9.5 ▲

1. In hyperbolic geometry, is the measure of an inscribed angle less than or greater than half the measure of its intercepted arc?
2. Why are any two given points on an ordinary circle corresponding points?
3. Prove that, for a limiting curve, a line perpendicular to a chord at its midpoint is a radius.
4. Prove that the segments of radii between any pair of limiting curves with the same ideal center are congruent.
5. Could a straight line intersect a limiting curve in three distinct points? Why?
6. Could a straight line intersect one branch of an equidistant curve in three distinct points? Why?
7. In hyperbolic geometry, if you are given three noncollinear points, what else would you need to know to decide whether the three points lie on a circle, a limiting curve, or one branch of an equidistant curve?
8. Show how to locate the baseline of an equidistant curve, given three points on one branch of the curve.
9. In Figure 9.30, if a tangent to the equidistant curve is drawn at B, what type of point of intersection would it have with the baseline?
10. In Figure 9.30, prove that if $ACFD$ is a Saccheri quadrilateral, the summit $\overline{AC}$ intersects $\overline{BE}$ at a point G such that $EG < EB$.
11. Prove that any segment connecting two points, one on each branch of an equidistant curve, is bisected by the baseline.
12. Explain how to locate the baseline of an equidistant curve, given three points, not all of which are on the same branch of the curve.
13. Show that three different equidistant curves pass through the vertices of a triangle, with two vertices on one branch and the third vertex on the other branch.
14. Compare the properties of $ABCD$ in Figure 9.31 with those of a parallelogram in Euclidean geometry.
15. Compare the various properties of the circle, limiting curve, and equidistant curve by preparing the information in the form of a table.
16. The topic of constructions in hyperbolic geometry is much more complex than it is in Euclidean geometry. Study sections on constructions in Wolfe, *Non-Euclidean Geometry*, and demonstrate a construction in hyperbolic geometry.

▲ 9.6 Elliptic Geometry

One type of non-Euclidean geometry—hyperbolic—has been introduced in the first five sections of this chapter. Two models for this geometry are discussed in Section 9.7.

Not long after the development of hyperbolic geometry, the German mathematician Riemann (1826–1866), who also contributed to many other areas of mathematics including topology and the geometry of metric spaces, suggested a geometry, now called *elliptic*, based on the alternative to the fifth postulate, which states that there are no parallels to a line through a point on the line.

Characteristic Postulate of Elliptic Geometry. Any two lines in a plane meet at an ordinary point.

Various finite geometries from Chapter 1, such as the three-point geometry and Fano's geometry, satisfy the requirement of this postulate, although they do not satisfy all the postulates of elliptic geometry.

It is also necessary to modify the postulational system of Euclid by replacing the statement about infinitude of a line with the milder statement that a line is *boundless* in elliptic geometry. An intuitive idea of the meaning of the word *boundless* is that the line cannot be enclosed by a circle lying in the same plane. That is, boundless means unbounded, as it was used in Chapter 3. It is also necessary to make further modifications including the statement that circles do not always have to exist, and that there may be infinitely many lines through some pairs of points. You should try to predict some of the consequences of the new postulate and these modifications before continuing.

If Figure 9.32 is in elliptic geometry, with $\overline{EH}$ and $\overline{FI}$ both perpendicular to $\overleftrightarrow{CG}$, these two lines meet at some point A because of the characteristic postulate that any two lines intersect. In elliptic geometry, it is customary to use curves to show straight lines. If $\overline{DE} \cong \overline{EF}$, then $\triangle AEF \cong \triangle AED$, and $\overline{AD}$ is also perpendicular to $\overline{CG}$. By an extension of this argument, it can be shown that every line through A intersects $\overleftrightarrow{CG}$ at right angles. Point A is called a *pole* of $\overleftrightarrow{CG}$ and the line is the *polar* for point A. Here, the distance from A to any point on $\overleftrightarrow{CG}$ is a constant. Recall other uses of the words *pole* and *polar* to show a relationship between a point and a line in various geometries encountered in this text.

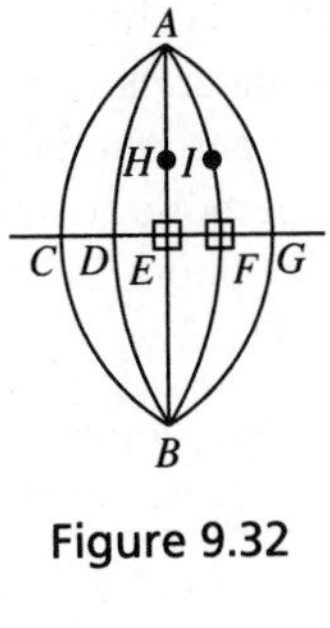

Figure 9.32

In Figure 9.32, if $\overline{AE} \cong \overline{EB}$, then $\triangle BEF \cong \triangle AEF$, and A, F, and B are also collinear. This means that B is also a pole of $\overleftrightarrow{CG}$, and that two lines intersect in two points. It is assumed here that A and B are distinct points, although it is also possible to consider them as identical (in *single elliptic geometry*). The common polar of the two points of intersection is the unique common perpendicular to the two lines. Interestingly enough, two straight lines, such as ADB and AFB, enclose a region in elliptic geometry. This region is called a *digon* or *biangle*.

Fortunately, a simple and familiar model exists for the elliptic geometry of the plane —the surface of the earth and lines of longitude on its surface. In Figure 9.32, think of A and B as the north and south poles, and $\overleftrightarrow{CG}$ as the equator. The Euclidean geometry on the surface of a sphere provides a model for two-dimensional elliptic geometry.

In Figure 9.33, let $\overleftrightarrow{HI}$ and $\overleftrightarrow{KJ}$ be lines in elliptic geometry such that H and I are the poles of $\overleftrightarrow{KJ}$, and K and J are the poles of $\overleftrightarrow{HI}$. In elliptic geometry, the distance from any line to its pole is constant—the same for all lines. Furthermore, a line is of finite length, and the length is four times the distance from the pole to the line.

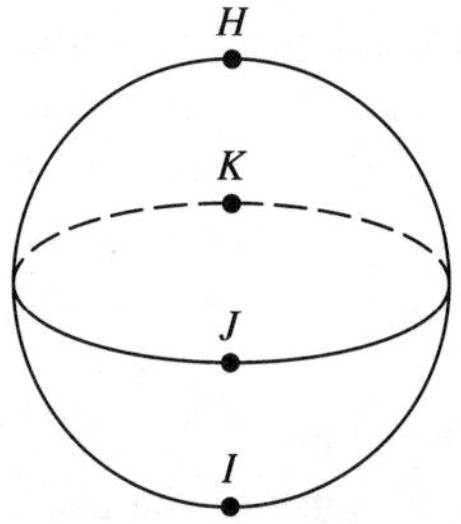

Figure 9.33

The use of a model also helps to explain what it means when a line is called boundless. A great circle on the sphere, representing a line in elliptic geometry, cannot be enclosed by a curve on the sphere. There is no way to "get around" the great circle from a point on one side of it to a point on the other without intersecting the great circle.

In elliptic geometry, there are no parallel or nonintersecting lines, since any two lines meet. Thus, concepts such as ideal points, omega triangles, nonintersecting lines and ultraideal points that are encountered in hyperbolic geometry do not exist in elliptic geometry. However, there are quadrilaterals and triangles that have some properties analogous to those encountered in hyperbolic geometry. The definition of Saccheri quadrilateral is the same here as in hyperbolic geometry. Of course, in elliptic geometry, since any two lines have two ordinary points in common, the base and summit of the Saccheri quadrilateral intersect, as do the two sides.

THEOREM 9.18 ➤ The segment joining the midpoint of the base and summit of a Saccheri quadrilateral is perpendicular to both the base and the summit.

The proof of this theorem is identical to the proof of Theorem 9.7 for hyperbolic geometry.

THEOREM 9.19 ➤ The summit angles of a Saccheri quadrilateral are congruent and obtuse.

Proof: In Figure 9.34, let $ABCD$ be any Saccheri quadrilateral, with O and O' the poles of $\overleftrightarrow{EF}$, the line joining the midpoints of the base and summit.

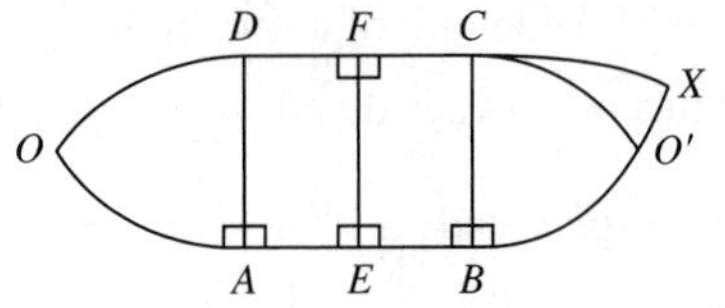

Figure 9.34

The fact that the summit angles are congruent comes from the congruent triangles used in the proof of Theorem 9.18.

To show that the summit angles are obtuse, it can first be established that their supplements are acute. If, as in Figure 9.34, X lies on BO' and is the pole of $\overleftrightarrow{BC}$, then $BX > BO'$, since $BO' < EO'$. Thus, $\angle BCX$ is a right angle, so $\angle BCO'$ is acute and its supplement, $\angle BCD$, is obtuse.

The following theorems are consequences of Theorem 9.19, and the proofs are left as Exercises 13–15 of Exercises 9.6.

THEOREM 9.20 ➤ In elliptic geometry, a Lambert quadrilateral has its fourth angle obtuse, and each side of this angle is shorter than the side opposite.

THEOREM 9.21 ➤ The sum of the measures of the angles of any triangle is greater than π.

THEOREM 9.22 ➤ The sum of the measures of the angles of any quadrilateral is greater than 2π.

You should find it worthwhile to return to the sections on hyperbolic geometry to see which of the concepts not already mentioned can be used in elliptic geometry. For example, there is no angle of parallelism in elliptic geometry. Circles do exist in elliptic geometry. They can be described as the set of all points a fixed (sufficiently small) distance from a given point. Since each point on a circle is also the same distance from the polar for the given point, a circle can be considered as an equidistant curve in elliptic geometry. Studying the model of the surface of the earth and lines of longitude can also result in the discovery of additional true statements about elliptic geometry.

▲ EXERCISES 9.6 ▲

1. Name two other geometries from Chapter 1 for which the characteristic postulate of elliptic geometry holds.
2. In ordinary Euclidean geometry, is a line boundless?

3. A line of latitude that is not the equator of a sphere represents what other concept in the model of elliptic geometry shown in Figure 9.33?
4. In elliptic geometry, if the distance from a line to its pole has a measure of two, what is the measure of the length of a line in that geometry?
5. When may two points not always determine a unique line in elliptic geometry?

For Exercises 6 and 7, for a digon, must it always be true that:

6. The pole of one of the lines forming the digon must be a point inside the digon?
7. There is exactly one common perpendicular to the lines forming the digon?
8. Explain why circles do not always exist in elliptic geometry for any point and any radius.
9. Do limiting curves exist in elliptic geometry? Why?
10. What is the maximum measure of the third angle in a triangle in elliptic geometry, if two of the angles are right angles?
11. Verify the fact that the proof of Theorem 9.18 is identical to the one for Theorem 9.7.
12. In elliptic geometry, where are two given lines farthest apart?
13. Prove Theorem 9.20.
14. Prove Theorem 9.21.
15. Prove Theorem 9.22.
16. Prove that similar but noncongruent triangles cannot exist in elliptic geometry.
17. ▷ Make a list of ten statements that are true for both elliptic and hyperbolic geometry.
18. ▷ Using the three dimensional model for elliptic geometry, develop the concept and definition for area that is analogous to the development in Section 9.4 for hyperbolic geometry.
19. ▷ Before studying the next section, try to discover a reasonable model for hyperbolic geometry.

▲ 9.7 Consistency of Non-Euclidean Geometry

Is the universe Euclidean or non-Euclidean? The question of which kind of geometry is the "right" geometry to fit the physical universe is one that may never be answered. While it might seem a simple problem to solve, there are formidable difficulties. For example, it might seem that physical measurement of the angles of triangles would easily settle the question of whether the sum is or is not equal to π. But physical measurement always involves errors. Besides, it is known that in non-Euclidean geometry the angle sum depends on the size of the triangle. The small part of the universe in which we live may not be big enough to contain triangles with a defect (or excess) large enough to be measured. Space travel may someday settle the question.

For most practical purposes, it makes little difference in our lives whether the universe is Euclidean or not. Euclidean geometry provides a simple model to use in most practical applications, such as engineering. Non-Euclidean geometry may have new applications as humans explore more of the universe. In many cases, it could be used instead of Euclidean geometry without making much difference. A modern theoretical model of the universe, based on Einstein's general theory of relativity, depends on a more complex non-Euclidean geometry than either hyperbolic or elliptic. The assumption is made that the universe is not exactly Euclidean, but that the nonuniform concentration of matter in space determines the extent of the deviation from the Euclidean model.

From a purely mathematical point of view, the truth of an axiomatic system is not what must be investigated. The important thing to determine is whether or not the system is *consistent*, as discussed briefly in Chapter 1. In other words, do the axioms of non-Euclidean geometry lead to *valid* conclusions, without any contradictions. Actually, mathematicians are interested in what is called *relative consistency*. They need to be sure that non-Euclidean geometry is as consistent as Euclidean geometry or as the algebra of real numbers.

Beltrami is given the credit for first proving the relative consistency of non-Euclidian geometry, in 1868. The proof of the relative consistency of a non-Euclidean geometry consists of finding a model within Euclidean geometry that, with suitable interpretations, has the same postulational structure as the non-Euclidean geometry. Then any inconsistency in the non-Euclidean geometry would mean there is also an inconsistency in Euclidean geometry.

A model for plane elliptic geometry was introduced in the previous section. Great circles on the surface of a sphere correspond to lines, and each line has two poles associated with it. Lines of latitude are models for equidistant curves.

Two-dimensional models in Euclidean geometry for hyperbolic geometry include those suggested by Felix Klein and Poincaré. The Klein model is illustrated in Figure 9.35. Points on the circle represent ideal points. Secants of the circle represent hyperbolic lines, with points on the chords interior to the circle representing ordinary points. In Figure 9.35, $\overleftrightarrow{AB}$ and $\overleftrightarrow{CD}$ are intersecting lines, $\overleftrightarrow{AD}$ and $\overleftrightarrow{CD}$ are parallel lines, and $\overleftrightarrow{AD}$ and $\overleftrightarrow{BC}$ are nonintersecting lines.

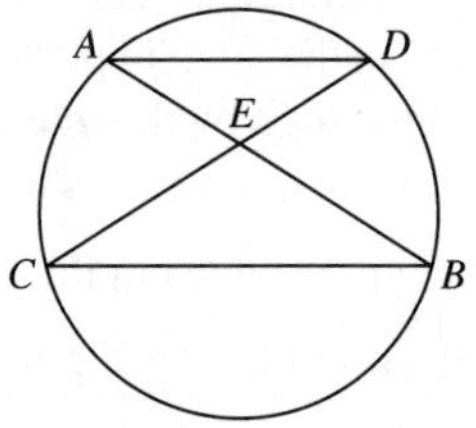

Figure 9.35

One advantage of the Klein model is that straight lines are represented by portions of straight lines. Two major problems in connection with the Klein model soon become apparent, however. One is that some interpretation of distance must be given so that a segment such as $\overline{AD}$ in Figure 9.35 has the characteristics of a line of infinite length. The second major problem is that some interpretation of angle must be given to avoid having the sum of the measures of the angles of a triangle equal to π.

The Poincaré model, developed by the great French mathematician Henri Poincaré (1854–1912), eliminates the second problem named in the previous paragraph, but not the first. The relationship between the Poincaré and the Klein models for hyperbolic geometry is illustrated in Figure 9.36. On a plane tangent to the sphere, Klein's model is shown as the interior of the circle C' congruent to the great circle C'' of the sphere.

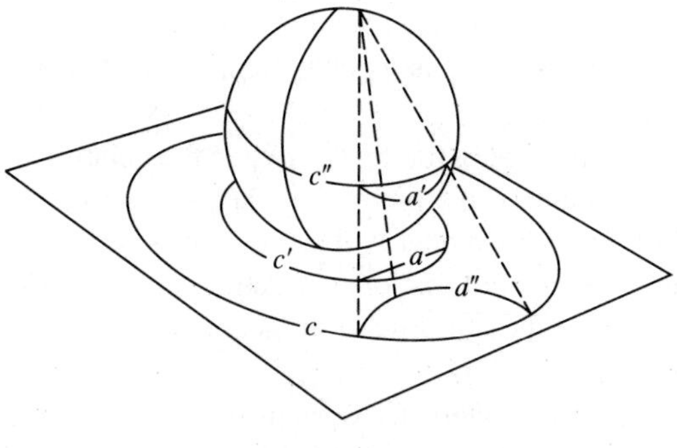

Figure 9.36

The points on C' and its interior are projected onto the bottom hemisphere of the sphere by a projection from an ideal point. Each chord, such as a, is projected into the arc a' of a circle orthogonal to C''. Now, the points on the bottom hemisphere are carried by a stereographic projection (considered in Section 6.4 as an application of inversive geometry) into the points on circle C and its interior. The arc a' is projected into an arc a'' orthogonal to circle C in the tangent plane. Poincaré's model for ordinary points in hyperbolic geometry is illustrated by the interior of C. The ordinary points on a line are illustrated by the points on an arc, such as a'', interior to C.

According to the development above, lines in Poincaré's model are represented by arcs of circles orthogonal to a given circle. In Figure 9.37, $\overset{\frown}{AB}$ and $\overset{\frown}{FG}$ represent intersecting lines, $\overset{\frown}{AB}$ and $\overset{\frown}{BC}$ represent parallel lines, and $\overset{\frown}{AB}$ and $\overset{\frown}{DE}$ represent nonintersecting lines.

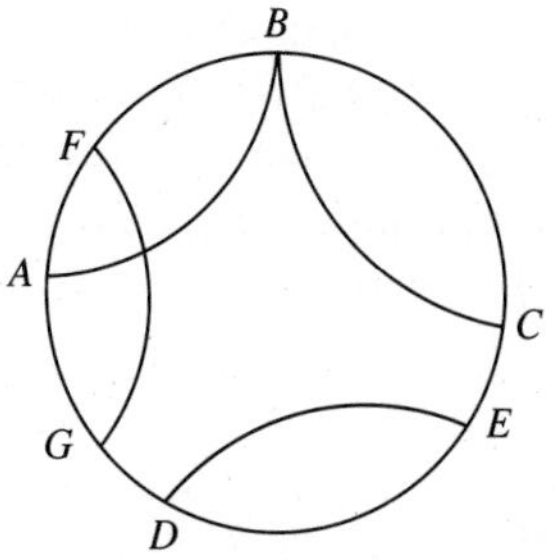

Figure 9.37

The ingenious way in which distance is defined in Poincaré's model is shown in the following formula, which refers to Figure 9.38:

$$CD = k \log_e \frac{(AC/CB)}{(AD/DB)}$$

where AC, CB, AD, and DB are lengths of segments, not arcs, and k is a parameter. Check to see that this formula does yield the correct results when the distance between two points is infinite (such as for D and B in Figure 9.38) or is zero—that is, if the two points coincide.

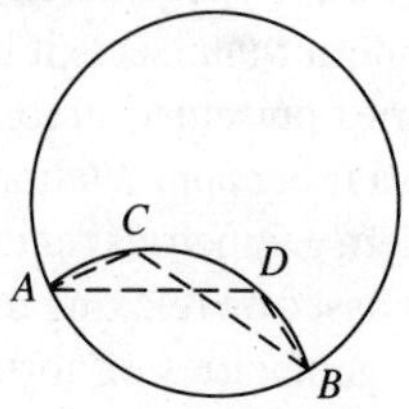

Figure 9.38

Fortunately, the measure of angles in Poincaré's model may be defined in the usual way for Euclidean geometry. This is true because angles are preserved under a stereographic projection. Inversion (see Chapter 6) is used as a key tool to show that the geometry of points and lines in Poincaré's model has the same postulational structure as the geometry of the hyperbolic plane. For example, a proof by inversion is used to prove that, in the Poincaré model, the sum of the measures of the angles of a triangle is less than π.

In Figure 9.39a, triangle ABC, which represents any triangle in the Poincaré model, is inverted into $A'B'C'$, shown in Figure 9.39b, by an inversion with center O, the inverse point of point B, with the fixed circle as the circle of inversion. It is apparent from the shape of $A'B'C'$, since arc $A'C'$ bends inward, that the sum of the angle measures is less than π. It is important to note that Figure 9.39 does not actually show the correct relative positions of the set of points in (a) and its inverse (b), since this relative position is unimportant in the proof. Actually the sets of points would overlap if drawn correctly, but such a drawing would be confusing, thus making it harder to interpret relationships within the inverted figure.

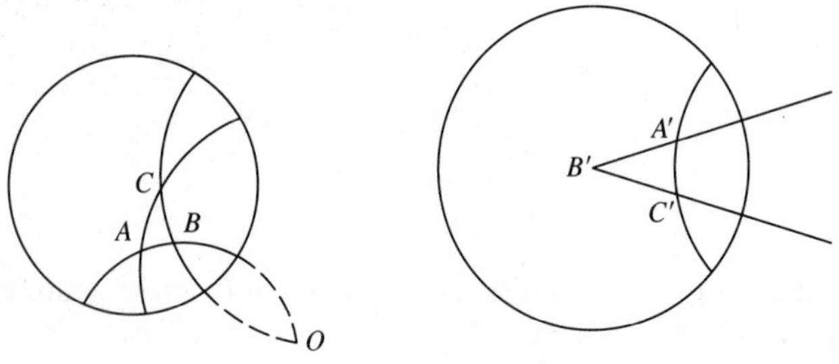

(a) (b)

Figure 9.39

Before leaving the Poincaré model, we should consider one other concept that helps to show differences among several geometries. The set of ideal points in a plane is called the *absolute*, and the nature of the absolute varies from one geometry to another. In elliptic geometry, the absolute is the empty set. In the geometry of the extended Euclidean plane (parabolic geometry), the absolute can be considered a straight line. In hyperbolic geometry, the absolute can be considered a conic (the circle in the Poincaré model) since a line through the interior intersects it in two distinct points.

Each of the modern geometries presented in this book has importance in its own right, but each also contributes to a fuller appreciation of the meaning of the expression "modern geometries." Other modern geometries remain to be explored by the interested reader. These include geometries based on relaxing the restrictions of Klein's definition from transformation to mapping, geometries of abstract spaces, differential geometry and geometries of more than three dimensions.

If, for example, we only need to have a mapping, then many sets of equations can be used as one element in defining the structure of a geometry—just so there is a unique value of x' and y' for a given value of x and y. We do not have to worry about showing that the inverse is a transformation. Here are several possibilities:

$$\begin{pmatrix} x' = x^2 + y^2 \\ y' = x^3 + y^3 \end{pmatrix} \qquad \begin{pmatrix} x' = x + 7 \\ y' = y^2 \end{pmatrix}$$

$$\begin{pmatrix} x' = \dfrac{9x}{x^2 - y^2} \\[2ex] y' = \dfrac{9y}{x^2 - y^2} \end{pmatrix}$$

Maurice Fréchet (1878–1973) is given credit for studying abstract spaces in 1906. The definition of Felix Klein was no longer adequate, and today we consider many structures geometries that have nothing to do with groups of transformations. An abstract space consists of a set of objects (often called *points*) and a set of statements of relationships involving these objects. The structure of a linear metric space includes a real number called the *metric* (distance function) of the space, which satisfies a list of postulates. Different metric spaces are specified by providing different definitions for the metric. In this setting, the theory of abstract spaces is involved in structures we might consider geometries.

A somewhat elementary example of a geometry with a different definition of metric than Euclidean geometry is called "taxicab" geometry. The grid in Figure 9.40 may be considered an arrangement of streets, on which a taxicab travels. The distance from one point on the grid to another is the shortest number of "blocks" the taxicab would take to go from one location to another.

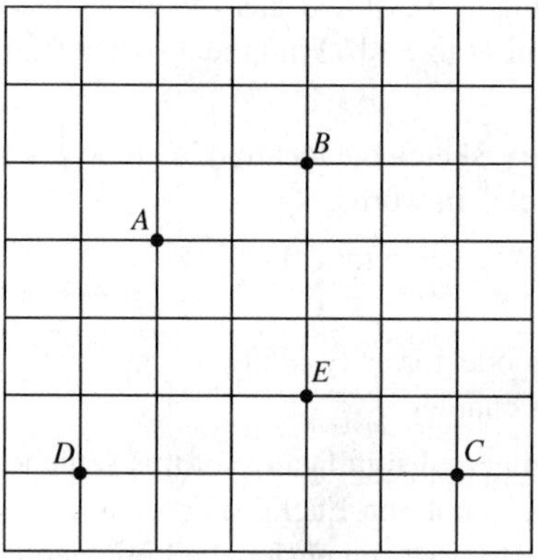

Figure 9.40

Example ➤ In Figure 9.40, find the distance from A to C and the distance from A to C by way of B. The distance from A to C is 7 units, and the distance from A to C by way of B is nine units.

▲ EXERCISES 9.7 ▲

Exercises 1–4 refer to the Klein model.

1. Draw a triangle with three ordinary points as vertices.
2. Draw an omega triangle.
3. Draw a triangle with exactly two ideal vertices.

4. Draw a triangle with three ideal vertices.

Exercises 5–8 refer to the Poincaré model.

5. Draw an example of an angle of parallelism.

6. Draw an omega triangle.

7. Draw a triangle with exactly two ideal vertices.

8. Draw a triangle with three ideal vertices.

9. Draw a Saccheri quadrilateral.

10. In Figure 9.38, find the distance AD, using the formula given.

11. In the proof by inversion about the measures of angles, explain in detail how Figure 9.39b was derived from Figure 9.39a.

12. Use the grid in Figure 9.40 and answer for taxicab geometry. a) Find the distance from point D to E. b) Find the distance from point B to D by way of point C.

13. In taxicab geometry, sketch the location of all points 4 units from a fixed point. Describe this "circle" in words.

▷ **14.** Study more about the Klein model and explain how distance and angle can be interpreted.

▷ **15.** Use the Poincaré model to verify additional theorems from hyperbolic geometry established in this chapter.

▷ **16.** Explain the meaning and significance of this sentence: "The geometry on the surface of a sphere is not non-Euclidean but instead provides a model in three-dimensional space for two-dimensional elliptic geometry."

▷ **17.** Investigate the use of a pseudosphere in providing a model for hyperbolic geometry.

▷ **18.** Use Chapter 19, "The Universe," from Weeks, *The Shape of Space*, or a similar reference to describe how modern scientists use non-Euclidean geometry to attempt to describe the physical nature of the universe.

▷ **19.** Give a definition for each type of geometry in the "geneology" of geometries on page 19 of Coxeter, *Non-Euclidean Geometry*.

▷ **20.** Make up an example of a mapping that might be used to define a geometry. Investigate to see if you can discover invariant properties in your geometry.

▷ **21.** List various definitions of the metric that will determine different metric spaces.

▷ **22.** Use a reference and carefully define "taxicab" geometry as a metric space.

▷ **23.** Study the definition of metric linear spaces and the theorem on the invariant norm, in Chapter 1 of Rolewicz, *Metric Linear Spaces*.

Appendix 1

Hilbert's Axioms

The axioms in Appendix 1 are from David Hilbert, *The Foundations of Geometry* (La Salle, Ill.: Open Court Publishing Company, 1950), by permission of Open Court Publishing Company. A new edition of *The Foundations of Geometry* was published by Open Court Publishing Company in 1980.

▲ Group I. Axioms of Connection

I, 1. Two distinct points A and B always completely determine a straight line a. We write $AB = a$ or $BA = a$.

I, 2. Any two distinct points of a straight line completely determine that line; that is, if $AB = a$ and $AC = a$, where $B \neq C$, then $BC = a$.

I, 3. Three points A, B, C not situated in the same straight line always completely determine a plane α. We write $ABC = \alpha$.

I, 4. Any three points A, B, C of a plane α, which do not lie in the same straight line, completely determine that plane.

I, 5. If two points A, B of a straight line a lie in a plane α, then every point of a lies in α.

I, 6. If two planes α, β have a point A in common, then they have at least a second point B in common.

I, 7. Upon every straight line there exist at least two points, in every plane at least three points not lying in the same straight line, and in space there exist at least four points not lying in a plane.

▲ Group II. Axioms of Order

II, 1. If A, B, C are points of a straight line and B lies between A and C, then B lies also between C and A.

II, 2. If A and C are two points of a straight line, then there exists at least one point B lying between A and C and at least one point D so situated that C lies between A and D.

II, 3. Of any three points situated on a straight line, there is always one and only one which lies between the other two.

II, 4. Any four points A, B, C, D of a straight line can always be so arranged that B shall lie between A and C and also between A and D, and furthermore, that C shall lie between A and D and also between B and D.

II, 5. Let A, B, C be three points not lying in the same straight line and let a be a straight line lying in the plane ABC and not passing through any of the points A, B, C. Then, if the straight line a passes through a point of the segment AB, it will also pass through either a point of the segment BC or a point of the segment AC.

▲ Group III. Axiom of Parallels

III. In a plane α there can be drawn through any point A, lying outside of a straight line a, one and only one straight line that does not intersect the line a. This straight line is called the parallel to a through the given point A.

▲ Group IV. Axioms of Congruence

IV, 1. If A, B are two points on a straight line a, and if A' is a point upon the same or another straight line a', then, upon a given side of A' on the straight line a', we can always find one and only one point B' so that the segment AB (or BA) is congruent to the segment $A'B'$. We indicate this relation by writing $AB \equiv A'B'$. Every segment is congruent to itself; that is, we always have $AB \equiv AB$.

IV, 2. If a segment AB is congruent to the segment $A'B'$ and also to the segment $A''B''$, then the segment $A'B'$ is congruent to the segment $A''B''$; that is, if $AB \equiv A'B'$ and $AB \equiv A''B''$, then $A'B' \equiv A''B''$.

IV, 3. Let AB and BC be two segments of a straight line a that have no points in common aside from the point B, and, furthermore, let $A'B'$ and $B'C'$ be two segments of the same or of another straight line a' having, likewise, no point other than B' in common. Then, if $AB \equiv A'B'$, and $BC \equiv B'C'$, we have $AC \equiv A'C'$.

IV, 4. Let an angle (h, k) be given in the plane α and let a straight line a' be given in a plane α'. Suppose also that, in the plane α', a definite side of the straight line a' be assigned. Denote by h' a half-ray of the straight line a' emanating from a point O' of this line. Then in the plane α' there is one and only one half-ray k' such that the angle (h, k) or (k, h) is congruent to the angle (h', k'), and at the same time all interior points of the angle (h', k') lie upon the given side of a'. We express this relation by means of the notation $\angle(h, k) \equiv \angle(h', k')$.
Every angle is congruent to itself; that is

$$\angle(h, k) \equiv \angle(h, k)$$

$$\angle(h, k) \equiv \angle(k, h)$$

IV, 5. If the angle (h, k) is congruent to the angle (h', k') and to the angle (h'', k''), then the angle (h', k') is congruent to the angle (h'', k''); that is to say, if $\angle(h, k) \equiv \angle(h', k')$ and $\angle(h, k) \equiv \angle(h'', k'')$, then $\angle(h', k') \equiv \angle(h'', k'')$.

IV, 6. If, in the two triangles ABC and $A'B'C'$, the congruences $AB \equiv A'B'$, $AC \equiv A'C'$, $\angle BAC \equiv \angle B'A'C'$ holds, then the congruences $\angle ABC \equiv \angle A'B'C'$ and $\angle ACB \equiv \angle A'C'B'$ also hold.

Group V. Axiom of Continuity

V. Let A_1 be any point upon a straight line between the arbitrarily chosen points A and B. Take the points $A_2, A_3, A_4, \ldots$ so that A_1 lies between A and A_2, A_2 between A_1 and A_3, A_3 between A_2 and A_4, etc. Moreover, let the segments $AA_1, A_1A_2, A_2A_3, A_3A_4, \ldots$ be equal to one another. Then, among this series of points, there always exists a certain point A_n such that B lies between A and A_n.

Note: An additional axiom, the Axiom of Completeness, appears in the French translation, but was not included in the five original groups of axioms. This axiom states that the elements of geometry form a system which cannot be extended, if we assume the five groups of axioms are valid.

Appendix 2

Birkhoff's Postulates

These axioms are reprinted from G. D. Birkhoff, "A Set of Postulates for Plane Geometry (based on scale and protractor)," *Annals of Mathematics*, Vol. 33, 1932, by permission of the *Annals of Mathematics*.

I. *Postulate of line measure*: The points $A, B, \ldots$ of any line l can be put into $(1, 1)$ correspondence with the real numbers x so that $|x_B - x_A| = d(A, B)$ for all points A, B.

II. *Point–line postulate*: One and only one straight line l contains two given points $P, Q(P \neq Q)$.

III. *Postulate of angle measure*: The half-lines $l, m, \ldots$ through any point O can be put into $(1, 1)$ correspondence with the real numbers $a \pmod{2\pi}$ so that, if $A \neq O$ and $B \neq O$ are points on l and m, respectively, the difference $a_m - a_l \pmod{2\pi}$ is $\angle AOB$. Furthermore, if the point B on m varies continuously in a line r not containing the vertex O, the number a_m varies continuously also.

IV. *Postulate of similarity*: If in two triangles, $\triangle ABC$, $\triangle A'B'C'$, and for some constant $k > 0$, $d(A', B') = kd(A, B)$, $d(A', C') = kd(A, C)$ and also $\angle B'A'C' = \pm \angle BAC$, then also $d(B', C') = kd(B, C)$, $\angle C'B'A' = \pm \angle CBA$, and $\angle A'C'B' = \pm \angle ACB$.

Appendix 3

Illustrations of Basic Euclidean Constructions

(a) (b)

Transferring a segment

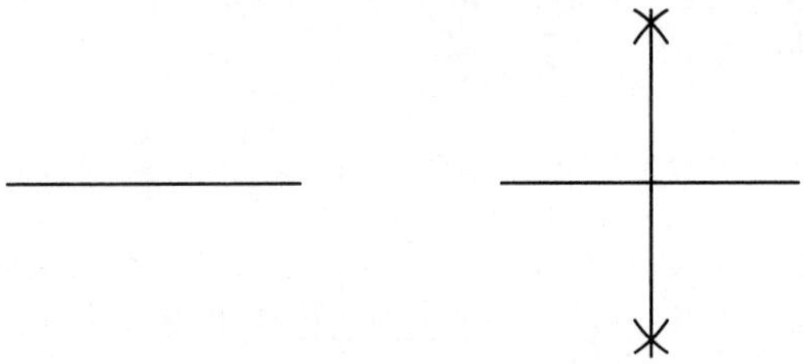

(a) (b)

Bisecting a segment

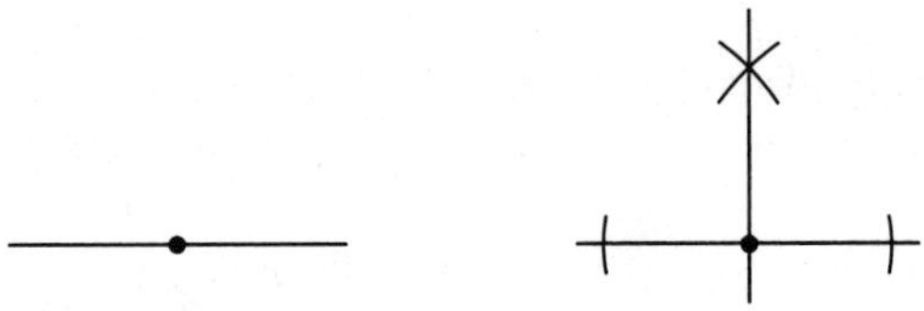

(a) (b)

Constructing a perpendicular to a line at a certain point

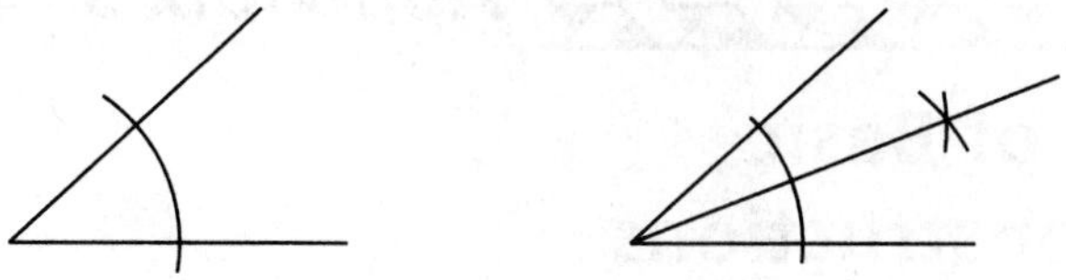

(a) (b)

Constructing an angle bisector

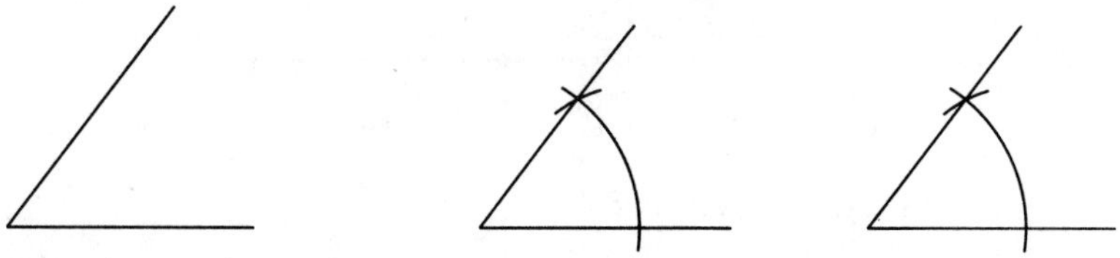

(a) (b)

Copying an angle

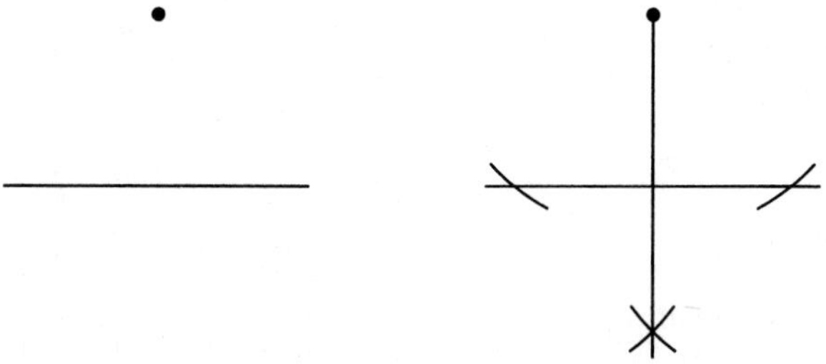

(a) (b)

Constructing a perpendicular
from a point to a line

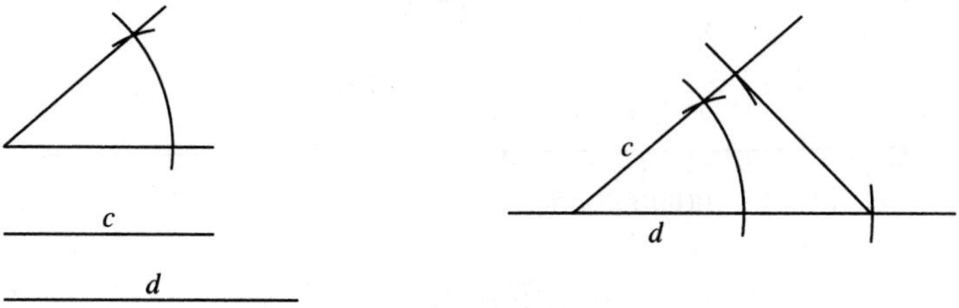

(a) (b)

Constructing a triangle, given an
angle and the two adjacent sides

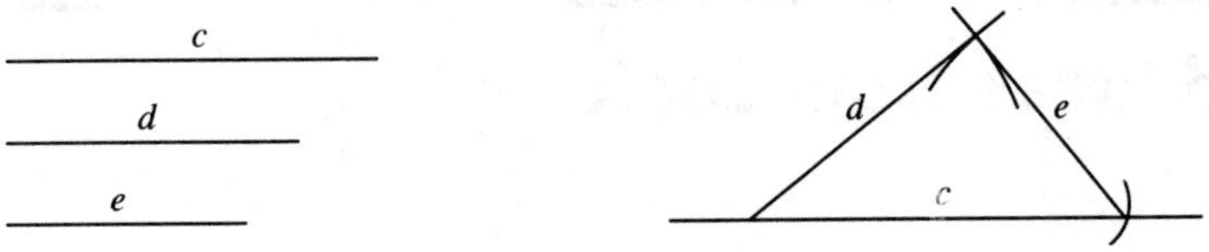

(a) (b)

Constructing a triangle, given the three sides

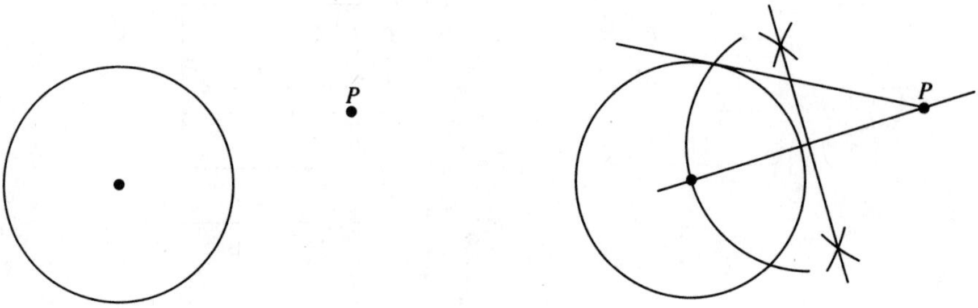

(a) (b)

Constructing a tangent from a point to a circle

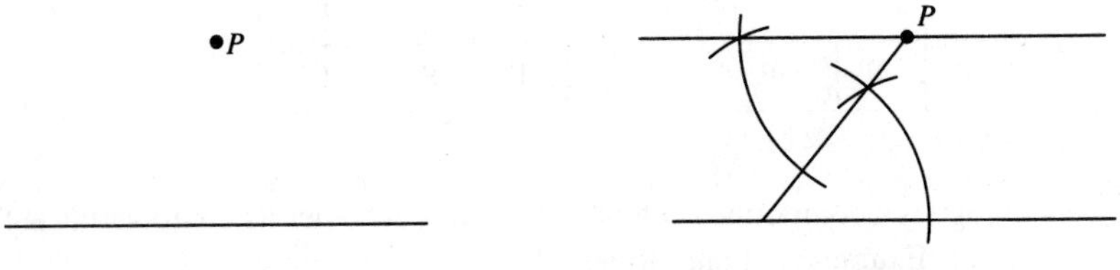

(a) (b)

Constructing a line through a point parellel to a given line

Appendix 4

Selected Ideas from Logic

1. Truth tables

Negation	
p	$\sim p$
T	F
F	T

Conjunction (and)		
p	q	$p \wedge q$
T	T	T
T	F	F
F	T	F
F	F	F

Disjunction (or)		
p	q	$p \vee q$
T	T	T
T	F	T
F	T	T
F	F	F

Implication		
p	q	$p \rightarrow q$
T	T	T
T	F	F
F	T	T
F	F	T

2. A tautology is a compound sentence that is true whether its components are true or not. Examples of tautologies:

$$p \vee \sim p$$

$$q \rightarrow (p \rightarrow q)$$

$$(p \wedge [p \rightarrow q]) \rightarrow q$$

3. A valid argument follows the pattern of a tautology. The conclusion is true if the argument is valid and if all the premises are true. A proof is a valid argument having all the premises true.

Appendix 5

Selected Theorems from Plane Euclidean Geometry

1. If two triangles have two sides and the included angle of one congruent, respectively, to two sides and the included angle of the other, the two triangles are congruent.
2. If two triangles have two angles and the included side of one congruent, respectively, to two angles and the included side of the other, the two triangles are congruent.
3. If two triangles have the three sides of one congruent, respectively, to the three sides of the other, the two triangles are congruent.
4. The segment joining the midpoints of two sides of a triangle is parallel to the third side and its length is equal to one-half the length of the third side.
5. The sum of the measures of the angles of a triangle is 180°.
6. The measure of an exterior angle of a triangle equals the sum of the measures of the two nonadjacent interior angles.
7. If two triangles have two angles of one congruent, respectively, to two angles of the other, the triangles are similar.
8. If two triangles have an angle of one congruent to an angle of the other and the including sides proportional, the triangles are similar.
9. If two triangles have their corresponding sides proportional, they are similar.
10. The altitude to the hypotenuse of a right triangle forms two right triangles similar to each other and to the original triangle.
11. The two diagonals of a parallelogram bisect each other.
12. The median of a trapezoid has a length equal to half the sum of the lengths of the bases.
13. The opposite angles of a parallelogram are congruent.
14. The two diagonals of a rhombus are perpendicular.
15. The sum of the measures of the interior angles of a polygon of n sides is $(n - 2)180^\circ$.
16. If two polygons are similar, then any two corresponding sides have the same ratio.

17. In a circle or in congruent circles, equal chords are equidistant from the center.
18. If a quadrilateral is inscribed in a circle, the opposite angles are supplementary.
19. The line joining the centers of two intersecting circles is the perpendicular bisector of the common chord.
20. An angle inscribed in a semicircle is a right angle.
21. A central angle in a circle is measured by its intercepted arc.
22. An inscribed angle is measured by half its intercepted arc.
23. An angle formed by a tangent and a chord to a circle is measured by half its intercepted arc.
24. The ratio of the measure of a sector of a circle to the measure of area of the entire circle is equal to the ratio of the degrees in the angle of the sector to 360°.
25. If two parallel lines are cut by a transversal, the alternate interior angles are congruent and the corresponding angles are congruent.

Appendix 6

The First Twenty-eight Propositions of Euclid

1. On a given finite straight line, to construct an equilateral triangle.
2. To place at a given point (as an extremity) a straight line equal to a given straight line.
3. Given two unequal straight lines, to cut off trom the greater a straight line equal to the less.
4. If two triangles have the two sides equal to two sides, respectively, and have the angles contained by the equal straight lines equal, they will also have the base equal to the base, the triangle will be equal to the triangle, and the remaining angles will be equal to the remaining angles, respectively, namely, those which the equal sides subtend.
5. In isosceles triangles the angles at the base are equal to one another, and, if the equal straight lines be produced further, the angles under the base will be equal to one another.
6. If in a triangle two angles be equal to one another, the sides which subtend the equal angles will also be equal to one another.
7. Given two straight lines constructed on a straight line (from its extremities) and meeting in a point, there cannot be constructed on the same straight line (from its extremities), and on the same side of it, two other straight lines meeting in another point and equal to the former two, respectively, namely, each to that which has the same extremity with it.
8. It two triangles have the two sides equal to two sides, respectively, and have also the base equal to the base, they will also have the angles equal which are contained by the equal straight lines.
9. To bisect a given rectilineal angle.
10. To bisect a given finite straight line.
11. To draw a straight line at right angles to a given straight line from a given point on it.
12. To a given infinite straight line, from a given point which is not on it, to draw a perpendicular straight line.
13. If a straight line set up on a straight line make angles, it will make either two right angles or angles equal to two right angles.

14. If with any straight line, and at a point on it, two straight lines not lying on the same side make the adjacent angles equal to two right angles, the two straight lines will be in a straight line with one another.

15. If two straight lines cut one another, they make the vertical angles equal to one another.

16. In any triangle if one of the sides be produced, the exterior angle is greater than either of the interior and opposite angles.

17. In any triangle two angles taken together in any manner are less than two right angles.

18. In any triangle the greater side subtends the greater angle.

19. In any triangle the greater angle is subtended by the greater side.

20. In any triangle two sides taken together in any manner are greater than the remaining one.

21. If on one of the sides of a triangle, from its extremities, there be constructed two straight lines meeting within the triangle, the straight lines so constructed will be less than the remaining two sides of the triangle, but will contain a greater angle.

22. Out of three straight lines, which are equal to three given straight lines, to construct a triangle: thus it is necessary that two of the straight lines taken together in any manner should be greater than the remaining one.

23. On a given straight line and at a point on it, to construct a rectilineal angle equal to a given rectilineal angle.

24. If two triangles have the two sides equal to two sides, respectively, but have the one of the angles contained by the equal straight lines greater than the other, they will also have the base greater than the base.

25. If two triangles have the two sides equal to two sides, respectively, but have the base greater than the base, they will also have the one of the angles contained by the equal straight lines greater then the other.

26. If two triangles have the two angles equal to two angles, respectively, and one side equal to one side, namely, either the side adjoining the equal angles, or that subtending one of the equal angles, they will also have the remaining sides equal to the remaining sides and the remaining angle to the remaining angle.

27. If a straight line falling on two straight lines make the alternate angles equal to one another, the straight lines will be parallel to one another.

28. If a straight line falling on two straight lines make the exterior angle equal to the interior and opposite angle on the same side, or the interior angles on the same side equal to two right angles, the straight lines will be parallel to one another.

Bibliography

ADLER, CLAIRE F. *Modern Geometry.* 2d ed. New York: McGraw-Hill, 1967.

ALBERT, ABRAHAM A., and SANDER, REUBEN. *An Introduction to Finite Projective Plane.* New York: Holt, Rinehart and Winston, 1968.

AMERICAN MATHEMATICAL ASSOCIATION. *Geometry.* Special Issue of the AMA Monthly, Vol. 97, No. 8, October, 1990.

ARTWICK, BRUCE A. *Applied Concepts in Microcomputer Graphics.* Englewood Cliffs, NJ: Prentice-Hall, 1984.

BANNON, THOMAS J. "Fractals and Transformations." *The Mathematics Teacher*, Vol. 84, March, 1991.

BARNETTE, DAVID. *Map Coloring, Polyhedra, and the Four-Color Problem.* Washington D.C.: Mathematical Association of America, 1984.

BARNSLEY, MICHAEL. *Fractals Everywhere.* Boston: Academic Press, Inc., 1988.

BARRY, EDWARD H. *Introduction to Geometrical Transformations.* Boston: Prindle, Weber & Schmidt, 1966.

BAULEC, YVES; BELLEMAIN, FRANCH; and LABORDE, JEAN-MARIE. *CABRI, The Interactive Geometry Notebook.* Pacific Grove, CA: Brooks/Cole Publishing Company, 1988–1992.

BENSON, RUSSELL V. *Euclidean Geometry and Convexity.* New York: McGraw-Hill, 1966.

BERGAMINI, DAVID, and editors of Time-Life Books. *Mathematics.* New York: Life Science Library, 1970.

BERGER, MARC. *Computer Graphics with Pascal.* Menlo Park, CA: The Benjamin/Cummings Publishing Co., Inc., 1986.

BEZUSZKA, et al. *Tessellations: The Geometry of Patterns.* Oak Lawn, IL: Creative Publications, 1977.

BLUMENTHAL, LEONARD M. *A Modern View of Geometry.* San Francisco: W.H. Freeman, 1961.

BULLARD, SIR EDWARD. "The Origin of the Oceans." *Scientific American*, September, 1969, Vol. 221 (3).

CAMP, DANE R. "A Fractal Excursion." *The Mathematics Teacher*, Vol. 84, April, 1991.

CAMPBELL, JOHN. *Introductory Cartography.* Englewood Cliffs, NJ: Prentice-Hall, Inc., 1984.

CHARTRAND, GARY, and LESNICK, LINDA. *Graphs and Digraphs.* 2d ed. Monterey, CA: Wadsworth and Brooks/Cole Advanced Books and Software, 1986.

CHINITZ, WALLACE. "Rotary Engines." *Scientific American*, February, 1969, Vol. 220 (2).

CHRESTENSON, H.E. *Mappings of the Plane*. San Francisco: W.H. Freeman, 1966.

COMAP. *Geometry's Future*. Conference proceedings sponsored by COMAP, Inc. Arlington, MA: COMAP, 1991.

COOPER, LEON, and STEINBURG, DAVID. *Methods and Applications of Linear Programming*. Philadelphia: W.B. Saunders, 1974.

COURANT, RICHARD, and ROBBINS, HERBERT. *What Is Mathematics?* London: Oxford University Press, 1951.

COXETER, H.S.M. *Introduction to Geometry*. 2d ed. New York: Wiley, 1969.

Edited by COXETER, H.S.M., EMMER, M., PENROSE, R., and TEUBER, M.L. *M.C. Escher, Art and Science*. Amsterdam: Ensevier Science Publishers, 1986.

COXETER, H.S.M. *Non-Euclidean Geometry*. 5th ed. Toronto: University of Toronto Press, 1965.

COXETER, H.S.M. *Projective Geometry*, 2d ed. Toronto: University of Toronto Press, 1974.

COXETER, H.S.M., and GREITZER, S.L. *Geometry Revisited*. New York: Random House, 1967.

DAVIS, DAVID R. *Modern College Geometry*. Reading, Mass.: Addison-Wesley, 1954.

DEMEL, JOHN T., and MILLER, MICHAEL J. *Introduction to Computer Graphics*. Monterey: Brooks/Cole Engineering Division, 1984.

DODGE, CLAYTON W. *Euclidean Geometry and Transformations*. Reading, MA: Addison-Wesley, 1972.

DORWART, HAROLD L. *The Geometry of Incidence*. Englewood Cliffs, NJ: Prentice-Hall, 1966.

EGGLESTON, H.G. *Convexity*. Cambridge: Cambridge University Press, 1966.

ESCHER, M.C. *The Graphic Work of M.C. Escher*. New York: Ballantine, 1971.

EVES, HOWARD. *An Introduction to the History of Mathematics*. 6th ed. Philadelphia: Saunders College Publishing, 1983.

EVES, HOWARD. *A Survey of Geometry*. 2 vols. Boston: Allyn and Bacon, 1964-65.

FABER, RICHARD L. *Foundations of Euclidean and Non-Euclidean Geometry*. New York: Dekker, 1983.

FALCONER, K.J. *The Geometry of Fractal Sets*. Cambridge: Cambridge University Press, 1985.

FISHBACK, W.T. *Projective and Euclidean Geometry*. New York: Wiley, 1962.

FISHER, J. CHRIS. "Curves of Constant Width from a Linear Viewpoint." *Mathematics Magazine*, Vol. 60, No. 3, June, 1987.

FLEGG, H. GRAHAM. *From Geometry to Topology*. London: The English Universities Press Limited, 1984.

FOLEY, JAMES D., and VAN DAM, ANDRIES. *Fundamentals of Interactive Computer Graphics*. Reading, MA: Addison-Wesley, 1982.

GANS, DAVID. *An Introduction to Non-Euclidean Geometry*. New York: Academic Press, 1973.

GANS, DAVID. *Transformations and Geometries*. New York: Appleton-Century-Croft, 1969.

GARNER, LYNN E. *An Outline of Projective Geometry*. New York: North Holland, 1981.

GILBERT, WILLIAM J. *Modern Algebra with Applications*. New York: John Wiley & Sons, 1976.

GOUDMAN, JACOB E. et al, eds. *Discrete Geometry and Convexity.* New York: New York Academy of Sciences, 1985.

GRAUSTEIN, WILLIAM C. *Introduction to Higher Geometry.* New York: Macmillan, 1930.

GRAY, JEREMY. *The Geometrical Work of Girard Desargues.* New York: Springer-Verlag, 1987.

GREENBERG, MARVIN J. *Euclidean and Non-Euclidean Geometries.* San Francisco: W.H. Freeman, 1980.

HILBERT, DAVID. *The Foundations of Geometry.* La Salle, Ill.: Open Court, 1980.

HOPKINS, E.J. and HAILS, J.S. *An Introduction to Plane Projective Geometry.* Oxford: Oxford University Press, 1953.

JOHNSON, DONOVAN A. *Paper Folding for the Mathematics Class.* National Council of Teachers of Mathematics, 1957.

JOHNSON, DONOVAN A., and GLENN, W.H. *Topology, the Rubber Sheet Geometry.* New York: McGraw-Hill, 1960.

KANDEL, ABRAHAM. *Fuzzy Mathematical Techniques with Applications.* Reading, MA: Addison-Wesley Publishing Co., 1986.

KÁRTESZI, P. *Introduction to Finite Geometries.* Amsterdam: North-Holland, 1976.

KLEE, VICTOR L., ed. *Convexity.* Proceedings of the Seventh Symposium in Pure Mathematics of the American Mathematical Society, Providence, RI, 1963.

KLINGENBERG, WILHELM. *Riemannian Geometry.* Berlin: Walter de Gruyter, 1982.

KRAUSE, EUGENE F. *Taxicab Geometry.* New York: Dover Publications, 1986.

LAY, STEPHEN R. *Convex Sets and their Applications.* New York: Wiley, 1982.

LEVI, HOWARD. *Topics in Geometry.* Boston: Prindle, Weber & Schmidt, 1970.

LEVY, LAWRENCE S. *Geometry: Modern Mathematics via the Euclidean Plane.* Boston: Prindle, Weber & Schmidt, 1970.

LICKORISH, W. B.R. and MILLETT, K.C. "The New Polynomial Invariants of Knots and Links." *Mathematics Magazine*, Vol. 61, No. 1, February, 1988.

LIETZMAN, WALTHER. *Visual Topology.* New York: American Ensevier, 1965.

LOCKWOOD, JAMES R., and RUNION, GARTH E. *Deductive Systems: Finite and Non-Euclidean Geometries.* Reston, VA.: NCTM, 1978.

LYUSTERNIK, L. *Convex Figures and Polyhedra.* New York: Dover, 1963.

MANDELBROT, BENOIT B. *The Fractal Geometry of Nature.* New York: W.H. Freeman, 1983.

MANDELBROT, BENOIT B. *Les Objects Fractals.* Paris: Flammarion, 1984.

MARSHALL, GEORGE R. *Computer Graphics in Application.* Englewood Cliffs, NJ: Prentice Hall, 1987.

MARTIN, GEORGE E. *The Foundations of Geometry and the Non-Euclidean Plane.* New York: Springer-Verlag, 1982.

MESERVE, BRUCE E., and IZZO, JOSEPH A. *Fundamentals of Geometry.* Reading, MA: Addison-Wesley, 1969.

MOISE, EDWIN E. *Elementary Geometry from an Advanced Standpoint.* Reading, MA: Addison-Wesley, 1963.

PATTERSON, E.M. *Topology*. New York: Interscience, 1959.

PEARSON, H.R., and SMART, J.R. *Geometry*. Boston: Ginn, 1971.

PEDERSEN, JEAN J. "Dressing Up Mathematics." *The Mathematics Teacher*. February 1968, Vol. 61 (2).

PEDOE, D. *A Course of Geometry for Colleges and Universities*. Cambridge: At the University Press, 1970.

PENNA, MICHAEL A. *Projective Geometry and its Applications to Computer Graphics*. Englewood Cliffs, NJ: Prentice-Hall, 1986.

PERFECT, HAZEL. *Topics in Geometry*. London: Pergamon Press, 1963.

PERVIN, W.J. *Foundations of General Topology*. New York: Academic Press, 1964.

POLYA, GEORGE. *Mathematical Discovery*. New York: Wiley, 1981.

PRATT, M.M. "Finite Geometries." Master's thesis, San Jose State College, San Jose, CA, 1964.

PRENOWITZ, WALTER, and SWAIN, HENRY. *Congruence and Motion in Geometry*. Boston: Heath, 1966.

RADEMACHER, H., and TOEPLITZ, O. *The Enjoyment of Mathematics*. Princeton: Princeton University Press, 1957.

REULEAUX, FRANZ. *The Kinematics of Machinery*. Translated and edited by Alex. B.W. Kennedy. London: Macmillan & Co., 1876.

RIETMAN, EDWARD. *Exploring the Geometry of Nature: Computer Modeling of Chaos, Fractals, Cellular Automata, and Neural Networks*. Blue Ridge Summit, PA: Windcrest, 1989.

ROGERS, DAVID F., and ADAMS, J. ALAN. *Mathematical Elements for Computer Graphics*. New York: McGraw-Hill, 1976.

ROLEWICZ, STEFAN. *Metric Linear Spaces*. Dordrecht, Holland: D. Reidel Publishing Company, 1985.

ROW, T.S. *Geometric Exercises in Paper Folding*. Chicago: The Open Court Publishing Company, 1941.

RYAN, PATRICK J. *Euclidean and Non-Euclidean Geometry*. Cambridge: Cambridge University Press, 1986.

SACCHERI, GIROLAMO. *Girolamo Saccheri's Euclides Vindicatus*. New York: Chelsea, 1986.

SNAPPER, ERNST, and TROYER, ROBERT J. *Metric Affine Geometry*. New York: Academic Press, 1971.

STASZKOW, RONALD and BRADSHAW, ROBERT. *The Mathematical Palette*. Philadelphia: Saunders College Publishing, 1991.

STEVENS, ROGER T. *Fractal Programming in C*. Redwood City, CA: M & T Publishing, Inc., 1989.

STRANG, GILBERT. *Linear Algebra and Its Applications*. 3rd Ed. San Diego: Harcourt, Brace, Jovanovich Publishers, 1988.

TOFFOLI, TOMMASO and MARGOLUS, NORMAL. *Cellular Automata Machines*. Cambridge, MA.: The MIT Press, 1985.

TORRETTI, ROBERTO. *Relativity and Geometry*. Oxford: Pergamon Press, 1983.

TSUZUKU, TOSHIRO. *Finite Groups and Finite Geometries.* Cambridge: Cambridge University Press, 1982.

TULLER, ANNITA. *A Modern Introduction to Geometries.* Princeton: Van Nostrand, 1967.

VALENTINE, F. *Convex Sets.* New York: McGraw-Hill, 1964.

WATTS BROTHERS TOOL WORKS. *How to Drill Square, Hexagon, Octagon, Pentagon Holes.* Witmerding, PA, 1966.

WEEKS, JEFFERY R. *The Shape of Space.* New York: Dekker, 1985.

WEINER, JONATHAN. *Planet Earth.* Toronto: Bantam Books, 1986.

WILSON, ROBIN J. *Introduction to Graph Theory.* 2d ed. New York: Academic Press, 1979.

WOLFE, HAROLD E. *Introduction to Non-Euclidean Geometry.* New York: Dryden Press, 1945.

WOODCOCK, ALEXANDER, and DAVIS, MONTE. *Catastrophe Theory.* New York: E.P. Dutton, 1978.

WYLIE, C.R. *Foundations of Geometry.* New York: McGraw-Hill, 1964.

YAGLOM, I.M., and BOLTYANSKII, V. G. *Convex Figures.* New York: Holt, Rinehart & Winston, 1961.

Answers to Selected Exercises

EXERCISES ➤ 1.1

1. $3^2 + 4^2 = 5^2$ **2.** 30 **3.** 36 **4.** Third side is not an integral length **5.** $\approx$ 12,525 km **6.** $\approx 9.1°$ **7.** 18,000 miles **8.** 18,000 miles **9.** False **10.** True **11.** False **12.** True **13.** True **14.** False

EXERCISES ➤ 1.2

2. All points are on the same line. **3.** Two distinct points are on more than one line. **4.** (1) There exist exactly three distinct books in the system. (2) Two distinct books are in exactly one library. (3) Not all the books in the system are in the same library. (4) Two distinct libraries have at least one book in common.
5. (1) There exist exactly three different students. (2) Two different students are on exactly one committee. (3) Not all the students are on the same committee. (4) Two different committees have at least one student in common.
6. Yes **7.** No **8.** Yes **9.** None **10.** Two **11.** No **12.** No **13.** No **15.** By Axiom 1, there are exactly 3 points in the geometry. By Axiom 2, the first and second of these are on one line, and by Axiom 3, the third point is not on this line. But by Axiom 2, this third point lies on a line through the first point, so the two lines through the first point contain all three points.

EXERCISES ➤ 1.3

1. (1) There exist exactly three distinct lines in the geometry. (2) Two distinct lines are on exactly one point. (3) Not all the lines of the geometry are on the same point. (4) Two distinct points are on at least one line. **3.** None **4.** (1) The total number of committees is four. (2) Each pair of committees has exactly one student in common. (3) Each student is on exactly two committees. **5.** No **6.** Four **7.** None
9. Consider any two lines of the geometry. By Axiom 2, these have one point in common, and by Theorem 1.4, they each have two other points. Since there are six points in the geometry, by Theorem 3, not all the points can lie on the two lines. **12.** Axion 2
13. (1) The total number of trees is four. (2) Each pair of trees has exactly one row in common. (3) Each row contains exactly two trees. **14.** No **16.** One

EXERCISES ➤ 1.4

1–5. For Exercises 1–5, numbers 1, 2, 3, and 4 are true. **6.** (1) There exists at least one library. (2) Every library has exactly three books in it. (3) Not all books are in the same library. (4) For two different books, there is exactly one library containing both of them. (5) Each two libraries have at least one book in common.

7. 1, 3, 4 **8.** $\begin{bmatrix} 4,1,2 & 4,2,3 & 4,3,5 & 4,6,7 \\ 4,1,3 & 4,2,5 & 4,3,6 & \\ 4,1,6 & 4,2,7 & 4,5,6 & \\ 4,1,7 & & 4,5,7 & \end{bmatrix}$

11. By Axioms 1 and 2, there is a line in the geometry with exactly three points, say A, B, C on it. By Axiom 3, there is a fourth point D not on this line. Consider the number of lines on D. By Axiom 4, there are at least three lines through D to points A, B, and C. The assumption of a fourth line violates either Axiom 4 or 5.

EXERCISES ➤ 1.5

1. (l) There exists at least one row. (2) Every row has exactly three trees in it. (3) Not all trees are in the same row. (4) There exists exactly one row containing a tree not on a row that contains no tree of the given row. (5) If a tree is not on a row, there exists exactly one different tree in the row such that the two trees do not have a row in common. (6) With the exception in Axiom 5, exactly one row contains each pair of distinct trees. **2.** 1, 3, 4 **3.** Two **11.** (a) A'; (b) B'
12. (a) $\overleftrightarrow{A'P}$; (b) $\overleftrightarrow{PB'}$ **13.** BRB' and ASA' **14.** 1, 3, 4 **15.** $\overleftrightarrow{A'T}$, $\overleftrightarrow{B'P}$, $\overleftrightarrow{B'R}$

EXERCISES ➤ 1.6

2. Four **3.** False **4.** True **5.** True **7.** a, b, and d **8.** (a) 21; (b) 31; (d) 57 **9.** Four **10.** Ten, five **11.** None **12.** Yes **13.** Two **14.** Four, four **15.** One **16.** No **17.** Twelve, six **18.** One **19.** No

EXERCISES ➤ 1.7

8. II, 5 **9.** I, 1 and I, 2 **10.** II, 1 – 4 **11.** Yes **12.** Yes **13.** Yes **14.** Yes

REVIEW EXERCISES ➤ CHAPTER 1

1. Elements. **2.** $\approx$ 30,361 km. The true statements are 3, 5, 6, 10, 11, 12, 13, 14, 16, 19, 20, 21, 22, 34, 46, 49, 50, 51, 52, 54, 55.

EXERCISES ➤ 2.1

2. c **3.** 13 **4.** -5 **5.** $-\frac{5}{2}$ **6.** $(7, -2)$ **7.** $(5, -3)$ **8.** $(2, -5)$
9. $\{(a, i), (c, j), (e, k)\}$ **10.** Not defined **11.** $\{(b, a), (d, c), (h, e)\}$
12. $\{(i, b), (j, d), (k, h)\}$ **13.** $(x, y) \to (x - 3, y - 1)$
14. $(x, y) \to (x - 3, y - 1)$ **15.** $(x, y) \to (x + 5, y - 2)$
16. $(x, y) \to (x - 2, y + 3)$ **17.** (26, 29) **18.** (262/5, 317/10) **19.** No
22. $a = 0, b = 0$

EXERCISES ➤ 2.2

3. R_1 **4.** R_2 **5.** $R(240)$ **6.** $R(120)$ **7.** No
8.

	R_1	I
R_1	I	R_1
I	R_1	I

10. Yes **11.** $R_1, I; I$

18. One subgroup with eight members; three subgroups with four members; five subgroups with two members; and one subgroup with one member

EXERCISES ➤ 2.3

1. Yes **2.** No **3.** No **4.** A rotation of zero degrees.
5. The angle of rotation could be 180°. **6.** $R(A, \alpha + \beta)$
8. A rotation of $360° - \theta$ degrees **9.** The segment is parallel or perpendicular to the line of reflection. **10.** Yes **11.** The same reflection followed by the inverse of the translation **12.** When the segment is parallel or perpendicular to the line of reflection
13. No

EXERCISES ➤ 2.4

1. $(3, -5)$ **2.** $(8, 2)$ **3.** $(2, -13)$ **4.** $(1, -\frac{13}{2})$ **5.** $(-5, -9)$
6. $(0, 0)$ **7.** $y' = 2x' + 1$ **8.** $7x' - y' - \frac{101}{5} = 0$ **9.** $2x' - 5y' - 16 = 0$
10. $3x' + 4y' + 2 = 0$ **12.** $(-\sqrt{2}/2, 7\sqrt{2}/2)$ **13.** $(\sqrt{3}, 1)$
14. $(\sqrt{2}/2, 3\sqrt{2}/2)$
15. $x - h = (x' - h)\cos\alpha + (y' - k)\sin\alpha$, $y - k = -(x' - h)\sin\alpha + (y' - k)\cos\alpha$
16. $\left(\dfrac{3 + 7\sqrt{3}}{2}, \dfrac{3\sqrt{3} - 7}{2}\right)$ **17.** $\left(\dfrac{17\sqrt{3} - 3}{2}, \dfrac{3\sqrt{3} - 3}{2}\right)$ **18.** $(-5, 10)$
19. $x = x' + 5$; $y = y' - 2$ **20.** $x = x' - 4$; $y = y' + 3$
21. $x = -x'$; $y = -y'$ **22.** $x = x'$; $y = -y'$
23. $x = y' \cos 30° + y' \sin 30°$; $y = -x' \sin 30° + y' \cos 30°$
24. $x = x' \cos 60° + y \sin 60°$; $y = x' \sin 60° - y' \sin 60°$ **25.** $x' = x + 4$; $y' = y + 7$
26. $x' = x - 1$; $y' = y - 5$

EXERCISES ➤ 2.5

1. $(14, 14)$ **2.** $(10, -25)$ **3.** $(-18, -1)$ **4.** $(-21, -11)$ **5.** $(9, 4), (6, 7)$
6. $(-4, -5), (1, -3)$ **7.** $(1, 4), (3, -2)$ **8.** $(0, 8), (16, -20)$
9. $(0, -3), (7, 0), (5, -5)$ **10.** $(1, 2), (-4, 0), (-3, 3)$
11. $(-2, -2), (-3, -3), (-4, -4), (-5, -5)$ **12.** $(15, 4), (5, 6), (10, 3), (45, 5)$
13. $(-25, 20), (-10, -15), (-25, -10), (-5, 15), (-30, 25)$
14. $(4, -5), (-3, -2), (-2, -5), (3, -1), (5, -6)$
21. $\begin{pmatrix} 0 & 1 \\ 1 & 0 \end{pmatrix}$ **22.** $\begin{pmatrix} 5 & 0 \\ 0 & 1 \end{pmatrix}$ **23.** $(-9, -3), (-2, -5), (-4, -7)$
24. $(-12, -15), (4, 40), (16, 70)$

EXERCISES ➤ 2.6

7. No **8.** No **9.** No **10.** 3 **11.** The result is that a and b in the general equations for the translation are zero, so the translation is the identity. **12.** Only the center of rotation is an invariant point. **14.** Let the lines be $y = 0$ and $y = mx$

EXERCISES ➤ 2.7

1. (a) $(-2, 1, 7)$; (b) $(0, 6, 10)$; (c) $(-4, -2, 2)$; (d) $(-2, -3, 8)$
2. (a) $(1, 3, -4)$; (b) $(3, 2, -1)$; **4.** $(-3\sqrt{2}/2, 4, 7\sqrt{2}/2)$ **5.** $(3, 8, -1)$
6. $(-2, -4, 3)$ **7.** $(-2, 3, 9)$ **8.** $(7, 4, -2)$ **12.** (a) Reflection, rotatory reflection, or glide reflection; (b) Reflection, rotatory reflection, or glide reflection
13. $x' = x,\ y' = y,\ z' = 4 - z$

EXERCISES ➤ 2.8

1. 4 cm **2.** $\frac{6}{5}$ **3.** Length and area **7.** The measure of area is multiplied by the square of the ratio of similarity. **8.** $\sqrt{51}/6$ **9.** They are equal.
10. $\left(\dfrac{9}{4}, \dfrac{15}{4}\right)$ **11.** $\left(\dfrac{50 + 9\sqrt{10}}{5}, \dfrac{20 + 13\sqrt{10}}{5}\right)$ **13.** Yes

EXERCISES ➤ 2.9

2. ≈ 1.4650 **3.** ≈ 1.3652 **4.** ≈ 1.1833 **5.** ≈ 1.3181

REVIEW EXERCISES ➤ CHAPTER 2

1. $fg = \{(3, 1), (7, 5), (6, 2)\}$; $gf = \{(2, 1)(1, 5), (5, 2)\}$ **2.** $(-1, 3, 3)$
6. Transformations **7.** Sample answer: I, R_1 **8.** 4 **9.** $\begin{pmatrix} 1 & 2 & 3 \\ 2 & 1 & 3 \end{pmatrix}$
10. $(x + 2)^2$ **11.** $x' + y' = 12$ **12.** $2x' + y' = 5$
13. Translations and glide reflections **14.** 3 **15.** $(1, \sqrt{3})$ **16.** $(3, -12, 5)$
17. $\frac{9}{4}$ **19.** ≈ 1.1674 **20–27.** For Exercises 20–27, numbers 20, 24, and 27 are groups.

EXERCISES ➤ 3.1

1. b, c, d, e, f, g **2.** a, b **4.** Yes; no **8.** The open one-dimensional neighborhood with radius r of a point P is the set of interior points of a segment of length $2r$ with P as midpoint. **9.** (e) The empty set, the set itself, the complement of the set. (f) The empty set, the entire plane, the empty set. **10.** (c) The empty set, $\{(x, y, 0) : x \geq y\}$, the complement of the boundary (d) The empty set, the set itself, the complement of the set **11.** (a) Neither (b) Closed (c) Closed; open; open; closed; closed; neither (d) Closed; closed; neither; closed (e) Neither
12. (a) a, d; (b) a, b **15.** It is neither closed nor open. **16.** (a) The interior of a circular region (b) A parabola and its interior

EXERCISES ➤ 3.2

1. No **2.** Yes **3.** Yes **4.** No **5.** Yes **6.** Meaningless because vertex is corner point **7.** No **8.** No **9.** All regular **10.** Vertices are corner points. **11.** Vertex is corner point. **12.** Vertices are corner points. **13.** (a) A triangular region is the intersection of three supporting half-planes determined by the sides of the triangle. (b) A convex polygonal region is the intersection of the n supporting half-planes determined by the sides of the polygon. **15.** If a line contains at least one boundary point of the set but no interior points, it is a supporting line for a convex set of points. If a line is not a supporting line for a convex set of points (but does contain at least one boundary point), then it contains interior points. If a line contains at least one boundary point of the set and also contains interior points, then the line is not a supporting line for a convex set of points.

EXERCISES ➤ 3.3

1. Yes **2.** No, not bounded **3.** No, not bounded **4.** No, not convex **5.** No, not bounded **6.** No, not always bounded **7.** No, not bounded **8.** Yes **9.** No, not closed **10.** No, might be empty **11.** One **12.** Yes **13.** No **17.** Line segment **19.** Three of the first and six of the second **20.** Two hundred Black Angus and one hundred Hereford

EXERCISES ➤ 3.4

1. Door handle and door, book on a shelf **2.** It fills an octant in space, with the vertex at the origin and three sides lying on the reference planes. **3.** No. It must pass through a boundary point of the convex set. **4.** Yes **5.** No **6.** No **7.** Yes **8.** Yes **9.** No **10.** No **11.** No **13.** A tetrahedral solid is the intersection of the four closed half-spaces containing the sides of the tetrahedron and the vertex not on the side.

EXERCISES ➤ 3.5

1. Circular region **2.** Triangular region **3.** The convex polygonal region **4.** Segment **5.** Entire plane **7.** Parabolic region **8.** Entire plane **9.** Angle and interior **10.** Triangular region or quadrangular region **11.** Packing a carton of books; getting a large, irregularly shaped package through a door **13.** Yes **14.** Yes **16.** A line and a point not on that line **18.** The angle and its interior **20.** A spherical region **21.** A tetrahedral region or other special cases **22.** No

EXERCISES ➤ 3.6

1. 1 cm, $\sqrt{2}$ cm **2.** 2 cm, $\sqrt{13}$ cm **3.** 1 cm, 1 cm **4.** $3\sqrt{187}/14$ cm, 7 cm **5.** True **6.** True **7.** False **8.** False **9.** True **10.** False **11.** False **12.** False **13.** True **14.** False **19.** $4\pi, 8\left(\pi - \sqrt{3}\right)$ **21.** $\sqrt{3}$

EXERCISES ➤ 3.7

3. Let $K = \{K_1, K_2, \ldots, K_N\}$ be N convex sets of points, $N \geq 3$, lying in two-space, so that every three sets have a nonempty intersection. Then the intersection of all the sets is not empty. **8.** Yes **9.** 6 **10.** Let $S = \{P_1, \ldots, P_n\}$ be any finite set of points in a plane. Then there is a point A such that every closed half-plane formed by a plane through A contains at least $n/3$ points of S.

REVIEW EXERCISES ➤ CHAPTER 3

1–10. For Exercises 1–4, number 1 is true. For Exercises 5–10, numbers 6 and 9 are correct. **11.** All the points of the plane. **12.** Vertices are corner points. Yes. **13.** Vertex is corner point. Ray is not a line. **14–20.** For Exercises 14–20, numbers 14, 16, and 19 are correct.
21.

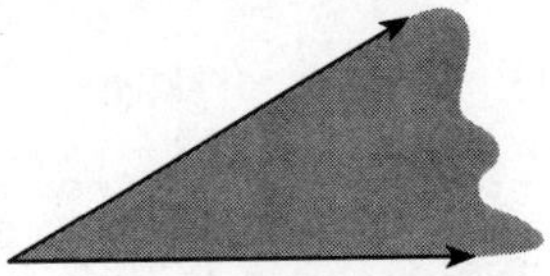

22. A triangular region with the three points as vertices **23.** The prism and its interior **24.** $2\sqrt{5}$ cm **25–28.** For Exercises 25–28, numbers 25 and 26 are correct. **29.** It is greater than the diameter of K. **30.** Yes **31.** $4P$ items and $12Q$ items.

EXERCISES ➤ 4.1

1. ABC, H; ACH, B; AHB, C; CHB, A **2.** No **3.** No **4.** Yes **5.** No **6.** At the vertex of the right angle **7.** For an obtuse triangle **8.** Three **9.** Yes **10.** ≈ 41 **11.** ≈ 8 **12.** ≈ 395 **13.** ≈ 430 **24.** This theorem is a consequence of the fact that the set of points at which a given segment subtends a given angle is a circle passing through the endpoints of the segment.

EXERCISES ➤ 4.2

1. No **2.** $\frac{AC}{CB} \cdot \frac{BD}{DF} \cdot \frac{FE}{EA} = -1$ **3.** $\frac{CA}{AB} \cdot \frac{BF}{FD} \cdot \frac{DE}{EC} = -1$ **4.** $\frac{FA}{AE} \cdot \frac{EC}{CD} \cdot \frac{DB}{BF} = -1$

5. $\frac{AF}{FE} \cdot \frac{ED}{DC} \cdot \frac{CB}{BA} = -1$ **10.** $\frac{AE}{EC} \cdot \frac{CF}{FB} \cdot \frac{BD}{DA} = 1$ **15.** An infinitude
16. An altitude

EXERCISES ➤ 4.3

1. (a) One; (b) One; (c) An infinite number **2.** 6 **3.** 6, 3

EXERCISES ➤ 4.4

1. A median **2.** Centroid **3.** One **5.** At the centroid **6.** No
7. Shorter **8.** $\frac{3}{4}$; $\frac{3}{5}$; $\frac{4}{5}$

EXERCISES ➤ 4.5

1. Yes **4.** The two Brocard points, circumcenter and symmedian point.
11. The four points are vertices of a square.
12. The four points are vertices of a square.

EXERCISES ➤ 4.6

1. 1.6180 **2.** $\frac{55}{34} \approx 1.618$ **3.** $1/x^4$ or $5 - 3x$ **5.** The interior angles have measures of 135, and this number is not a factor of 360. **11.** 28 **12.** 18
13. $\frac{45}{2}$ **14.** 21 **17.** 45°

EXERCISES ➤ 4.7

12. $\frac{3}{8}$ **13.** $\frac{1}{6}$ **14.** $\frac{\pi}{4+\pi}$ **15.** $\frac{3\sqrt{3}}{4\pi}$ **16.** (a) $\frac{6}{7}$ (b) $\frac{1}{3}$ (c) $\frac{99}{100}$
17. (a) $\frac{1}{10}$ (b) $\frac{1}{2}$ (c) $\frac{3}{4}$

REVIEW EXERCISES ➤ CHAPTER 4

1. Incenter **2.** Circumcenter **3.** 3 **4.** 6 **5.** 6 **6.** $\frac{CK}{KF} \cdot \frac{FA}{AB} \cdot \frac{BD}{DC} = -1$
7. One **9.** One **10.** Angle bisector **11.** Incenter and symmedian point
12. Centroid, orthocenter **13.** $\frac{73}{2}$ **15.** $\frac{1}{2}$

EXERCISES ➤ 5.2

1. Yes **2.** Yes **3.** Yes **4.** No **5.** Yes **6.** Yes **7.** Yes **8.** No

20. $$\frac{(c-a)(fg-eh)-(g-e)(bc-ad)}{(c-a)(h-f)-(g-e)(d-b)},$$

$$\frac{(d-b)}{(c-a)}\left[\frac{(c-a)(fg-eh)-(g-e)(bc-ad)}{(c-a)(h-f)-(g-e)(d-b)}\right]+\frac{bc-ad}{c-a}$$

EXERCISES ➤ 5.3

3. There are 4, 3, 2, 1, or no solutions **4.** 1 solution **5.** 1 or no solution
6. 1 or no solution **7.** 1 or no solution **8.** 2, 1, or no solutions
9. 4 solutions **10.** 4, 3, 2, 1, or no solutions **11.** 2, 1, or no solutions
12. 1 solution **13.** 2, 1, or no solutions **14.** 1 solution **15.** 1 or no solution
16. 1 or no solution **17.** 1 solution **18.** 1 or no solution
19. 1 or no solution

EXERCISES ➤ 5.4

1. None **2–7.** For Exercises 2–7, numbers 2, 3 and 7 are algebraic.
8. x is a solution of $x^4 - 4x^2 + 2 = 0$. **9.** Yes, it is a solution of $x^3 - 2 = 0$.
10. ≈ 2.5198 **12.** The only possibilities are $\pm 1, \pm\frac{1}{2}, \pm\frac{1}{4}, \pm\frac{1}{8}$, and each of these can be checked by synthetic division to show it is not a solution. **15.** None

16. This makes the cosine of an acute angle equal to 1. **17.** The equation has no rational solution.

REVIEW EXERCISES ➤ CHAPTER 5

8. The solver assumes that the construction has been made, then examines the completed picture of the solution to find the needed connections between the unknown elements in the figure and the given facts in the original problem. **9.** A triangle similar to the required triangle can be constructed with the given data. **10.** A preliminary triangle that can be constructed immediately from the given information. **13.** Yes **14.** $x^4 - 6x^2 + 4 = 0$ **15.** No **17.** Fold the focus over the directrix.

EXERCISES ➤ 6.1

1. $\frac{16}{3}$ **2.** 3 **3.** $\frac{4}{3}$ **4.** $\frac{1}{3}$ **5.** $\frac{8}{49}$ **6.** $\frac{\sqrt{5}}{3}$ **7.** $\frac{189}{50}$ **8.** $\frac{n^2}{m}$
9. No **10.** Outside the circle of inversion **12.** When the circle does not pass through the center of inversion; when the circle does pass through the center **13.** The image is a circle containing the center of inversion. **14.** (a) No (b) no **16.** The identity transformation **17.** Two distinct intersecting lines, a circle and an intersecting line, or two intersecting circles **18.** A circle inside the circle of inversion and concentric to it **19.** On the circle of inversion

EXERCISES ➤ 6.2

1. Two orthogonal circles, a circle and a line through the center
12. A line not passing through the center of inversion.

EXERCISES ➤ 6.3

1. $\left(\frac{3}{25}, \frac{4}{25}\right)$ **2.** $\left(\frac{12}{25}, \frac{16}{25}\right)$ **3.** $(2, 2)$ **4.** $\left(\frac{9}{2}, \frac{9}{2}\right)$ **5.** $(0, 3)$

6. $\left(0, \frac{48}{9}\right)$ **7.** $\left(\frac{56}{25}, \frac{8}{25}\right)$ **8.** $\left(\frac{7}{2}, \frac{1}{2}\right)$ **9.** $\left(\frac{50}{17}, \frac{47}{17}\right)$ **10.** $\left(\frac{70}{17}, \frac{42}{17}\right)$

11. $\left(\frac{53}{13}, \frac{57}{13}\right)$ **12.** $\left(\frac{38}{13}, \frac{47}{13}\right)$ **14.** $2x^2 + 2y^2 - 9x = 0$

15. $5x^2 + 5y^2 - 9x = 0$
16. $2x^2 + 2y^2 + 9x + 9y = 0$ **17.** $36x^3 + 36xy^2 - 81y^2 = 0$

18. $9y^3 + 9x^2y - 81x^2 = 0$ **19.** $x^2 + y^2 = \frac{81}{16}$ **21.** $\frac{2 - 5i}{29}$ **22.** $\frac{8 - 20i}{29}$

23. $\frac{3 + 7i}{58}$ **24.** $\frac{12 + 28i}{58}$ **25.** $\frac{4 - 8i}{5}$ **26.** $\frac{9 - 18i}{5}$ **27.** $\frac{27 + 99i}{130}$

28. $\frac{48 + 176i}{130}$

EXERCISES ➤ 6.4

1. The other point would be the ideal point. **3.** A tangent to a circle passes through the center of a circle orthogonal to the circle at the point of tangency.
4. A circle circumscribed about a right triangle has the hypotenuse as a diameter.
15. The same plane **16.** $\left(\frac{147}{14}, \frac{49}{14}, \frac{98}{14}\right)$ **17.** $\left(\frac{16}{19}, \frac{40}{19}, \frac{24}{19}\right)$

REVIEW EXERCISES ➤ CHAPTER 6

1. Those on the circle of inversion **2.** When it is the circle of inversion or orthogonal to the circle of inversion **3.** $\frac{25}{4}$ **4.** $\sqrt{6}$ **5.** No
6. The identity **7.** The same two circles **9.** No **10.** Both are orthogonal to the circle of inversion. **13.** $\left(\frac{8}{13}, \frac{12}{13}\right)$ **14.** $3x'^2 + 3y'^2 + 9y' = 0$
15. $x'^2 + y'^2 = \frac{1}{9}$ **16.** $\frac{8}{3} + \frac{8}{3}i$ **17.** If two orthogonal circles intersect at a point on a third circle, and if one of them is tangent to the third circle at that point, then the second one of them is orthogonal to the third circle. **18.** If a circle passes through the points of intersection of two circles, then it is orthogonal to a circle that is orthogonal to both of the intersecting circles. **19.** $\left(-\frac{81}{14}, \frac{243}{14}, \frac{162}{14}\right)$

EXERCISES ➤ 7.1

1. $A'B' = AB\cos\theta$ **2.** Top center **3.** Off to the left **4.** No **6.** Not invariant **7.** Invariant **8.** Not invariant **9.** Not invariant **10.** Not invariant **11.** Not invariant **12.** Invariant **13.** Not invariant **14.** Not invariant **15.** Not invariant **16.** Not invariant

EXERCISES ➤ 7.2

1. There are fewer invariant properties. **2.** 1, 3, 4 **3.** 1, 2 **4.** 1, 3
5. 2, 3 **6.** 1, 2, 3 **7.** 3 **8.** 3 **9.** 3 **11.** Two **12.** Two
13. Two of them would be ideal points. **14.** Yes

EXERCISES ➤ 7.3

1. 3 **2.** 3 **3.** If A and B are distinct points of a plane, there is at least one line on both points. **4.** On a line are at least four points.
5. A triangle (trilateral) consists of three nonconcurrent lines and the points of intersection of each pair.
6. The three diagonal points of a complete quadrangle cannot be collinear.
7. Three points not on the same line determine a plane.
8. A point is determined by two intersecting lines.
9. A plane and a line on the same point might not be incident.

EXERCISES ➤ 7.4

3. A point at infinity **4.** Yes **5.** Yes **6.** Yes **7.** Yes **8.** Yes

EXERCISES ➤ 7.6

1. The last coordinate is zero. **2.** (a) (3, 8, 1); (b) $(2, \frac{3}{4}, 1)$; (c) (1, −4, 1)
3. (a) (2, 5) (b) $\left(-\frac{1}{2}, \frac{3}{2}\right)$; (c) $\left(\frac{2}{3}, -\frac{5}{3}\right)$ **4.** $\left(\frac{5}{3}, \frac{2}{3}, 1\right)$; (10, 4, 6); (20, 8, 12)
5. No **6.** (1, −1, 0) **7.** $(b, -a, 0)$ **8.** $3x_1 + x_2 - 2x_3 = 0$
9. $3X_1 + X_2 - 2X_3 = 0$ **10.** $x_1 - x_3 = 0$ **11.** (1, 0, −1) **12.** $x_1 - x_3 = 0$
14. (0, 2, 1)

EXERCISES ➤ 7.7

1. −2, 3, 1 **2.** (5, −4) **3.** (−4, −4) **4.** $-\frac{3}{2}$ **5.** $\frac{5}{2}$ **6.** $\frac{3}{28}$
7. $\frac{28}{25}$ **8.** (1, 1, −1); (1, 1, 1); (1, 1, 0); (4, 4, −1) **9.** $\frac{5}{3}$ **11.** Similarity
12. None **13.** Affine **14.** Translation
15. Unequal scaling; (3, 0, 4, 1), (0, 2, 4, 1), (0, 0, 0, 1), (3, 2, 0, 1)
16. Translation; (4, 4, 3, 1), (3, 5, 3, 1), (3, 4, 2, 1), (4, 5, 2, 1)
17. Three-point perspective; (1, 0, 1, 8), (0, 1, 1, 7), (0, 0, 0, 1), (1, 1, 0, 6)
18. Projection; (1, 0, 4, 1), (0, 1, 4, 1), (0, 0, 4, 1), (1, 1, 4, 1)
19. Reflection; (1, 0, 1, 1), (0, −1, 1, 1), (0, 0, 0, 1), (1, −1, 0, 1)
20. Shear; (5, 2, 4, 1), (4, 1, 1, 1), (0, 0, 0, 1), (1, 3, 3, 1)
21. Overall scaling; (1, 0, 1, 2), (0, 1, 1, 2), (0, 0, 0, 2), (1, 1, 0, 2)
22. Two-point perspective; (1, 0, 1, 3), (0, 1, 1, 4), (0, 0, 0, 1), (1, 1, 0, 6)
23. Projection; (1, −2, 1, 1), (0, −2, 1, 1), (0, −2, 0, 1), (1, −2, 0, 1)
24. Reflection; (1, 0, −1, 1), (0, 1, −1, 1), (0, 0, 0, 1), (1, 1, 0, 1)

EXERCISES ➤ 7.8

1. The product of a reflection and the same reflection is the identity. **2.** No
3. Rotation of 180° **4.** Rotation of 120° **5.** Rotation of 90°
8. A unique one-dimensional involution is determined by two pairs of corresponding points.
9. A two-dimensional projectivity that leaves the four vertices of a complete quadrangle invariant is the identity transformation. **11.** No

EXERCISES ➤ 7.9

1. It could be a degenerate conic. **2.** Figures 7.6, 7.23, 7.25, 7.26, 7.27, 7.29
3. The lines of the ranges of points in the projectivity defining the line conic are also lines of the conic. **4.** The points corresponding to the common point of the two ranges of points determining a line conic are the points of contact of the two ranges of points.

5. $3x_1 + x_2 + x_3 = 0$ **7.** Yes **9.** There are 720 permutations of six things taken six at a time. Half of these are duplicates caused by reversing the order of points. Of the remaining 360, there are six different names for each of 60 hexagons.

REVIEW EXERCISES ➤ CHAPTER 7

1–3. For Exercises 1–3, 2 and 3 are invariant. **7.** 3
8. A complete quadrilateral has 3 pairs of opposite vertices. **9.** 3 **10.** $(-17, 5, 1)$
11. $(-4, 3, 0)$ **12.** $\frac{4}{3}$ **13.** $\frac{28}{45}$ **14.** 2 **15.** $(9, 8, 1)$
16. $2x_1 - 5x_2 + 7x_3 = 0$ **17.** 9 **18.** Yes
19. Unequal scaling; $(4, 3, 3, 0)$, $(2, 6, -1, 1)$
20. Translation; $(2, 1, 3, 0)$, $(-3, 7, -2, 1)$ **21.** Two-point perspective; $(2, 1, 3, 7)$, $(1, 2, -1, 2)$ **22.** $2X_1 + 3X_2 + 4X_3 = 0$

EXERCISES ➤ 8.1

1. A rotation; $x \rightarrow$ cotangent x **2.** The inverse also has an inverse that is a continuous transformation. **4.** Length of a segment; area of a region; number of sides of a polygon **5.** A convex set is a connected set for which the curve joining two points can always be a segment. **6–21.** For Exercises 6–15, answers are 6, 7, 8, 9, 11, 13. For Exercises 16–21, answers are 16, 18, 19, 20. **22.** Penny
23. Washer **24.** Button **26–29.** For Exercises 26–29, answer is 26.

EXERCISES ➤ 8.2

1–9. For Exercises 1–4, answers are a and 4. For Exercises 6–9, answers are 6, 7, 8.
11. Vertex C is not counted. Vertex E is counted. **12.** Odd
13. Multiply connected **16.** Yes

EXERCISES ➤ 8.3

1. No. Images would not always be in the circular region. **3.** No **4.** No
5. Yes **6.** No **7.** Yes **8.** Yes **9.** No **10.** No **11.** No
12. No **13.** Yes **14.** No **15.** Yes **16.** Yes **17.** No **18.** No
19. No **20.** No **21.** Yes **22.** No **23.** No **24.** Yes **25.** No
26. No **27.** No **28.** No **29.** No **30.** No **31.** No **32.** No
33. No **34.** No

EXERCISES ➤ 8.4

2. Zero **3.** Prism **4.** Teacup **5.** Dining room chair with arms
6. Three-ring binder pages **7.** Button **9.** Four; four **12.** The theorems only apply to countries topologically equivalent to a circular region. **13.** More than four countries with a common point would each require a separate color.

EXERCISES ➤ 8.5

1. By rows; 6, 12; 20, 30; 12, 20, 30 **5.** 0 **6.** 2 **7.** 0 **8.** -2 **9.** -4
10. 5 **11.** 7 **12.** Impossible **13.** Impossible **16.** Two interlocking pieces
17. No. Two interlocking pieces **18.** No. One twisted piece **19.** $n \geq 3$

REVIEW EXERCISES ➤ CHAPTER 8

1. No **2.** No **3.** Yes **4.** Yes **5.** Yes **6.** 2 **7.** Yes **8.** One **9.** 20 **10.** 20 **11.** 2 **12.** 5 **13.** No **14.** 48 cm^2 **15.** Curve **16.** Not a curve **17.** Curve **18.** Closed curve **19.** Not a curve

EXERCISES ➤ 9.1

3. If a straight line intersects one of two parallel lines, it will not always intersect the other. **4.** Straight lines parallel to the same straight line are not always parallel to each other. **5.** There does not exist one triangle for which the sum of the measures of the angles is π radians. **6.** There does not exist a pair of similar but noncongruent triangles. **7.** There does not exist a pair of straight lines the same distance apart at every point. **8.** It is not always possible to pass a circle through three noncollinear points. **9.** a **10.** AGJ; AIL; AKM **12.** Two

EXERCISES ➤ 9.2

1. Otherwise, two distinct lines could be both parallel and intersecting.

EXERCISES ➤ 9.3

1. Two **2.** The sum of the measures of the angles would be 2π. **4.** Adjacent side **5.** Not the same at every point. **6.** Summit **9.** 70.371° **10.** The defect of the original triangle is equal to the sum of the defects of the two smaller triangles.

EXERCISES ➤ 9.4

1. There would exist a rectangle. **2.** At their common perpendicular **3.** An infinite number **4.** No **12.** The sum of the measures of the summit angles is equal to the sum of the measures of the three angles of the original triangles.

EXERCISES ➤ 9.5

1. Less than **2.** The angles determined by the radii and the segment joining the points are congruent. **5.** No **6.** No **7.** The kind of point of intersection of perpendicular bisectors of segments joining the points **9.** Ultra-ideal point **12.** Connect the midpoints of two of the segments joining pairs of points.

EXERCISES ➤ 9.6

1. Four-line geometry; $PG(2, 3)$ **2.** Yes **3.** Circle **4.** 8 **5.** When they are the two poles of a line **6.** No **7.** Yes **9.** No, there are no ideal points. **10.** Less than 2π **12.** At the polar of their common points

EXERCISES ➤ 9.7

10. Infinite **11.** $\overleftrightarrow{A'B'}$ is the image of circle A, B; $\overleftrightarrow{B'C'}$ is the image of circle B, C. Circle A', C' is the image of circle A, C. Each of the three images is orthogonal to the image of the fundamental circle. **12.** (a) 4 (b) 11

REVIEW EXERCISES ➤ CHAPTER 9

1. A, B, C; B, D, F; B, E, I **4.** 2
5. That one of the pairs of corresponding angles at a vertex was congruent **7.** 2; an infinitude **8.** Summit **9.** Nonintersecting **10.** 2 **11.** The sum of the measurements is 130.315°. **12.** Yes **14.** 3 **16.** Equidistant curves **17.** 4

Index

D

E

F

I

J

K

L

M

N

O

P

T

U

V

W

Y

Z